D0504123
19-75
WITHDRAWN
FROM STOCK
L LIBRARY

Introduction to

Electron Microscopy

Third Edition

Pergamon Titles of Related Interest

Dummer *ELECTRONIC INVENTIONS AND DISCOVERIES, 2nd Edition*
Roddy *INTRODUCTION TO MICROELECTRONICS, 2nd Edition*
Bradbury *THE MICROSCOPE, Past and Present*

Related Journals*

MICROELECTRONICS AND RELIABILITY
MICRON

*Free specimen copies available upon request.

Introduction to
Electron Microscopy

Third Edition

Saul Wischnitzer, Ph.D.
Professor, Department of Biology
Yeshiva University, New York City

Pergamon Press
New York ☐ Oxford ☐ Toronto ☐ Sydney ☐ Paris ☐ Frankfurt

Pergamon Press Offices:

U.S.A.	Pergamon Press Inc., Maxwell House, Fairview Park, Elmsford, New York 10523, U.S.A.
U.K.	Pergamon Press Ltd., Headington Hill Hall, Oxford OX3 0BW, England
CANADA	Pergamon Press Canada, Ltd., Suite 104, 150 Consumers Road, Willowdale, Ontario M2J 1P9, Canada
AUSTRALIA	Pergamon Press (Aust.) Pty. Ltd., P.O. Box 544, Potts Point, NSW 2011, Australia
FRANCE	Pergamon Press SARL, 24 rue des Ecoles, 75240 Paris, Cedex 05, France
FEDERAL REPUBLIC OF GERMANY	Pergamon Press GmbH, Hammerweg 6, Postfach 1305, 6242 Kronberg/Taunus, Federal Republic of Germany

Library of Congress Cataloging in Publication Data

Wischnitzer, Saul.
Introduction to electron microscopy.

Bibliography: p.
Includes indexes.
1. Electron microscopy. I. Title.
QH212.E4W57 1980 502'.8'25 80-15266
ISBN 0-08-026298-8
ISBN 0-08-028038-2 (pbk.)

Printed in the United States of America

Dedicated to my parents whose
many sacrifices over the years made this book possible.

PREFACE

THE AIM of this new edition is to provide an updated introduction to the foundations of electron microscopy, to outline some practical aspects of instrument operation, and to discuss the rationale of the methodology of biological specimen preparation. It now covers scanning electron microscopy (SEM) on an equal footing with transmission electron microscopy (TEM) and updates some material on TEM.

Although the electron microscope is an instrument which has been designed and developed by physicists and engineers, the majority of those now making use of this instrument are in the fields of biomedical research. Unfortunately, such workers often lack the preparation for a comprehensive understanding of the mathematical and physical framework upon which the design of this instrument is based. This volume was written in the belief that it is both possible and desirable for investigators in the biomedical sciences to attain a useful understanding of the theoretical and operational aspects of the electron microscope. They may thus be able to appreciate the potentials as well as the limitations of this research tool and should be in a more favorable position to pursue their own research problems successfully when using this instrument.

Other presentations of this subject are often either on an elementary level for the layman or else complex and directed almost exclusively toward those who have a substantial background in physics and mathematics. This book seeks to bridge the gap between these two levels of presentation, with the average biomedical investigator especially in mind. None of its mathematical treatments goes beyond the use of algebra and trigonometry.

A discussion of the electron microscope should have wide applicability irrespective of the particular instrument that the microscopist is, or contemplates, using. Therefore, this book has not been limited to any particular current model. Where particular examples may be useful, one or another is mentioned specifically.

At the time of writing the first edition of this book in 1962, specimen preparation methodology was still undergoing radical change. By the appearance of the second edition in 1970, the stage was reached when a useful set of preparatory procedures became relatively stabilized and offered the microscopist a considerable degree of flexibility. Moreover, specialized approaches such as the application of histochemistry, autoradiography, and immunochemistry, in combination with electron microscopy, were introduced. All of these developments made it appropriate to discuss some theoretical and practical aspects of methodology.

While improvements in instrumentation and methodology have occurred during the 1970s, these have been essentially in the realm of refinements rather than breakthroughs. They are discussed in detail in the several series of reviews published over the past few years (see p. 364). The most prominent advance that has taken place has been in the area of scanning electron microscopy. This instrument has now achieved a significant status as a microscopic investigatory tool.

This third edition therefore now consists of two sections—the first dealing with transmission electron microscopy (TEM), and the second with scanning electron microscopy (SEM). Each section is in turn divided into two similar parts, one dealing with instrumentation and the other with methodology. As a result, this book merits, both by content and approach, the continued use of its title as an introduction to both major areas of electron microscopy.

Five additional features of this edition are: a discussion of high voltage electron microscopes; Appendicies on beam intensity and on vacuum pumps; an updating of the General References; and a revision of the Bibliography. The new third section of this book contains 16 figures, many of which are scanning electron micrographs demonstrating some of the techniques used in conjunction with the SEM.

The author wishes to express once again his gratitude to Dr. L. Ornstein, Professor of Pathology (Biology), Mt. Sinai School of Medicine, and Dr. A. Wachtel, Professor of Biology, University of Connecticut for their valuable advice and critical reading of the manuscript, and acknowledges the contributions of electron micrographs by various investigators, often originators of the techniques described in this text or of some of their must useful variants. The secretarial assistance of Mrs. Ceceil Levinson is also deeply appreciated.

New York City — SAUL WISCHNITZER

CONTENTS

SECTION ONE: TRANSMISSION ELECTRON MICROSCOPY

Part One: Instrumentation

1. Introduction 1

2. Historical Review 4

3. Basic Theory of Electron Microscopy 6
   - A. Nature of Light Beams 6
   - B. Resolution 8
     - 1. Diffraction 8
     - 2. Limit of Resolution 10
   - C. Nature of Electron Beams 14
   - D. Electron Emission 15
   - E. Electron Optics 19
     - 1. Introduction 19
     - 2. Magnetic Field 20
     - 3. Action of Magnetic Fields as Lenses 24
     - 4. Magnetic Focusing 28
     - 5. Evolution of Magnetic Lenses 31
     - 6. Electrostatic Lenses 34
   - F. Analogy Between the Light and Electron Microscopes 37

4. The Electron Microscope 38
   - A. Illumination System 38
     - 1. Electron Gun 38
       - a. Filament 39
       - b. Shield 40
       - c. Anode 40
       - d. Non-biased and Biased Emitters 40
       - e. Self-biased Gun 42
       - f. Operation of the Self-biased Gun 44
       - g. Brightness of Electron Source 45
     - 2. Condenser Lens 46
       - a. Aperture Angle of Illumination 48
       - b. Intensity of Illumination 48
       - c. Condenser Lens Operation 49

B. Imaging System 50
1. Objective Lens 50
a. Pole Pieces 51
b. Lens Aberrations 53
c. Limits Set by Objective Lens Aberrations 57
d. Contrast and Image Formation 59
e. Depth of Field 61
f. Objective Lens Operation 62
2. Projector Lens 62
a. Depth of Focus 63
b. Magnification and Final Image Formation 65
c. Range of Magnification 66
C. Image Translating System 68
1. Fluorescent Observation Screen 69
2. Photographic Recording 69
3. General Considerations in Image Translation 70
a. Beam Intensity Level 70
b. Choice of Photographic Emulsion 70
D. Other Electron Microscope Components 72
1. Specimen Chamber and Holder 72
2. Objective Aperture 73
3. Photographic Apparatus 75
4. Vacuum System 76
5. Power Supply 80
a. Filament Current Supply 80
b. High Voltage Supply 80
c. Lens Current Supply 81
E. Operational Requirements 81
1. Alignment 81
a. Electron Gun—Condenser Alignment 84
b. Illumination—Objective Lens Alignment 85
c. Adjustment of Illumination Tilt 85
d. Objective Lens—Projector Lens Alignment 87
e. Objective Lens—Intermediate Lens Alignment 87
f. Objective Lens—Screen Alignment 87
g. Objective Aperture Alignment 87
2. Detection and Correction of Lens Asymmetry 89
3. Disturbances 93
a. Stray Magnetic Fields 93
b. Mechanical Disturbances 93
c. Electrical Disturbances 94
d. Specimen Instability 94
e. Contamination 95

4. Vacuum Leaks 95
5. Specimen Damage 96
a. Exposure to Vacuum 96
b. Exposure to Electron Beam 97
c. Exposure to Hydrocarbons 97
d. Exposure to Non-Hydrocarbon Residual Gases 99
F. Operation of the Electron Microscope 100
1. Operating Adjustments 100
2. Photography 100
3. Determination of Magnification 101
4. Test of Resolution 102
G. Accessories for the Electron Microscope 102
1. Double Condenser 102
2. Image Intensifiers 103
3. Stereomicroscopic Accessories 104
4. Electron Diffraction Accessories 104
H. Other Types of Electron Microscopes 105
1. Electrostatic Electron Microscopes 105
2. Scanning Electron Microscopes and Microprobes 105
3. High Voltage Electron Microscopes 107
I. Differences Between the Light and Electron Microscopes 111

Part Two: Methodology

5. Introduction 115

6. Historical Review 116

7. Obtaining of the Specimen 120
A. Preparation of Excised Specimens 120
B. *In Situ* Fixation 121
C. Perfusion 121
D. Handling of Small Specimens 123
E. Handling of Human Material 124
F. Discussion 124

8. Fixation 125
A. Osmium Tetroxide Fixation 125
1. *p*H of the Fixative and Buffering Media 126
2. Tonicity of the Fixative 126
3. Temperature of the Fixative 126
4. Duration of Fixation 127
5. Modifications of the Standard Osmium Tetroxide Medium 127

B. Other Fixation Media 129
1. Aldehyde Fixatives 129
a. Formalin 129
b. Glutaraldehyde 129
2. Permanganate 131
C. Discussion 133

9. Dehydration 137
A. Dehydration Media 137
B. Duration of Dehydration 138

10. Embedding 139
A. Methacrylate 139
B. Water-Soluble Embedding Media 143
C. Polyester Resins 145
D. Epoxy Resins 147
1. Araldite 151
2. Epon 151
3. DER–334 152
4. Maraglas 152
E. Discussion 152

11. Microtomy 154
A. Principles of Microtomy 154
B. Ultramicrotomes 157
C. Knives 160
1. Glass Knives 161
2. Diamond Knives 162
D. The Trough 163
E. Section Thickness 164
F. Discussion 166

12. Staining 169
A. Principles of Staining 169
B. Positive Staining 170
C. Negative Staining 171

13. Specialized Electron Microscope Techniques 175
A. Ultrastructural Nucleoprotein Localization 175
B. Ultrastructural Enzyme Cytochemistry 177
C. Tracers in Electron Microscopy 180
D. Ultrastructural Immuno-Electron Microscopy 181
E. Electron Autoradiography 186
F. High Resolution Electron Microscopy 189
G. Electron Microscopic Analysis of Cell Fractions 195

H. Cryofixation for Electron Microscopy 197
1. Freeze-Substitution 199
2. Freeze-Etching 201
I. Surface Examination of Small Specimens 203
1. Dispersion Methods 203
2. Replica Technique 204
3. Surface Spreading Technique 204
J. Quantitative Electron Microscopy 207

SECTION TWO: SCANNING ELECTRON MICROSCOPY

Part One: Instrumentation

14. Introduction 213

15. Historical Review 216

16. Basic Theory 217

17. Design of the SEM 219
A. Electron Optical Column 219
1. Electron Gun 219
a. Tungsten Filament Cathode 219
b. Lanthanum Hexaboride Rod Cathode 223
c. Field Emission Gun 224
d. Discussion 225
2. Electron Lenses 226
a. Image Demagnification 227
b. Lens Aberrations 228
c. Final Condenser Lens 231
B. Image Translating System 233
1. Scanning System 233
2. Signal Processing 234
a. Beam–Specimen Interaction 234
b. Electron Collection 239
c. Signal Amplification 239
d. Signal Display 240
e. Signal Selection 241
C. Other Instrument Components 242
1. Apertures 242
2. Specimen Stage 243
3. Magnetic Shielding 243
4. Vacuum System 244

18. Image Characteristics 246
A. Contrast 246
1. General Considerations 246
2. Secondary Electron Emission Mode 246
3. Enhancement in Final Image 248
B. Resolution 249
1. Microscope Operating Conditions 249
2. Beam-Specimen Interactions 251
3. Induced Interference Problems 251

19. Current SEM Developments 252
A. Low Voltage Microscopy 252
B. Scanning Transmission Electron Microscopy 253
C. Scanning Electron Microprobe Analysis 254

20. Operational Considerations 255
A. Instrument Testing 255
B. Instrument Settings 255
C. Signal Monitor 257

21. Comparative Value of the SEM 258

Part Two: Methodology

22. Introduction 263

23. Historical Review 264

24. Prefixation Handling 265
A. Mild Pretreatment 265
B. Dissection 265
C. Chemical Pretreatment 267

25. Fixation 269
A. Fixation Parameters 269
B. Fixation Methods 271
1. Immersion 271
2. Dripping 271
3. Injection 271
4. Vascular Perfusion 271

26. Postfixation Handling 272
A. Dehydration 272
B. Drying 272
1. Air Drying 272
2. Critical Point Drying 273
3. Freeze-Drying 273

C. Specimen Mounting 274
D. Specimen Coating 274
1. Coating Methods 275
2. Coating Materials 275

27. Special Processing 277
A. Sectioning 277
B. Freeze-Fracturing 277
C. Freeze-Etching 279
D. Ion Beam Etching 279
E. Replication 281
F. Cell Surface Labeling 281
1. Antibody-Latex Sphere Coupling 283
2. Silica Sphere Labeling 283
3. Antigen-Sandwich Labeling 284
4. Ferritin Labeled Antibody 284
5. Other Antibody Markers 284
6. Other Markers 284
G. Living Material 285

28. Complementary TEM and SEM Studies 286

Appendices
A. The Nature of Light 291
B. Wavelength of an Electron 294
C. Summation and Resolution of Vectors 297
D. Fundamentals of Electronics 299
E. Dependence of Condenser Aperture Angle on Focal Length 305
F. Depth of Field and Depth of Focus 307
G. Spherical Aberration 311
H. Chromatic Aberration 313
I. Optical Theory of Cylindrical Lenses 315
J. Theory of Fresnel Diffraction Fringes 317
K. Electron Diffraction 323
L. Shadow Casting 329
M. Intensity Relations in Electron Beams and Images 332
N. Vacuum Pumps 335
Literature Cited 343
References 348
Bibliography 361
Glossary 391
Author Index 393
Subject Index 397

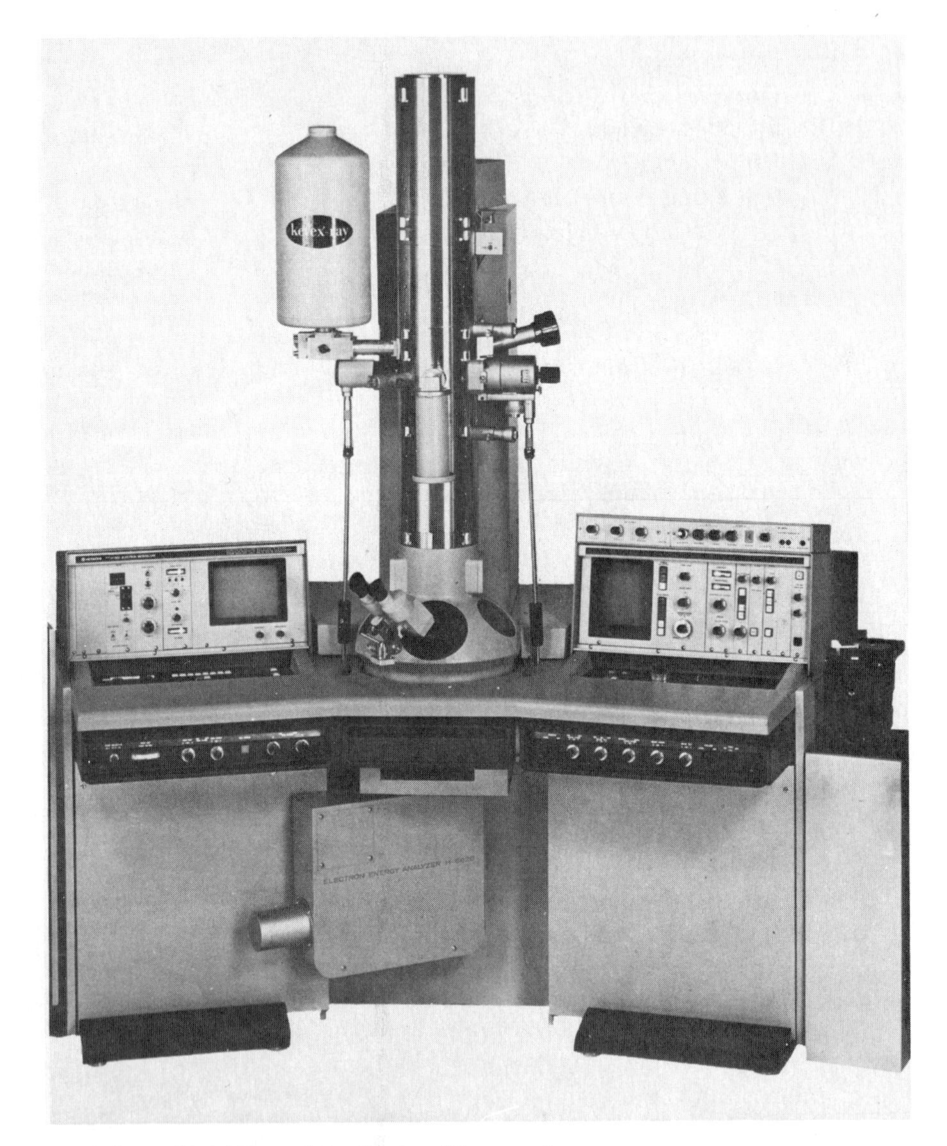

A Recent Model Transmission Electron Microscope*

---

*Hitachi H-600

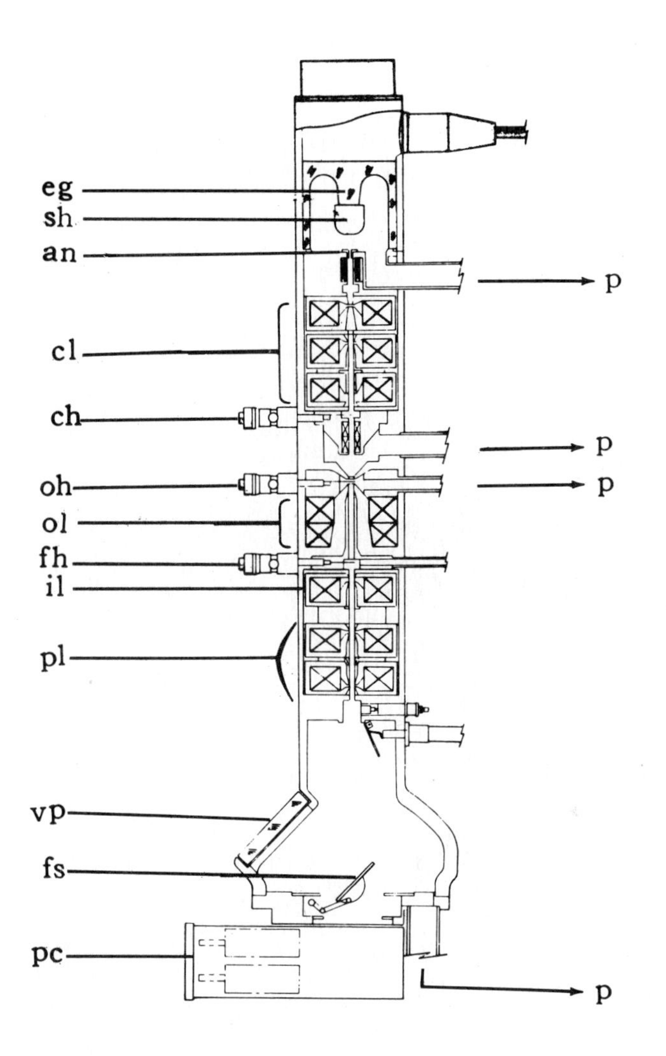

H-600 Column

eg — electron gun
sh — shield
an — anode
cl — condenser lenses (triple)
ch — condenser aperture holder
oh — objective aperture holder
ol — objective
fh — field limiting aperture holder
p — to pumping system
il — intermediate lens
pl — projector lens
vp — viewing port
fs — fluorescent screen
pc — photographic chamber port

A Recent Model Scanning Electron Microscope*

*Hitachi S-450

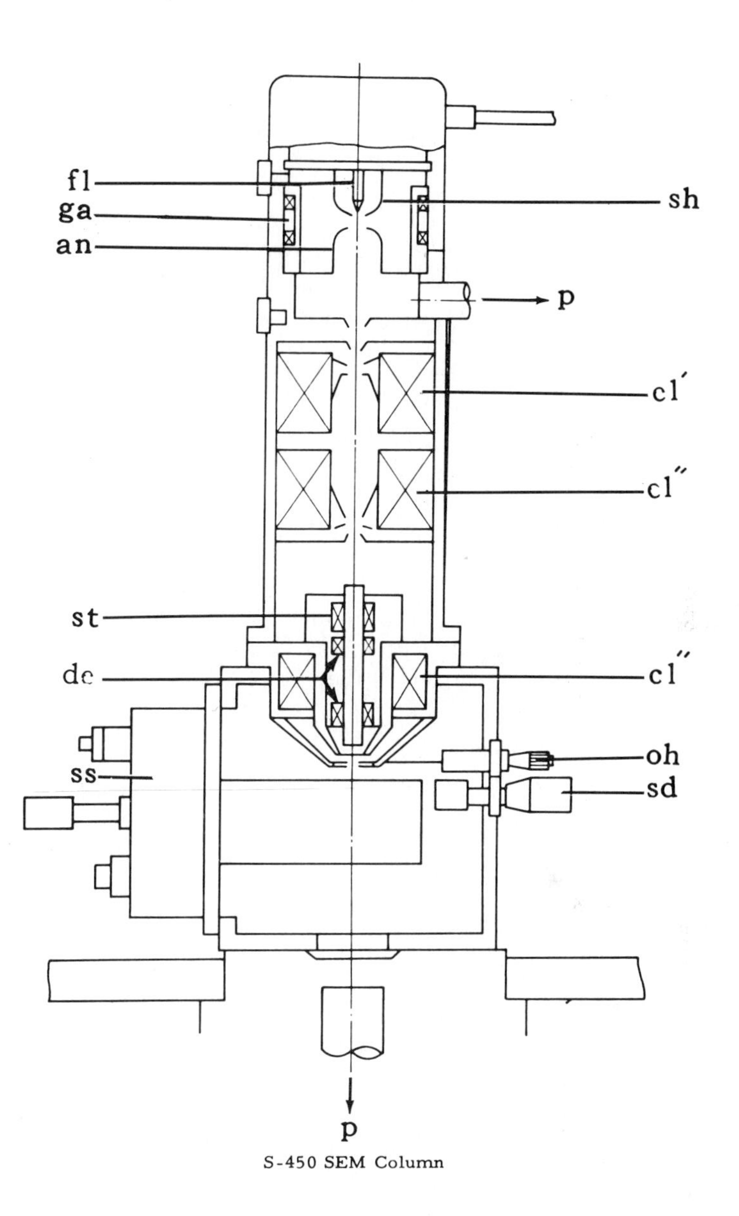

S-450 SEM Column

fl — filament
ga — gun alignment
an — anode
st — stigmator
dc — deflection coils
ss — specimen goniometer stage

sh — shield
cl' — first condenser lens
cl'' — second condenser lens
cl''' — final condenser lens
oh — objective aperture holder
sd — secondary electron detector

p — to pumping system

Section One

TRANSMISSION ELECTRON MICROSCOPY

Part One

INSTRUMENTATION

Chapter 1

INTRODUCTION

IN THIS, the twentieth century, the term "explorer" can more readily be applied to the scientist than to the relatively few who are engaged in conquering and mapping the remaining unknown areas of this earth. In this era the predominant courses of scientific exploration are in opposite directions. Some seek to uncover the mysteries of the universe beyond the earth, while other scientists are expending their energies to reveal fine details of the world of matter. To successfully expose the grandeur of the microcosmos of matter, in the last analysis, depends on the development of adequate instruments that, so to speak, are capable of extending our capacity for vision beyond its natural limitation. Or to present it more usefully, gaining an insight into the structural organization of matter is considerably facilitated by our sense of vision, which is man's most critical faculty for the understanding of his environment.

The compound microscope was a magnificent contribution of the 16th century toward the goal of the exploration of matter. The electron microscope represents, thus far, one of the twentieth century's most significant contributions to this never-ending pursuit. For by this means, more than a thousand-fold extension of our vision beyond the limits of the light microscopic level can be attained. This leap into the unknown is of special significance because it brings man's ability to "directly" observe and study matter down to the molecular realm. At this level it is possible to visualize the simplest forms of life, the viruses, as well as the major building blocks of biological matter, the macromolecules.

A remarkable feature of the present technological "explosion" is that while the light microscope reached its peak of perfection after more than two centuries of development, the electron microscope reached a comparable level in the short span of two decades. On the other hand, while in the last fifty years it has been possible to examine and evaluate a vast cellular and histological realm by means of the light microscope, it will probably take a far greater time to record and interpret all the information made visible on the ultrastructural level by means of this relatively new instrument.

TABLE I

DIMENSIONAL LEVELS IN THE ANATOMICAL SCIENCES

DIMENSIONS OF BASIC UNITS	CLASSIFICATION	METHOD OF DIRECT OBSERVATION OF BASIC UNIT	BASIC UNITS
1 mm, 1,000 µ, 10,000,000 A 0.1 mm, 100 µ, 1,000,000 A	Gross Anatomy	Eye and simple lens	Organs Small organisms
10 µ, 100,000 A	Histology	Light microscope	Tissues
1 µ, 10,000 A 0.1 µ, 1,000 A	Cytology	MICROSCOPES: Light Phase contrast U-V	Cells Bacteria
.01 µ, 100 A .001 µ, 10 A	Submicroscopic Morphology	Electron microscope	Cell organelles Viruses Macromolecules
.0001 µ, 1 A	Molecular Level	(None)	Molecules
	Atomic Level	(None)	Atoms

1 micron (1 μ) = 1×10^{-3} = 1,000 nm = 10,000 Å

1 nanometer (1 nm) = 1×10^{-6} mm = 10 Å

1 Angstrom (1 Å) = 1×10^{-7} mm = 0.1 nm

The electron microscope, by providing visual access to the world between the cellular and molecular levels, has presented a great challenge to the biologist. He now has a span extending approximately over three orders of magnitude (10^{-4} to 10^{-7} cm) to explore and to correlate with adjacent levels (see Tables I and II). The technical advances that have been made in an attempt to meet this challenge and the results thus far obtained indicate that, in time, man will also succeed in "conquering" this largely unexplored world.

TABLE II

DIMENSIONAL RELATIONSHIPS ON THE MICRO- AND SUBMICROSCOPIC LEVELS

Total Mag.	Dimensional Relationships	Structure	Actual Size
O	Unmagnified ·	HUMAN EGG	0.1 mm
600 X	Magnified 600x	HUMAN EGG	100 μm
600 X		LIVER CELL	25 μm
600 X		ERYTHROCYTE	7 μm
600 X		BACILLUS	2μm × 0.5μm
6,000 X	Magnified again 10x	BACILLUS	2000 nm × 500 μm
6,000 X		BACTERIOPHAGE	50 nm = 500 Å
600,000X	Magnified again 100x	BACTERIOPHAGE	500 Å
600,000X		FERRITIN MOL. CORE	50 Å
600,000X		LIPID MOLECULE	10 Å

An enormous expansion of our understanding of the ultrastructure of matter has taken place during the last fifteen years as a result of development and application of the electron microscope. Instrumentation and many techniques have by now become routine. Thus the potential to investigate the hitherto unexplored region that lies between the lower limit of resolution of the light microscope and the upper limits of the molecular level has become a reality. In this gap lie the answers to many fundamental problems of biological and medical significance; these are now within our grasp. Thus by increasing the "visual potential" of investigators in the life sciences it has become possible, by means of electron microscopy, to establish a link between the world of the cytologist and pathologist on one hand, and that of the biochemist and biophysicist on the other.

Chapter 2

HISTORICAL REVIEW*

AMONG the most outstanding achievements of the sixteenth century was the construction of the first crude light microscope. With this instrument it was possible for man to begin his study of the fine structure of matter. Along with other improvements, the introduction, two centuries later, of lenses that were corrected for the intrinsic optical defects present in earlier systems brought the light microscope near its peak of perfection. Although these developments opened a vast field for exploration, the factor that restricted the study of the ultrastructure of cells and tissues was the fundamental nature of light which imposes a limit to the resolution even of optical systems completely free of optical defects (see p. 11). The introduction of dark-field, phase contrast and interference microscopy during the first half of the twentieth century broadened the potentialities for morphological investigations. Nevertheless, these advances in microscopy did not contribute to the possibility for the direct visualization of biological matter smaller than about 0.2 μ.**

The fundamental physical concepts upon which electron microscopy is based can be traced back to the latter part of the nineteenth century. But two essential developments lead directly to the building of the electron microscope. The first was the de Broglie theory (1924)[1] that implied that a moving electron could be considered to have the properties of a light-like wave (see p. 14). The second major development was the demonstration by Busch (1926)[2] that suitably shaped magnetic or electrostatic fields could be used as true lenses to focus an electron beam to produce an image.

The milestones in the development and perfection of the TEM, whereby its use blossomed in the 1950s can be summarized as follows:

1932 Knoll and Ruska published[3] a description of the first transmission electron microscope which can be considered to be the prototype of modern magnetic instruments.† Their instrument consisted of an electron source and two

* The numerical superscripts used in this chapter and in other places further in the text refer to literature citations on p. 343.

** Microscopic dimensional units are defined on p. 8.

† Rüdenberg[4], in 1931, took out German patents describing an electron microscope. His priority over Knoll and Ruska in the development of the electron microscope is, however, open to serious question[5].

magnifying lenses. A condenser lens was not used. The resolution obtained with this instrument was below that attained with the light microscope. Nevertheless they secured the first electron micrographs of an illuminated specimen.

— Brüche and Johannson[6] built the first electrostatic electron microscope. This instrument was functionally very limited since it was designed for producing enlargements of the emitting source (self-luminous objects).

1934 Ruska[7] described an improved version of his electron microscope to which a condenser lens was added. The micrographs obtained indicated that the potential existed to surpass the resolving power of the light microscope.

1935 Driest and Muller[8], who modified Ruska's instrument somewhat, were able to attain resolutions beyond the level of the light microscope.

1938 von Borries and Ruska[9] constructed an electron microscope of advanced design which was capable of resolving 100 Å.

1939 Commercial production of the microscope developed by von Borries and Ruska[10] was begun in Germany by Siemens and Halske.

— Mahl[11] constructed the first successful electrostatic electron microscope (capable of resolving 70 Å) that made use of non-self-luminous specimens. Limited commercial production of this miscroscope was initiated in Germany by AEG.

— Burton, Hillier and Prebus[12] constructed an operational electron miscroscope in Canada.

1941 Hillier and Vance[13] and others constructed the RCA type B electron microscope (capable of resolving 25 Å) which was the first commercially produced electron microscope in North America.

1944 von Ardenne[14] demonstrated resolving powers of 12–15 Å.

1946 Hillier[15] attained a resolution of 10 Å, which is at the border of the practical limit of resolution of these instruments as determined by spherical aberration (see p. 58) and comes quite close to the theoretical limit of resolution which is about 2.0 Å.

1947 Hillier and Ramberg[16] developed the "compensated objective" lens (see p. 9).

Many subsequent refinements consisting of more stable power supplies, shorter focal length objectives, greater mechanical and thermal stability and multiple conveniences which make the required meticulous attention to operational detail relatively easy have led to routine instrument resolution near 5 Å and occasionally near 2 Å.[17,18]

Chapter 3

BASIC THEORY OF ELECTRON MICROSCOPY

As WILL be seen shortly, electron microscopy is based on very similar theoretical principles to those of light microscopy. The new instrument became technically feasible as a result of the development of suitable "lenses" capable of establishing and directing the path for the type of radiation used in the system, namely, the electrons.

A. NATURE OF LIGHT BEAMS*

As is well known, light is most conveniently looked upon as having both a *corpuscular* and a *wave nature* in order to satisfactorily explain the results of various physical experiments. In the former interpretation, a light source emits fine particles which travel along straight lines in a homogeneous medium and are received by the sensitive medium (e.g., the retina of the eye, photographic emulsion, etc.). However, an example which emphasizes the limitations in interpreting the nature of light only in terms of the corpuscular theory is apparent when we try to explain the following simple experiment by means of this theory. A screen with a pinhole aperture is placed in the path of a wide beam of light which, in terms of the corpuscular theory, is expressed as a bundle of parallel lines, each being made up of fine particles. Were this interpretation sufficient, then we would expect a narrow straight beam to emerge through the aperture which would be observable only when a detector is placed in line with the aperture (Fig. 1). In reality, the small amount of light which passes through is observable from points off axis because it spreads out from the aperture in a spherical manner which is best understood and illustrated in terms of the wave theory of the nature of light.

According to electromagnetic theory, the light source initiates a "vibrational motion" which transmits energy in the direction of propagation. The wave motion of light is analogous to that produced by dropping a

* For a more detailed discussion see Appendix A.

pebble into water (Fig. 2) and is illustrated in Figure 3. The horizontal distance from any point on one wave to the corresponding point in the next wave is known as the *wavelength* (Fig. 4). Each of the various colors of the spectrum is characterized by its individual wavelength (Fig. 5).

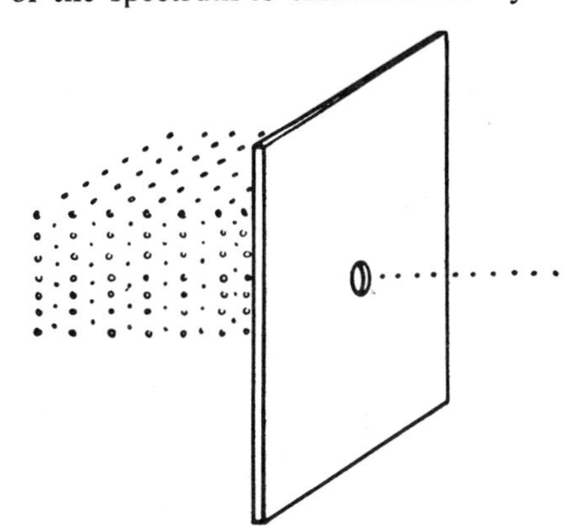

FIG. 1. Corpuscular theory of light (compare with Fig. 7).

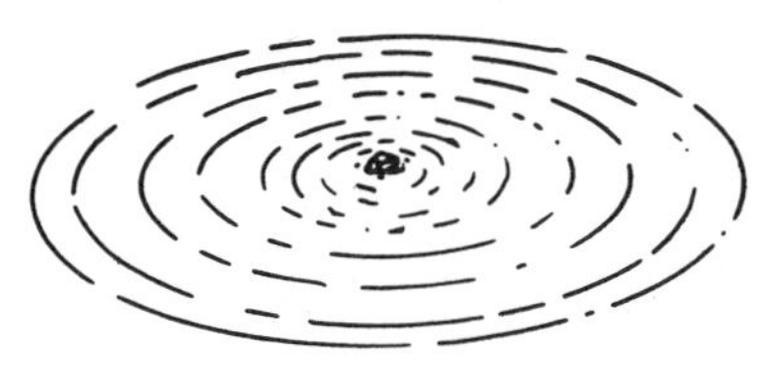

FIG. 2. Water waves radiating from a disturbance.

FIG. 3. Light waves in transverse representation.

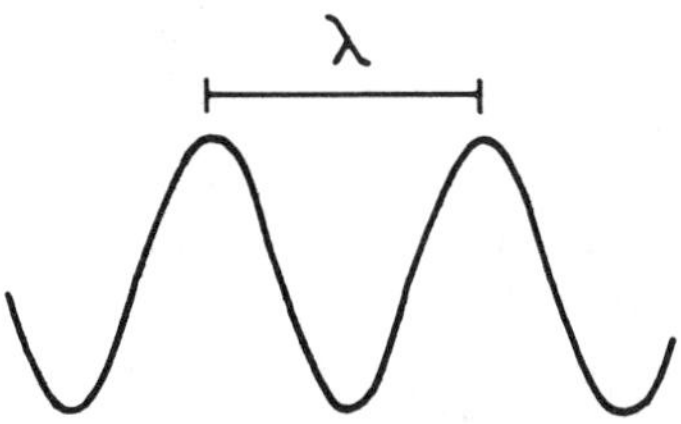

λ

FIG. 4. Wavelength.

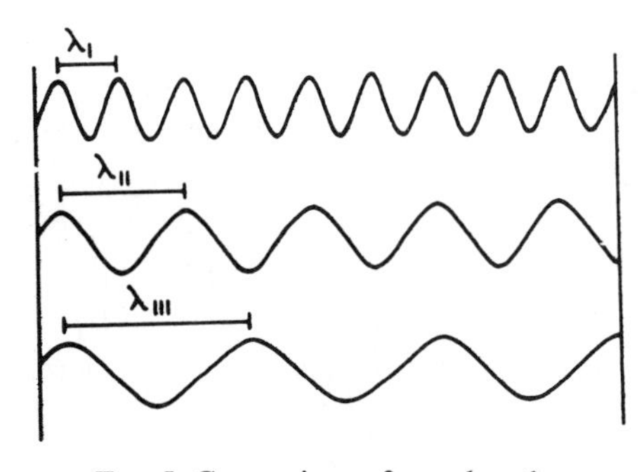

FIG. 5. Comparison of wavelengths.
e.g.: $\lambda_{I} = 2000$ Å (ultraviolet).
$\lambda_{II} = 4000$ Å (violet).
$\lambda_{III} = 6000$ Å (red).

B. RESOLUTION*

A function of a lens system might be to obtain an exact image of an object. However, this goal is unattainable because of a phenomenon called diffraction, which is a consequence of the interaction of wave-like illuminating beams and objects.

1. Diffraction

Consider the example of the wave motion generated by dropping a pebble into a pool of water. If a barrier which has a small opening in its center is placed across the middle of the pool surface, the wave motion observed can be described as follows (Fig. 6). The point where the pebble hits the water would serve as a focal point for the emanating wavelets, all of which would be reflected back from the barrier except at the opening. There a small arc or *wave front* from each of the waves passes through the opening which acts as new focal point for the origin of wavelets. To establish an analogous situation for light, we must then conceive of light in terms involving outward propagation as concentric *spherical* waves from a source. Thus, in the aforementioned experiment, when a beam of light is intersected by a screen which contains a pinhole aperture, the wave fronts from the light source pass through the aperture which serves as a focal point for the origin of new waves. Because of the concentric spherical nature of their mode of propagation, they can be seen off axis beyond the screen, as well as axially (Fig. 7). This phenomenon, known as *diffraction*, is defined as bending or spreading of light into the region behind an obstruction that the waves have passed. [Diffraction-producing obstructions in the path of a beam of light from a point source suitable for demonstration of this phenomenon may be a pinhole aperture, a narrow slit, a fine wire, or a sharp-edged object (see p. 317)].

It is usually operationally cumbersome to illustrate optical phenomena in terms of waves or wave fronts. It has been found to be much more convenient when dealing with optical instruments to indicate the motion of wave fronts by means of rays. Thus the path taken by a series of wave fronts

* *Resolution* can be defined as the minimum separation which produces a detectable change in contrast in an area of an image of two points, between the points, and is expressed in units of length.

The various units have the following values:

1 micron (μ) = 0.001 mm = 10,000 Å.
1 millimicron (1 mμ) = 0.001 μ = 10 Å.
1 Angstrom (1 Å) = 0.1 mμ = 1×10^{-7} mm.

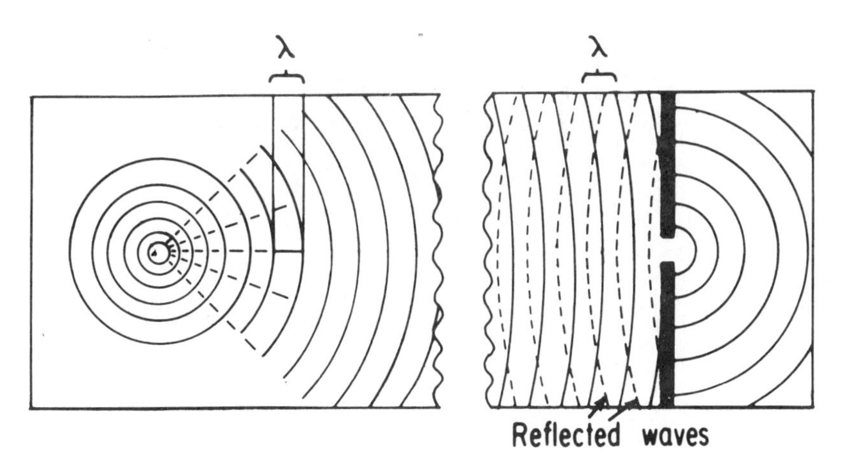

FIG. 6. Diffraction of water waves.

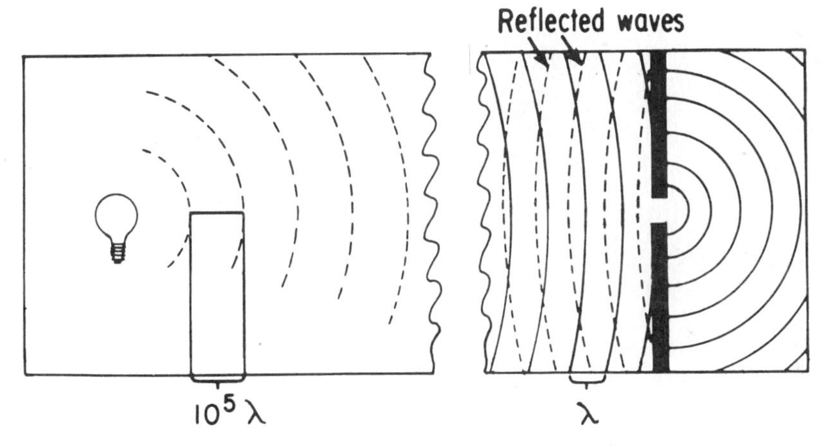

FIG. 7. Wave theory of light. (In this figure the right hand portion represents a 7×10^4 magnification relative to the left side.)

can be expressed as a series of rays each being perpendicular to the wave front (see Fig. 8).

Diffraction is very significant so far as the formation of images by lenses is concerned. Thus, if a strong beam of light illuminates a pinhole in a screen and thus the pinhole serves as a point source and the light passing through is focused by an apertured "perfect" lens on a second screen (Fig. 8), the image obtained is not a pinpoint of light but rather a bright central disk surrounded by a diffuse ring of light. Even if monochromatic light were used to illuminate the point source and were to pass through a "perfect" lens, the image *cannot* be a sharp one but rather a

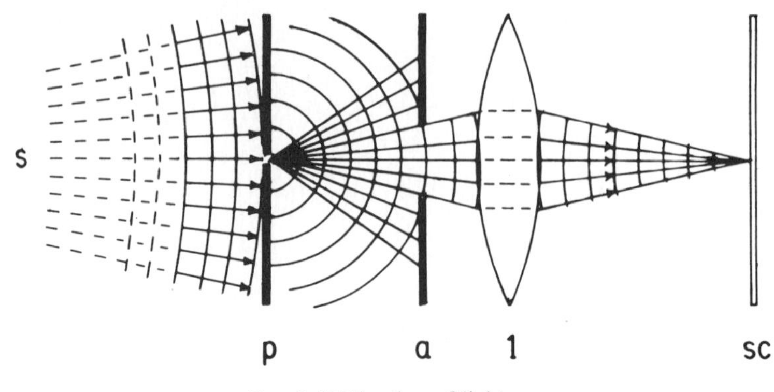

FIG. 8. Diffraction of light waves.

s— light from distant source.
p— pinhole aperture.
a— circular limiting aperture.
l — biconcave lens.
sc— viewing screen.
d — Airy disk.

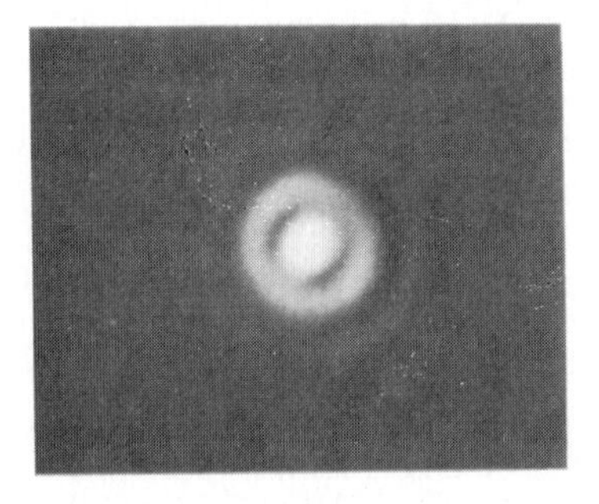

FIG. 9. Airy disk. An actual photograph of the diffraction pattern resulting from light passing through a pinhole in an aluminum film and imaged by a "perfect" lens. The distribution of light in Airy disk is shown in the adjacent graph. Airy found that 84% of the light is concentrated in the central bright disk. The remainder of the light is distributed among the surrounding rings.

diffuse disk composed of concentric rings (Fig. 9). This type of image is known as an "Airy disk," being named after the scientist, Sir George Airy, who first carefully investigated this phenomenon. The reason for this is that diffraction of light occurs whenever the free passage of a wave front is impeded by an obstacle. These diffraction effects occur around the circular edge of the limiting aperture, which produces new waves and thus alters the final distribution of light. Each element of the object under view which removes or impedes the passage of light produces the same kind of effect and the limiting margins of the lens itself produce a related effect. This indicates how diffraction limits our ability to produce a perfect image.

2. Limit of Resolution

The study of the diffraction of light was put on a quantitative basis by the work of Sir George Airy, Lord Rayleigh and Ernst Abbe during the nineteenth century.

The aforementioned experiment demonstrated how a light emitting point is imaged by a "perfect" lens as an Airy disk whose diameter, *d*, is expressed in *Abbe's equation** as:

$$d = \frac{0.612\lambda}{n \sin \alpha} \quad (1)$$

where:

d = the radius of the first dark ring measured at the minimum.
λ = wavelength of image-forming radiation.
n = the index of refraction of the medium between the point source and the lens, relative to free space.
α = half the angle of the cone of light from the specimen plane accepted by the front lens of the objective. This angle is also called the half-aperture angle and it is measured in radians.

This equation, strictly speaking, refers to light-emitting points. It, however, also applies to points in the object plane of a transmission light microscope if the aperture angle of the illumination reaching the specimen is about equal to the aperture angle of the lens** (Fig. 10).

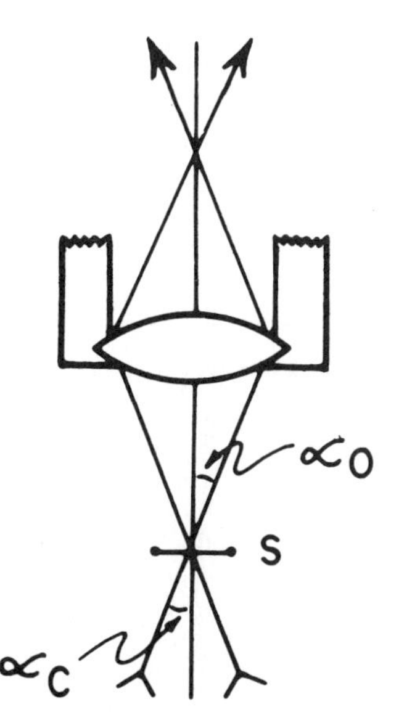

FIG. 10. Light path in a light microscope.
α_c — aperture angle of the illumination which reaches the specimen plane.
s — specimen plane.
α_0 — aperture angle of light which travels from the specimen plane to the objective lens.

* In another form, the above equation can be given as: $d = \frac{0.612\lambda}{N.A.}$ where: N.A. (numerical aperture) = $n \sin \alpha$ and represents the light gathering power of the lens aperture (i.e., a measure of the obliquity of the rays which it can accept and utilize).

** The constant in Abbe's equation varies between 0.61–1.22 as the illuminating aperture ($n \sin \alpha_c$) varies from 1 to 0.

The significance of Abbe's equation, which *mathematically expresses the magnitude of the diffraction effect*, is that it provides a numerical value for the limit of resolution of any aberration free optical system. This can be appreciated when we consider the case of two point sources located a short distance apart in the object plane. Each of these point sources will produce an Airy disk in the image plane. If the object points are sufficiently far apart to produce Airy disks that do not overlap (to any considerable extent), the points are easily resolvable; that is, they are distinctly seen as two separate points (Fig. 11A). If, however, the disks overlap so that the first dark ring of each falls on the central bright disk of the other, the object points are not seen as separate entities (Fig. 11B). Specifically, the two points will just be resolvable if the centers of the disks are separated by a distance slightly greater than d. Since d in Abbe's equation is expressed in terms that are relative to the object plane rather than image plane, it follows that the object points are resolvable if they are separated by a distance greater than d.

From Abbe's equation it is apparent that to attain the maximum resolution with a light microscope, as indicated by a minimum value for d, the optical conditions should be such that λ have a minimal value, while n and $\sin \alpha$ should be maximal. If we establish these three factors then we

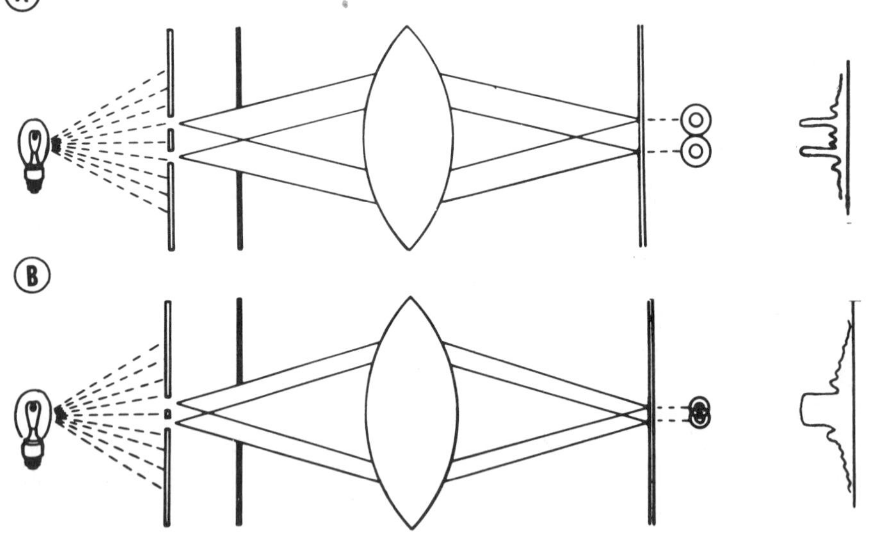

FIG. 11. Resolution of two point sources.
(A) Airy disks do not overlap. Points are completely resolvable. Intensity peaks remain distinct.
(B) Airy disks overlap. Points are not resolvable. Intensity peaks fused.

can obtain the ultimate limit of resolution of the light microscope (as limited by diffraction). This can be done by obtaining the following conditions:

(a) n must be increased as far as possible beyond the value for air which is unity. This can be done by the use of an immersion oil and front lens which have high values (*circa* 1.5).

(b) α must be as close as possible to 90° (a limiting angle which can be reached only when the specimen is placed "within" the lens).

(c) λ must be near 4000 Å (violet light) which is the shortest wavelength to which the average eye is sensitive.

For example, with $n = 1.5$ and $\sin \alpha = 0.87$ (that is, typical for an oil immersion lens of N.A. = 1.3) and $\lambda = 4000$ Å, the theoretical resolution attainable is about 0.2 μ*. In other words, two points in the specimen which are not separated by at least this distance will not be seen as two distinct points but will be observed as a single blurred image (Fig. 12).

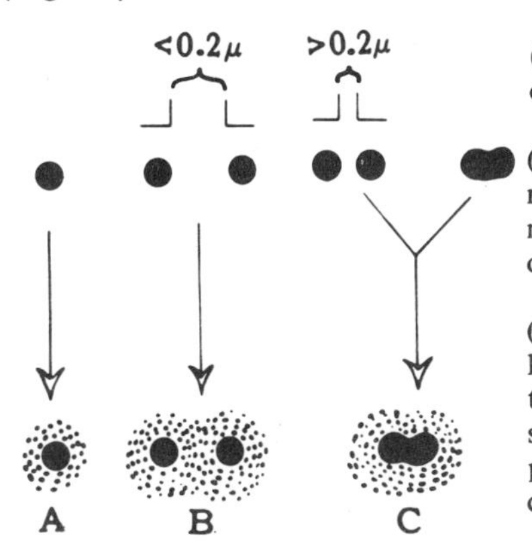

FIG. 12. Meaning of resolution.
(A) If a particle is examined with a light microscope under high magnification, its periphery appears unsharp.
(B) If two adjacent particles which are separated by about 0.2 μ are observed under the microscope their images will infringe upon each other but nevertheless their individual character can still be seen.
(C) If two adjacent particles are separated by less than 0.2 μ then the microscope images of the particles fuse and an oblong structure is seen. The same figure would be evident if the particle in reality was oblong. Under these conditions such adjacent particles are not microscopically resolvable.

Since the value of $\sin \alpha$ and n cannot be significantly increased beyond the stated values, the only factor that can be readily altered to attain a decrease in d and thereby improve resolving power is to decrease the value of λ. This has been done by the use of ultraviolet light which lies beyond the lower end of the visible spectrum. With this form of radiation, used with a quartz lens system, the resolution that can be attained is about 0.1 μ (1000 Å).

---

* More exactly the limit of resolution of the light microscope when we substitute the values discussed above in Abbe's equation we get:

$$d = \frac{(0.61)(4000)}{(1.5)(0.87)} = 1870\text{Å or } 0.187\,\mu$$

It is apparent, therefore, that a source of radiation with even a smaller wavelength must be employed in order to attain substantially higher resolution. Such a suitable source is a beam of *electrons* whose properties, as will be discussed in the next section, have the prerequisite needed to attain this goal.

C. NATURE OF ELECTRON BEAMS

The fact that moving electrons might be used as a kind of "illumination" was implicit in the theory, advanced by de Broglie, that moving particles have wave-like properties. By combining some of the principles of classical physics with the quantum theory (see Appendix B), he concluded that a wavelength can be assigned to moving particles which can be calculated from the equation:

$$\lambda = \frac{h}{mv} \tag{2}$$

where:

λ = wavelength of particles.
h = Planck's constant.
m = mass of the particle.
v = velocity of the particle.

which, when the known values* are substituted, becomes:

$$\lambda = \frac{12.3}{\sqrt{V}}\ \text{Å} \tag{3}$$

---

* When an electron with the charge e passes through a potential difference V its kinetic energy is:

$$\tfrac{1}{2}mv^2 = eV$$

(when V approaches the velocity of light (c), m must be replaced by $\dfrac{m}{\sqrt{1-\left(\dfrac{v}{c}\right)^2}}$

from which it follows that:

$$v = \sqrt{\frac{2eV}{m}} \text{ and } \lambda = \sqrt{\frac{h^2}{2meV}}.$$

Substituting the known values in this equation:

$$h = 6.6 \times 10^{-27};\quad m = 9.1 \times 10^{-28};\quad e = 4.8 \times 10^{-10},\ e.s.u.$$

and expressing V in volts, we obtain:

$$\lambda = \sqrt{\frac{150}{V}} \cdot 10^{-8} cm = \frac{12.3}{\sqrt{V}}\ \text{Å}$$

This formula tells us that the wavelength of a beam of electrons is dependent upon the potential, V, through which it has been accelerated. Thus, for example, if the latter were 60,000 V (60 KV), the wavelength of the electron beam would be about 0.05 Å, which is 1/100,000 as long as the wavelength of ordinary green light.

In 1927, (G. P.) Thomson and Reid in England, and Davison and Germer in the United States, experimentally measured the wavelength of an electron beam and found it to be in agreement with that which can be calculated from de Broglie's formula. Thus it became apparent that high velocity electrons, as a result of their small wavelength, might provide a suitable form of radiation to fulfill the requirements for high resolution microscopy.

Abbe's equation (1; see p. 11) can therefore be rewritten, after substituting for λ with equation (3), as:

$$d = \frac{(0.61)(12.3)}{n \sin \alpha \sqrt{V}} \tag{4}$$

Since electron microscope aperture angles are always very small (see p. 58) $\sin \alpha \simeq \alpha$, and since both the object and image are usually in field-free space in an electron microscope, the refractive index $n = 1$. Thus (4) can be written as:

$$d \approx \frac{7.5}{\alpha \sqrt{V}} \text{Å} \tag{5}$$

Thus, in the electron microscope, the level of resolution is basically determined by the accelerating potential and angular aperture of the objective lens.

Substituting specific realistic values in (5) (where $\alpha = 10^{-2}$ radians, $V = 10^5$ volts), we get:

$$d \approx 2.4 \text{Å}$$

D. ELECTRON EMISSION

Electrons are negatively charged particles which, from a simplified point of view, can be considered to be located in planetary orbits around atomic nuclei (Fig. 13). They are prevented from escaping from these orbits by the strong attractive force created by the positively charged nucleus. This force is least for a valence electron (which is located in an

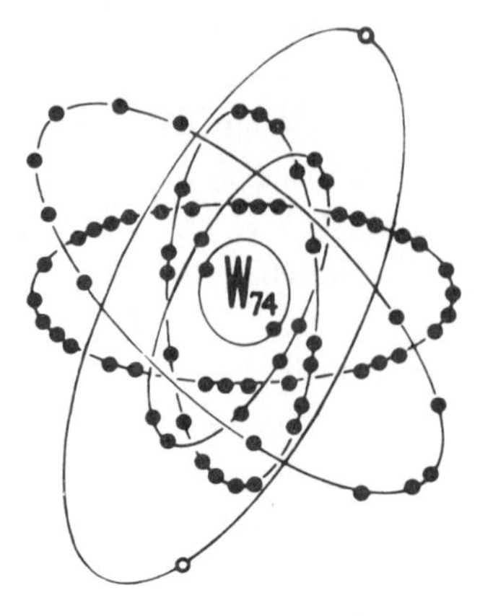

FIG. 13. Electron configuration of a tungsten atom.

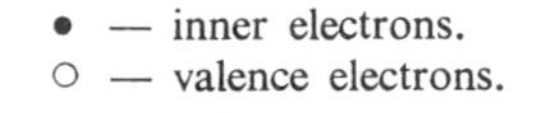
• — inner electrons.
○ — valence electrons.

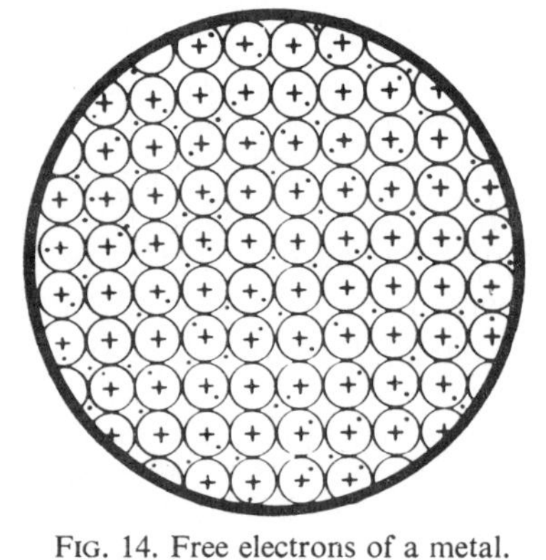

FIG. 14. Free electrons of a metal.
+ — ion with positive atomic core.
• — valence electrons some of which are potentially free.

outer orbit), because it is screened by the neutralizing charges of the inner electrons. As a result the valence electrons can be relatively easily detached.

In a metal, the valence electrons can be detached from one atom by the attractive force of the positive charge of its neighboring atomic cores or ions so easily that such electrons are designated as *free electrons* (Fig. 14) (since they do not belong to any particular atom). Each of these free electrons "collides" frequently with other electrons and ions producing continual changes in their velocity and direction of motion. Thus some such electrons near the surface of the metal may escape, leaving a positive net charge on the surface at the point of electron emission. Energy must be expended by the free electron to overcome the positive attractive force which tends to pull it back to the metal. The potential which must be overcome for an electron to escape the surface is called *work function**. Thus work function is a measure of the work required to overcome attractive forces within the solid, plus that which is required to overcome the positive image charge on the surface of the metal after the electron escapes. Each substance has its characteristic work function (see p. 40). A metal which has a low work function is capable of emitting a more copious supply of electrons at a given temperature than one with a high work function.

---

* The theory of the movements of charged particles in electrostatic (and magnetic fields) is known as *electron ballistics*. The movements of such particles depend upon the charges and masses of the particles, the strength of the fields and the laws of motion.

Since electric potential, E, is defined by the work, W, done in moving a unit charge, it follows that for a charge Q

$$E = \frac{W}{Q} \qquad (1)$$

or

$$W = EQ \qquad (2)$$

W is also referred to as the *potential energy* of the charge in the field before it moves. The kinetic energy is equal to $\frac{1}{2}mv^2$, where v is the velocity of the particle *after* the work, W, has been performed, and kinetic energy gained is equal to the potential energy lost, therefore:

$$\tfrac{1}{2}mv^2 = EQ \qquad (3)$$

or

$$v = \sqrt{2\frac{Q}{m}E} \qquad (4)$$

where:

v = speed

m = mass

(This equation shows that both the speed and the kinetic energy acquired by a charged particle moving in an electric field is determined solely by the total potential, E, through which the particle has moved).

The work function are defined as the potential, E, which an electron of charge e must overcome to escape from a solid. We may substitute e for Q, and equation (4) can be written as:

$$v = \sqrt{2\frac{e}{m}E} \qquad (5)$$

This equation, therefore, also permits us to calculate the minimum initial velocity of emission that an electron must have in order to overcome the potential of, for example, a tungsten filament by substituting the known values as follows:

$$e = 1.6 \times 10^{-19} \text{ coulomb}$$

$$m = 9.1 \times 10^{-31} \text{ kilogram}$$

thus

$$\frac{e}{m} = 1.76 \times 10^{11} \text{ coulomb per kilogram}$$

E = work function (4.52 volts for tungsten; see p. 40), therefore:

$$v = \sqrt{(2)(1.76 \times 10^{11})(4.52)}$$

$$v = 1.26 \times 10^{6} \text{ meters per second}$$

$$v = 783 \text{ miles per second}$$

To provide the necessary energy for substantial numbers of free electrons to overcome the positive surface charge, the emitting material is heated, for example, by passing an electron current through it. This results in increased numbers of electron collisions which, in turn, increase their velocity and energy and enables more of them to escape the surface. This mechanism of creating an electron source is called *thermionic emission* and provides the basis for the design of the usual electron gun.

An equation representing electron emission:

$$I = AT^2\varepsilon^{-(We/kT)} \qquad (6)$$

where:

I = emission current density (amps per cm^2 of emitting surface).
A = constant (60.2 amps per cm^2 deg^2 for tungsten).
T = absolute temperature (degrees C + 273).
ε = natural base of logarithms.
W = work function (volts) (4.52 volts for tungsten).
e = charge of the electron (1.6×10^{-19} coulombs).
k = Boltzmann's constant (1.39×10^{-9} joules per degree absolute).

The experimental curve for tungsten represented by equation (6) is shown in Figure 15.

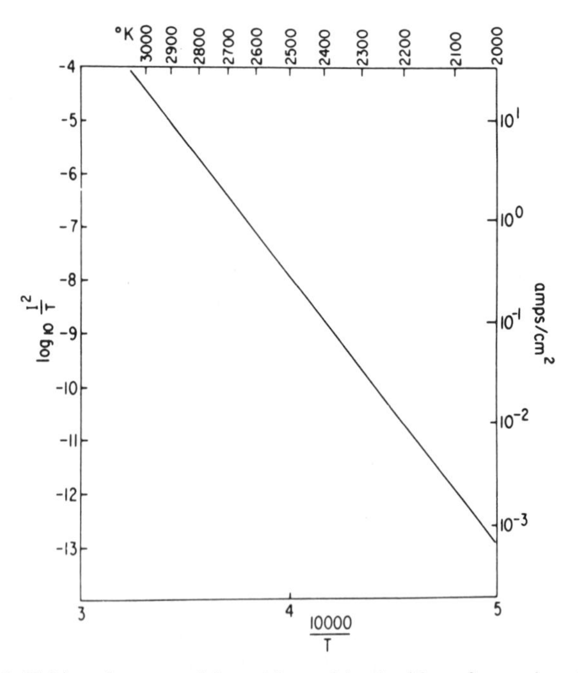

FIG. 15. Taking the natural logarithm of both sides of equation (6) we get log I = 2 (log AT) – We/KT. So long as 2 log AT is small compared to We/KT, log I will be proportional to 1/T as indicated in the above plot.

E. ELECTRON OPTICS

1. Introduction

In the light microscope, glass lenses are used to obtain a magnified image of a specimen. The action of a lens is depicted in Figure 16. When parallel rays of light, such as those coming from a distant point source, strike or leave the surface of the lens at an angle other than 90°, their direction is changed. This phenomenon, which is known as *refraction,* is due to the change in the velocity of the light waves as they pass the boundary between media of different densities. With a properly shaped lens, the rays representing the direction of propagation of these waves are brought to a focus at a single point. If a screen were interposed at the focal point at right angles to the optical axis of the lens, the point would appear as a small brilliant spot, which represents an *image* of the source.

In the electron microscope, where magnetic lenses (see below) are usually used, the situation is only partially analogous to that described above. Such lenses do produce a deviation in the trajectory of the electrons from a point source which causes them to converge at a single focal point. This effect, however, does not involve sharp changes in the velocity of the electrons along the optical axis.

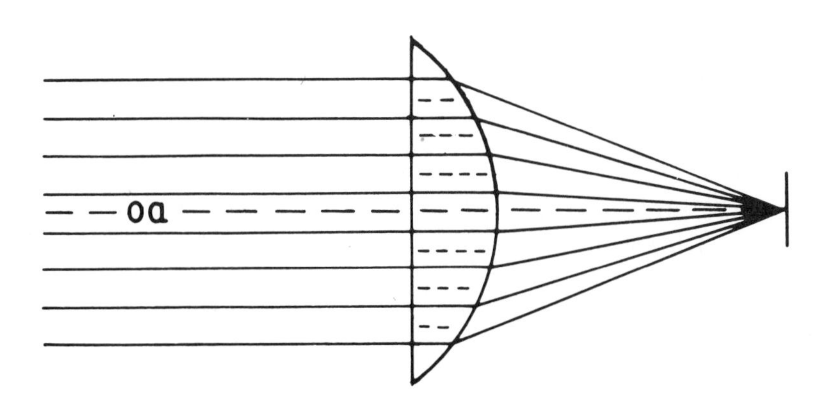

FIG. 16. Focusing of light rays by refraction.
The focusing action of the lens is partly explicable by considering it to be made up of a large number of sections of prisms each of which bends, that is, refracts the rays of light passing through so that all cross the optical axis (*oa*) at the common focal point. Parallel rays of light are brought to the common focal point and are visible on a screen as a small brilliant spot.

2. Magnetic Field

It is common knowledge that magnets have the ability to attract or repel some kinds of bodies that are located some distance from them. The space around a magnet contains the *magnetic field* which can be represented graphically by *lines of force* connecting north to south poles (Fig. 17).

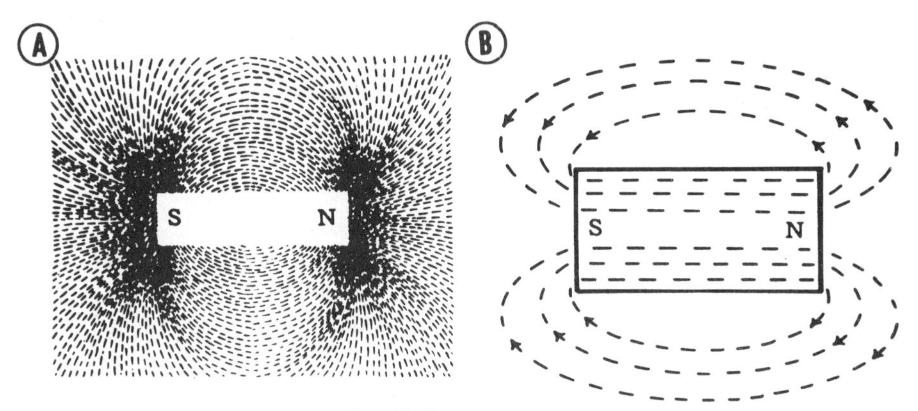

Fig. 17. Bar magnet.
(A) Pattern of iron filings around bar magnet.
(B) Course of the lines of force of the magnetic field.

The collection of lines of force (flux) representing a magnetic field are known collectively as *magnetic field intensity*. The *magnetic flux density* is a measure of the number of lines of flux per unit area.

In 1819, Oersted noted that a compass needle will be deflected when brought near a conductor carrying an electric current (Fig. 18). This was the first direct evidence of what had been long suspected, namely, a link between electricity and magnetism. Oersted determined that the compass deflection was due to a magnetic field established around the conductor by the current in the conductor.

Shortly after Oersted's discovery, Ampere determined the shape of the magnetic field about a conductor carrying a current. This can be demonstrated by passing a copper wire vertically through the center of a stiff cardboard that is in a horizontal position. When the ends of the vertical conductor are connected to a dry cell and iron filings are sprinkled over the surface of the cardboard, the iron filings form a pattern of concentric circles around the conductor (Fig. 19). If a small compass is placed at various points on a circle of filings, the needle always comes to rest tangent to the circle. If the direction of current in the vertical conductor is reversed,

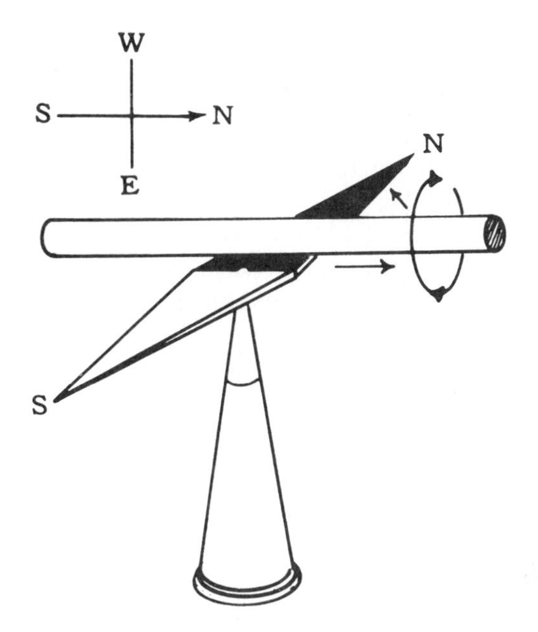

FIG. 18. Oersted's experiment. A current in a wire located above a compass needle (a movable magnet) causes deflection of the needle.

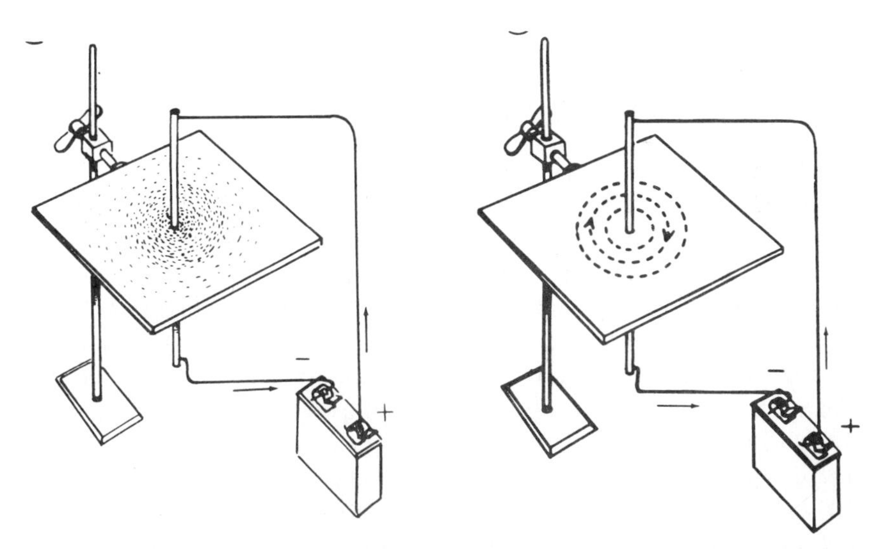

FIG. 19. Magnetic field around a conducting wire. When the circuit is closed and the cardboard stage tapped gently, the iron filings become arranged in a pattern which shows that a magnetic field surrounds the straight conductor (wire).

FIG. 20. Magnetic field around a conducting wire. The lines of force are in concentric circles with the wire in the center; the plane of the circles is perpendicular to the wire.

the compass needle again becomes aligned tangent to the circle of filings, but the N pole points in the opposite direction. Thus it can be concluded that *a magnetic field encircles the path of an electric charge in motion*. The lines of flux (force) are closed concentric circles lying in a plane perpendicular to the conductor, with the axis of the conductor as their center (Fig. 20).

If the wire carrying a current is bent into a loop, then the lines of flux from all segments of the loop must pass through the inside of the loop in the same direction and these circles are arranged concentrically around the conductor (Fig. 21A). Such a loop can be considered a magnet, since the lines of flux are arranged symmetrically around an axis passing through the center of the loop (Fig. 21B), much as in the case of a very short bar magnet.

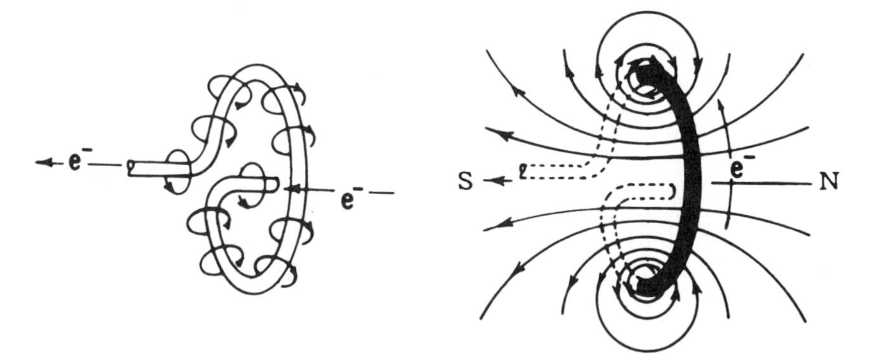

FIG. 21. The magnetic field induced by current passing through a conductor shaped as a circular loop.

There are two ways to increase the pole strength of such a magnet. One is to increase the magnitude of the current in the conducting loop. The other is to form additional loops in the conductor similar to the original one and aligned with it. A linearly arranged group of conducting loops would assume the form of a helix, and a long helically wound coil is known as a *solenoid* (Fig. 22). The cylindrical volume inside the loops and extending the length of the coil is called the *core*. When a current is passing through the solenoid, the core of each loop, or turn, becomes a magnet, and the core of the solenoid is a magnetic tube through which practically all the magnetic flux passes. The solenoid itself thus exhibits a field like that of a long bar magnet but since the field is established by the flow of electric current it is known as an *electromagnet*.

A solenoid having a core of air, wood or some other non-magnetic material does not produce a very strong electromagnet, for the permeability (μ) of all non-magnetic substances is essentially equal to that of air, i.e.,

$\mu = 1$. Substitution of such materials for air does not appreciably change the flux density. Soft iron, on the other hand, has a high permeability, and if an iron bar is placed near or in the core of the solenoid (Fig. 23), the potential magnetic field strength, i.e., flux density, is greatly increased.* Strong electromagnets thus usually have a ferromagnetic core with a high permeability. The strength of the electromagnet depends on the magnitude of the current in amperes and the number of turns in the coil.**

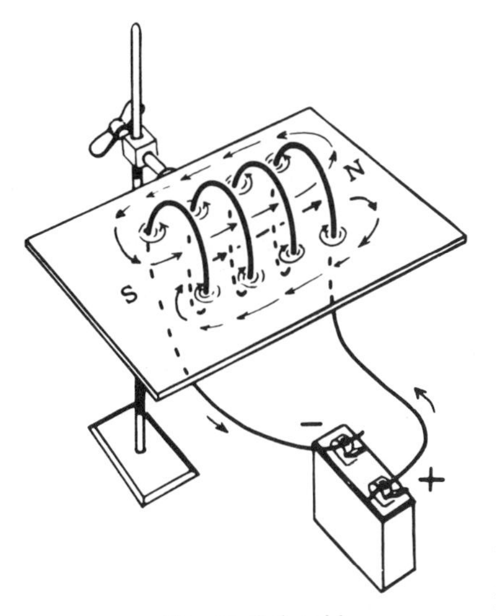

FIG. 22. Solenoid.

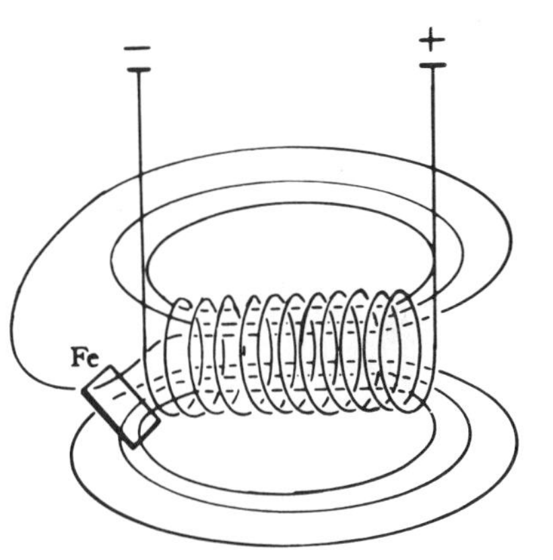

FIG. 23. Solenoid and soft-iron bar. The presence of a piece of soft iron (*Fe*) increases the strength of the magnetic field by attracting the adjacent lines of force.

---

* As a ferromagnetic material is subjected to a magnetizing force, the flux density increases until the material is saturated. If the magnetizing force is then reduced to zero, the magnetization does not return to zero but lags behind the magnetizing force. This phenomenon is known as *hysteresis*. Another advantage of soft iron is the fact that when used in an electromagnet its hysteresis is low.

** The flux density in the center of a solenoid whose length is large in comparison to its diameter, is given by the expression

$$B = \mu\left(\frac{NI}{l}\right) \tag{1}$$

where:

B = flux density.
μ = permeability.
N = number of turns.
I = strength of current.
l = length of solenoid.

since $NI/l = H$, the magnetic field intensity, i.e., the strength which the magnetic field generates, equation (1) can be written as:

$$B = \mu H \tag{2}$$

Electromagnets are the basic units used in the magnetic lenses of most electron microscopes. They are more versatile than permanent magnets because their magnetic fields can be easily varied by adjusting the magnitude of the current passing through the solenoid.

3. Action of Magnetic Fields as Lenses

Oersted, as we noted in the previous section, discovered that a current flowing through a wire established a magnetic field. He demonstrated that a fixed conductor bearing a current deflects a movable magnet (see Fig. 18). The converse of the above discovery was also found to be true. Namely, it has been shown that a fixed magnet deflects a movable conductor bearing a current (Fig. 24). It was this observation that implied that a stream of electrons (which is equivalent to the flow of electricity in a movable conductor) moving through a magnetic field would be deviated from its original trajectory in a direction at right angles to both the direction of the field and the direction of motion of the electrons (Fig. 25).

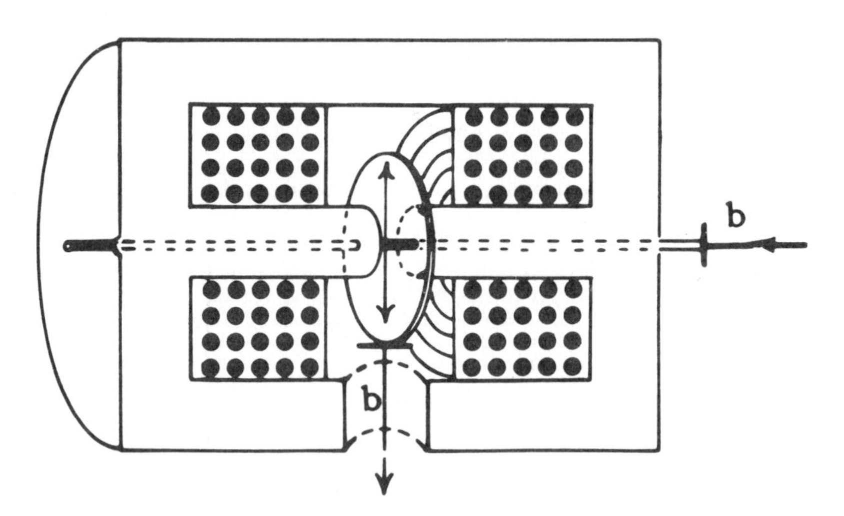

FIG. 24. An example of Faraday's dynamo (in longitudinal section). A wheel which makes contact with two brushes (*b*) is located between the poles of an electromagnet (the stationary magnet). If current is passed through the coils and the wheel is rotated, current will flow through a circuit connecting the brushes. If a current is fed through the coils and the brushes, the wheel rotates. This then is a Faraday motor. The motion of the conductor (the wheel) is perpendicular to both the direction of the current and the direction of the magnetic lines of force.

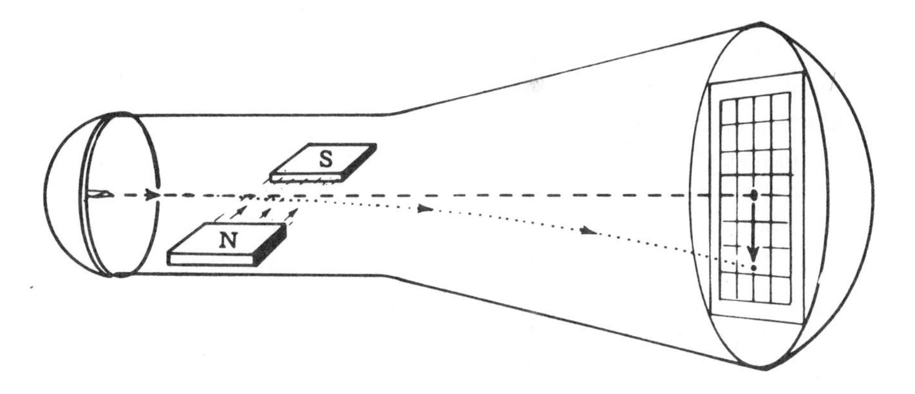

FIG. 25. Thompson's experiment. A stream of electrons originating from a source and passing, *in vacuo*, through a magnetic field produced by a pair of magnets will be deflected. The direction of deflection demonstrates that electrons are negatively charged particles of matter.

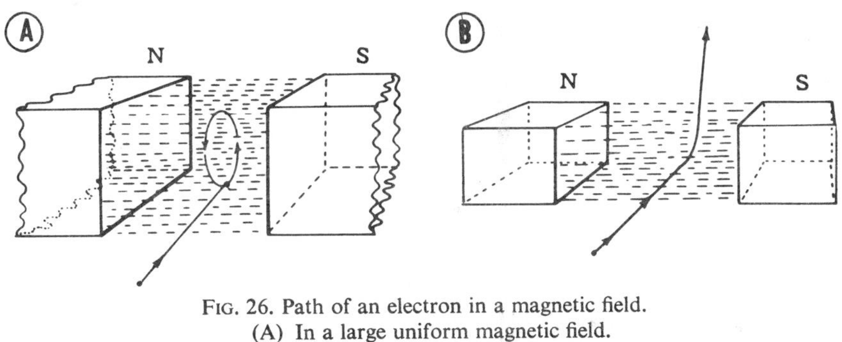

FIG. 26. Path of an electron in a magnetic field.
(A) In a large uniform magnetic field.
(B) In a small uniform magnetic field.

Having noted that an electron stream can be affected by a magnetic field, we can proceed to examine the details of what takes place during such an interaction. In the case of a *large uniform* field at *right* angles to the direction of the stream of moving electrons (Fig. 26A), the force of the magnetic field would cause each electron to inscribe a helical path of a definite radius through the field. However, if the electron is moving *parallel* to the field, no force is exerted upon it and it is not deviated from its course. On the other hand, if the field occupies only a small volume (Fig. 26B), an electron entering *perpendicular* to the field may escape from the field before it completes a full circle. In the field it will follow a path which is an arc of a circle. Its course upon leaving the field is along a straight line tangent to the circle at the "point" of departure from the field. It can be shown that the speed *along the path* of the electron in this deviated course remains the same as before entering and after leaving.

In the electron microscope, the situation is much more complex than described in the simple cases just given. For in actuality, from a single object point, a whole cone of electrons moves at many different angles in or into the magnetic field (lens). The directions of motion of each electron can, however, be understood in terms of its vector components (see Appendix C). Thus, we can graphically portray an electron moving through a uniform magnetic field, with the vector V, which represents its speed and direction (Fig. 27A). If the vector makes an acute angle with the field, the trajectory which the electron will follow can be determined by dividing the vector into its components V_1 and V_{11}. That is, we examine the motion as if the electron moves first with a velocity V_1 at right angles to the field and then with a velocity V_{11}, parallel to the field. The path of the electron when V_1 and V_{11} act simultaneously can thus be computed.

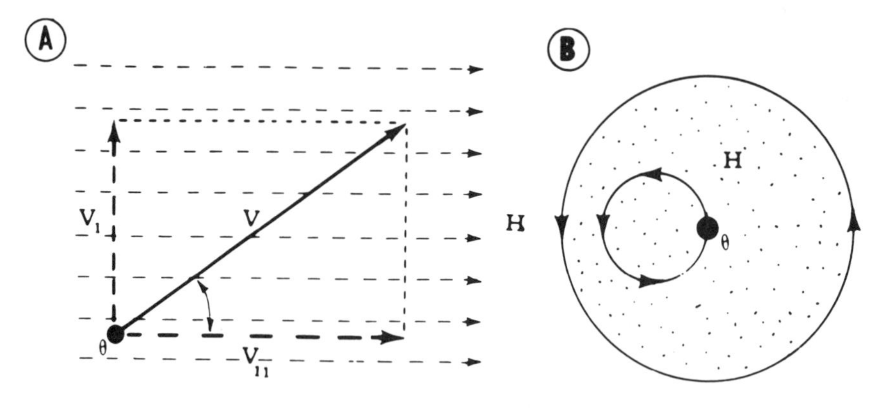

FIG. 27. Electron vectors in a magnetic field.
(A) An electron at point θ is moving with a velocity, V, through a uniform magnetic field, H, at an angle to that field. To determine the trajectory of the electron, the vector quantity V is resolved into its components V_1 and V_{11}. (B) A cross-section or end-on view to demonstrate the circular path of the vector component V_1 which is at right angles to the field.

As discussed above, the electron following the course of V_1 at right angles to the field will inscribe a complete circle returning to its point of origin (Fig. 27B). On the other hand, when the electron follows the course of V_{11}, parallel to the field, it is not deviated and continues along a straight line. The path that such an electron will take when it has both velocities simultaneously is, therefore, a helix; that is, a circle drawn out by movement along an axis parallel to its plane (Fig. 28A).

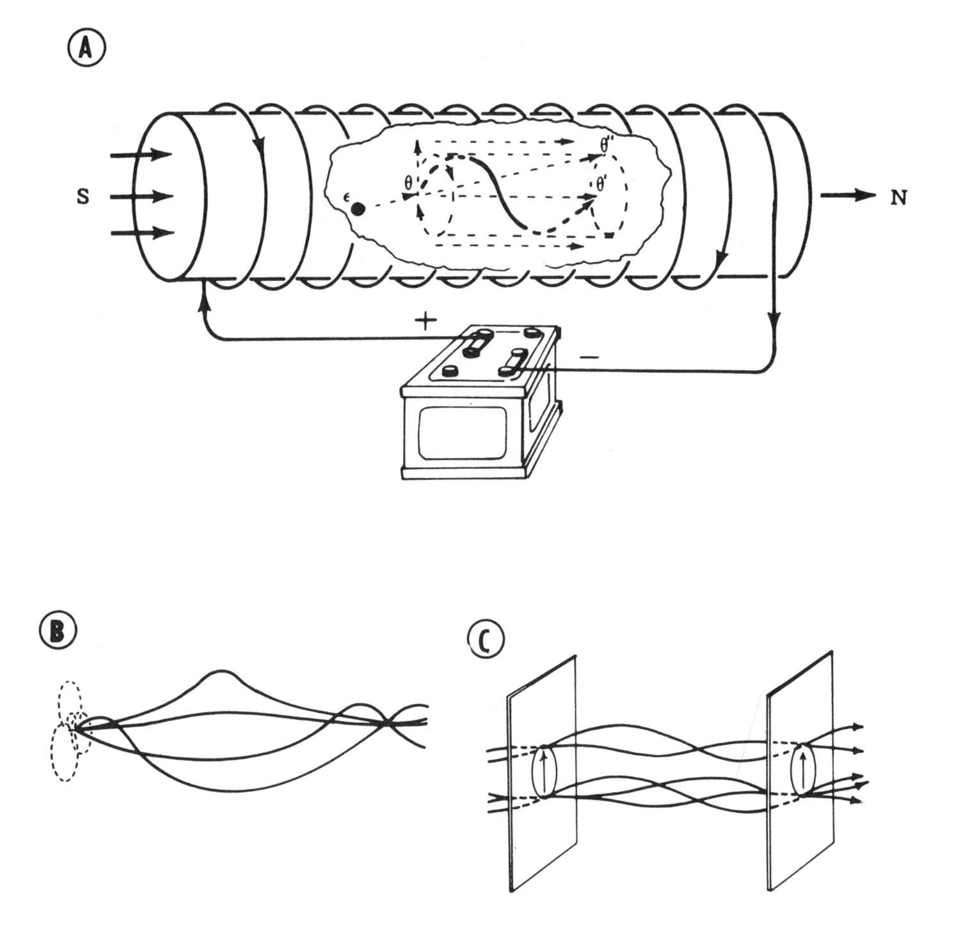

FIG. 28. Focusing action of a uniform magnetic field.

(A) An electron ε enters the uniform portion of the magnetic field of a long solenoid (the lens) at an oblique angle. When it reaches the point θ which lies on the axis of the coil, its trajectory will be determined by its vector components (V_1 and V_{11}). These simultaneously active forces produce a circular forward moving motion (i.e., a helical path which is tangent to the axis) which is completed at θ'. If the magnetic field were not initiated when the electron arrived at θ, then its trajectory would continue along a straight line to point θ''. Thus we see that a magnetic lens, like its light optical counterpart, functions by forcing an electron which is diverging from the axis, back onto the axis at the point further along.

(B) If a number of electrons have the same axial velocity, the pitch of their helices will be the same, and all electrons passing through a given point will, after one revolution, come together at a second point. The latter is located at a distance equal to the pitch from the first point in the direction of the magnetic field.

(C) By means described in the preceding magnetic field illustrations, a weak magnetic field can produce an erect image of an object placed in its path which can be rendered visible on a fluorescent screen.

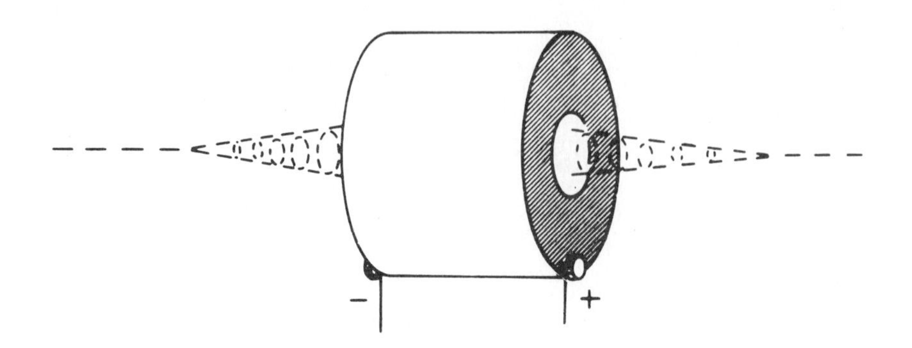

FIG. 29. Magnetic lens action. The focusing effect of a magnetic lens is schematically illustrated.

By these means it becomes apparent how an electron which is diverging from the axis of a uniform magnetic field can be forced back onto the axis some distance away. When a number of electrons travelling with the same velocity originate from the same point, but are directed along different trajectories at small angles to a uniform magnetic field, they will follow helical courses which make them all eventually converge on another point further along the axis of the magnetic field (Fig. 28B, C). Thus the magnetic field acts as a lens (Fig. 29).

4. Magnetic Focusing

If the distance, q, between a screen and a glass lens is varied (the source-to-lens distance remaining fixed), there is only one distance, i.e., value for q, which provides an in-focus image on the screen. This distance, i, is known as the *image distance*. At all other distances, the image produced is out-of-focus and is, therefore, larger and diffuse (Fig. 30A, B). This is due to the fact that the rays come to their focal point either before or (in principle) behind the screen. In practice, the in-focus position with a light microscope is found by changing the distance between the objective lens and the specimen.

The action of a glass lens has already been described (see p. 19). When the source is at infinity, the rays that strike the lens are parallel, and an in-focus image will be obtained when the image distance, i, is *equal* to the focal length, f, (Fig. 30C). This operationally defines the *focal length* which is a *fixed* parameter of a glass lens from which its imaging properties can be computed.

To arrive at a situation which is more analogous to the mechanism of focusing with magnetic lenses, we must consider a system where the focal

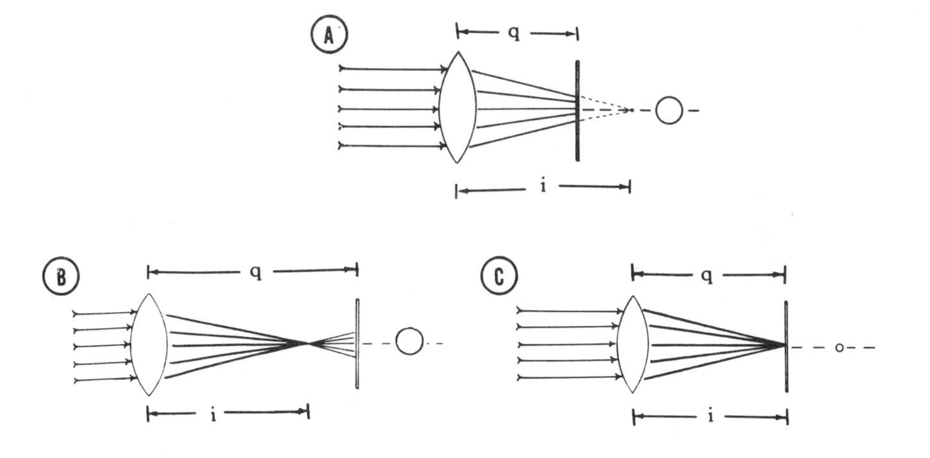

FIG. 30. Focusing an image with a glass lens where the focal length is fixed.
(A) Where screen distance q is smaller than the distance i, an out-of-focus, i.e., large and diffuse image is produced on the screen.
(B) Where the screen distance q is larger than the image distance i, an out-of-focus image is produced.
(C) Where the screen distance q equals the image distance i, an in-focus, i.e., small, brilliant image is produced on the screen.

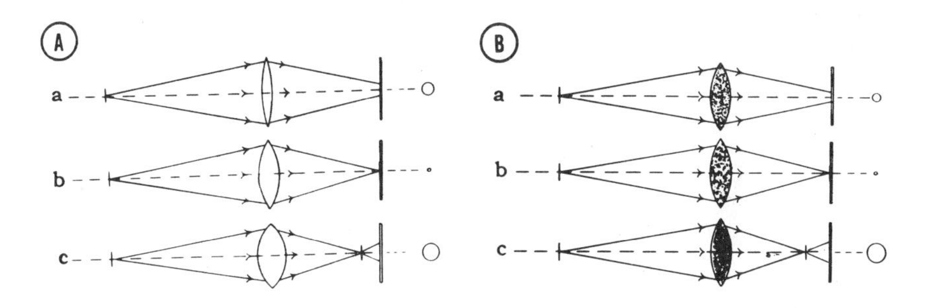

FIG. 31. (A) Focusing an image with glass lens where the screen distance is fixed.—I.
(a) Where a thin lens has a focal length which is too long it produces an out-of-focus image.
(b) Where a lens of medium thickness has the appropriate focal length it forms an in-focus image.
(c) Where a thick lens has too short a focal length it forms an out-of-focus image.
(B) Focusing a glass lens where the screen distance is fixed.—II.
(a) Where the index of refraction of a glass lens produces a focal length which is too long, the image is out-of-focus.
(b) Where the index of refraction of a glass lens produces the correct focal length, the image is in-focus.
(c) Where the index of refraction of a glass lens produces a focal length which is too short, the image is out-of-focus.

length is *not fixed*, as in the above case, but can be varied. The focal length of a lens depends upon the curvature of its surfaces and on its index of refraction. Thus by altering either (or both) of these factors we can obtain different focal lengths. If a source is imaged by a series of lenses of varying surface curvatures (Fig. 31A) or indices of refraction (Fig. 31B) and if the object distance and screen distance are constant, only the lens with the appropriate focal length will produce an in-focus image on the screen while the others will be out-of-focus.

We can now consider focusing with the electron microscope where the positions of the magnetic lenses are usually fixed. (This requires specified distances between the respective lenses for each lens combination.)

The focal length of a magnetic lens is defined as the distance behind the lens at which the image of a point electron source located at infinity would be focused. If a screen were to be placed in the plane of this image and the focal length increased or decreased, an out-of-focus image would result. In a like manner, for a source located at a finite distance and a screen in the appropriate image plane, as in light optics (Fig. 31 A, B), a decrease or increase in focal length will also produce an out-of-focus image (Fig. 32). For magnetic lenses, focusing is usually achieved by varying the current which passes through the electromagnet. This, in turn, changes the

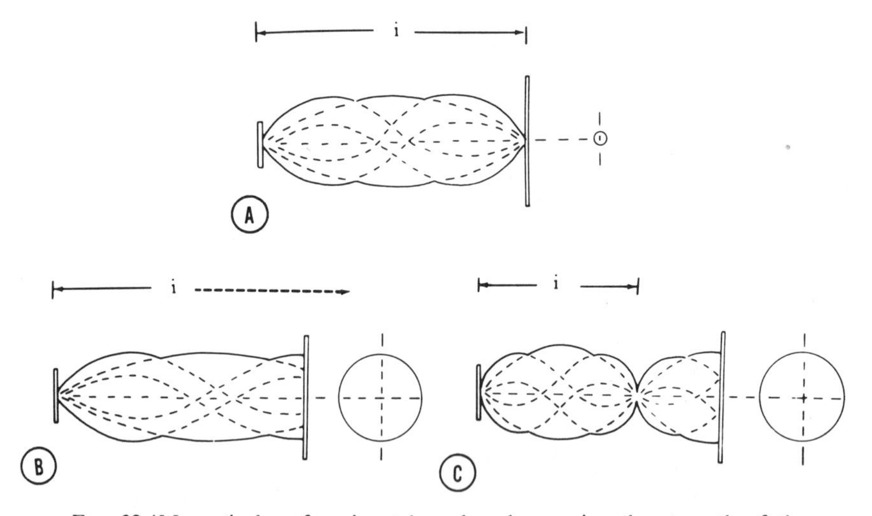

FIG. 32. Magnetic lens focusing takes place by varying the strength of the current flowing through the lens coil which varies the focal length and, therefore, image distance.
(A) Appropriate image distance which produces an in-focus image.
(B) Image distance is too long; image is out-of-focus.
(C) Image distance is too short; image is out-of-focus.

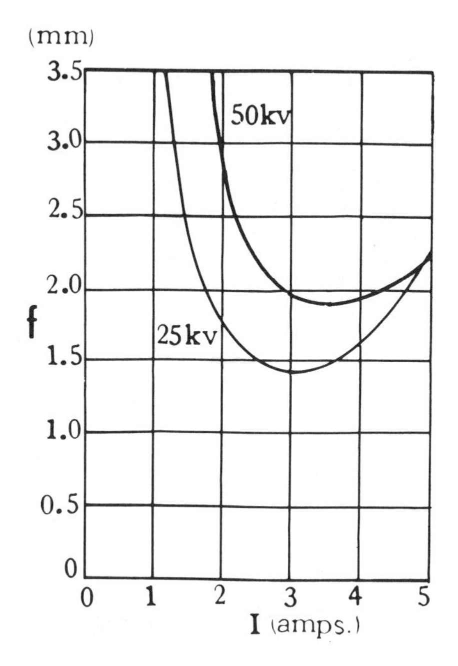

FIG. 33. Variation of focal length, f, with coil current, I, for two operating voltages for a 1000 turn lens coil. (Note: these curves are based on measurements obtained with a magnetic lens.)

strength of the magnetic field and thereby alters the focal length of the lens and is equivalent to a combined change in both "refractive index" and "curvature of surface" (Fig. 33).

Permanent magnets may also be used as lenses in electron microscopes. In such a case, since coil currents are not involved, another mechanism for changing the focus must be provided. For example, the focal length of permanent magnetic lenses can be altered by changing the gap length between the pole pieces (see p. 51) or the specimen position can be changed as in the light microscope.

5. Evolution of Magnetic Lenses

The first demonstration of magnetic focusing took place in 1899. However, an understanding of the magnetic focusing of an electron beam came about as a result of the work of Busch. In 1926–27, he published a series of classical papers wherein he proved both mathematically and experimentally that a short solenoid had the properties of a *lens*; namely, that the well-known formulae of light optics were applicable to such units. This work laid the foundation for the field of electron optics.

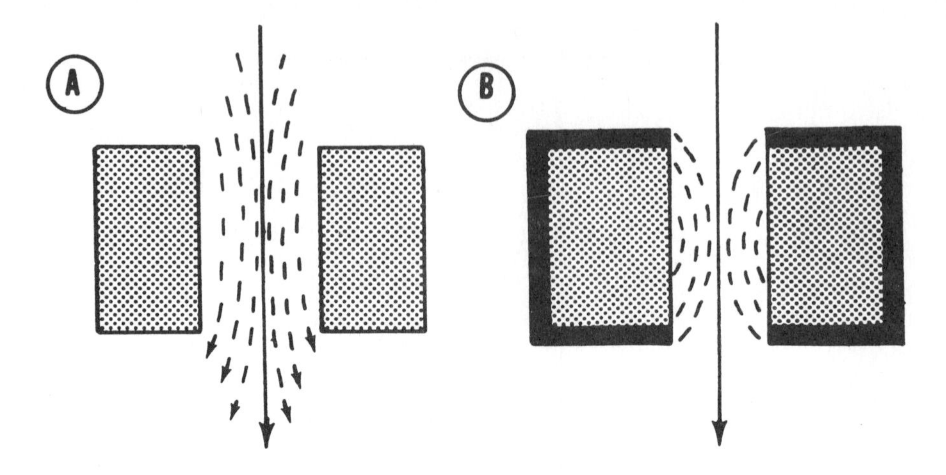

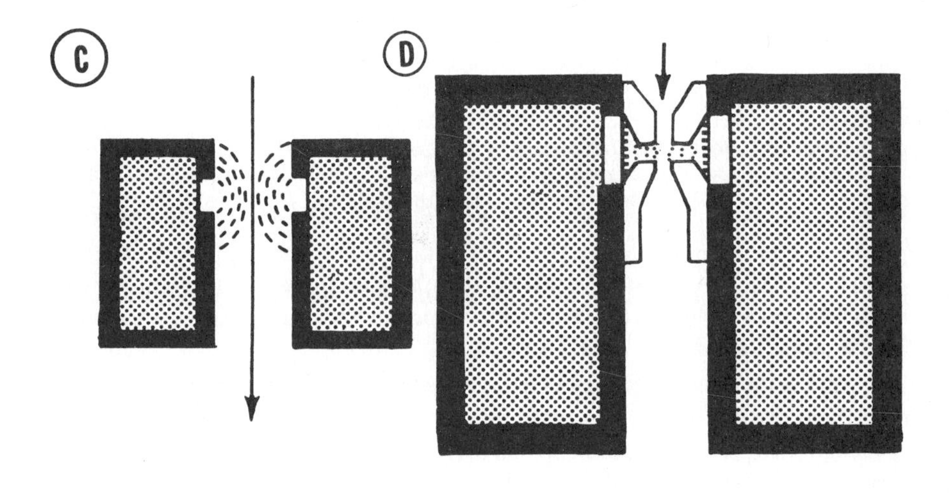

FIG. 34. Evolution of magnetic electron lenses.
(A) Short solenoid used as a magnetic lens (Busch, 1925).
(B) Soft-iron casing enclosing outer surface of the solenoid thus concentrating the magnetic field (Gabor, 1927).
(C) Soft-iron encasing the solenoid except at a narrow annular gap thereby reducing the magnetic field to a very short region along the lens axis (Knoll and Ruska, 1931).
(D) Modern objective lens consisting of a soft-iron encased solenoid and soft-iron pole pieces so as to have an enormously concentrated field at the level of the annular gap (Ruska, 1934).

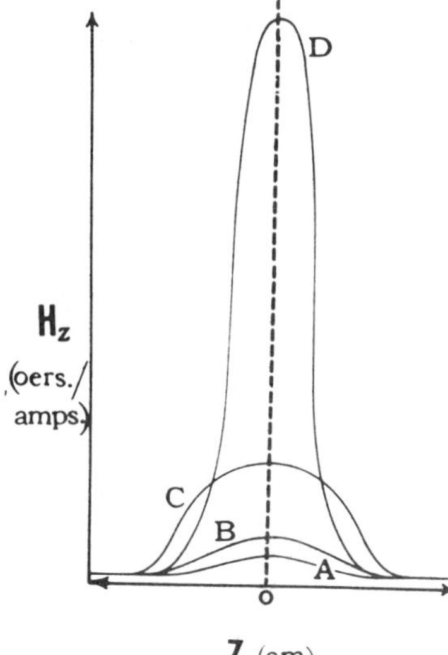

FIG. 35. Field strength distribution curves. The curves A-D correspond to the respective kinds of lenses described in Figure 34. Each represents the field strength along the long axis of the lens. The changes in the shape of the curves represent the shortening or concentration of the field over a shorter axial distance.
H_z — longitudinal magnetic field.
Z — distance along the lens axis of symmetry.

It was thus established that a diverging electron beam originating at a point on the electron optical axis, when passing through a magnetic field produced by a short solenoid, can be brought to a focus at a second point on the same axis. Magnetic lenses, therefore, were shown to be capable of forming images of objects, point for point. The use of a short solenoid as a lens had the drawback of producing a magnetic field which was spread out for a considerable distance along the optical axis with only a small volume, located at the center of the coil, having maximum intensity (Figs. 34A, 35). Gabor, in 1927, succeeded in concentrating the magnetic field of the lens. He accomplished this by encasing the coil in a soft-iron collar except on its inner wall (Fig. 34B). In such a lens the magnetic field is increased in intensity since most of the lines of force run through the iron casing.

In 1931, Ruska and Knoll extended Gabor's idea by completely encasing the coil with soft iron except at a narrow annular gap in the inside of the coil (Fig. 34C). As would be expected, this produced an even greater concentration of the field along a very short axial distance (Fig. 35). With such a coil, considerably less current is needed to obtain a magnetic field whose strength is equal to that of an unencased coil.

In order to obtain large magnifications it is necessary that the focal length of a lens should have a small value corresponding to a large flux density. This could be accomplished by reducing the physical size of the lens and *keeping the coil current constant*. This would require a reduction in the wire diameter of the coil and would ultimately result in overheating.

In 1934, Ruska overcame this difficulty by introducing soft-iron pole pieces with open axial bores at the position of the annular gap (Fig. 34D). This design set the pattern for subsequent high power lenses by providing a very concentrated magnetic field and, therefore, an extremely short focal length with currents and coil resistances that would not lead to overheating.

6. Electrostatic Lenses

Most lenses incorporated in currently manufactured electron microscopes are of the magnetic type, both because spherical and chromatic aberration coefficients of such lenses (k_s *and* k_c; pp. 54, 55), are somewhat smaller than the corresponding electrostatic lens coefficients and the performance of such lenses is much less sensitive to the quality of the vacuum (i.e., level and composition of residual gas phase) and the cleanliness of the components. Nevertheless, it is desirable to discuss electrostatic lenses because the parts of the electron gun, as will be seen, exert electrostatic lens-type effects. Also the accumulation of charge on non-conducting contamination on physical apertures can transform them into such lenses (see p. 98).

A very special electrostatic "lens" can be established by placing two identical plane metal plates near each other and by connecting them to the terminals of a battery. By these means a difference in potential between the plates is established so that, for example, a negatively charged particle would be attracted to the positive plate. When such an experiment is carried out under appropriate conditions, the movements of particles which enter the space perpendicular to the plates are seen to continue to follow paths which are defined as *lines of force* and are perpendicular to the plates (Fig. 36A).

The work done in moving from one plate to another can be divided into equal parts which are marked off along the lines of force. If the corresponding points on the neighboring lines of force are themselves connected by surfaces, we would demark zones of equal work. The latter surfaces are called *equipotential surfaces* (Fig. 36A). That this particular electrostatic field is *uniform* is reflected by the plane equipotential surfaces (represented by straight lines) which subdivide it.

The interval of the field represented by the distance through which a unit charge must be moved to consume a unit of work (that is, the distance between two consecutive equipotentials) defines the unit of *potential difference*, and the rate of change of potential difference along a line of force is called the *potential gradient.*

More pertinent is the path of electrons in a *non-uniform field.* This is represented by curved and/or non-uniformly spaced surfaces subdividing

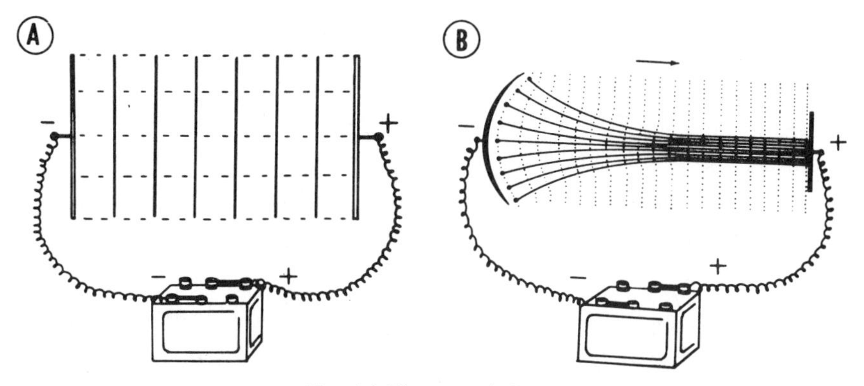

FIG. 36. Electrostatic lens.
(A) Such lenses can be made by attaching two parallel metal plates to the terminals of a battery. The horizontal *lines of force* are intersected by the vertical *equipotential surfaces* which subdivide the uniform electrostatic field. (B) The equipotential surfaces subdividing a non-uniform field are the paths of electrons which travel from the cathode to the anode and intersect all equipotential surfaces at right angles.

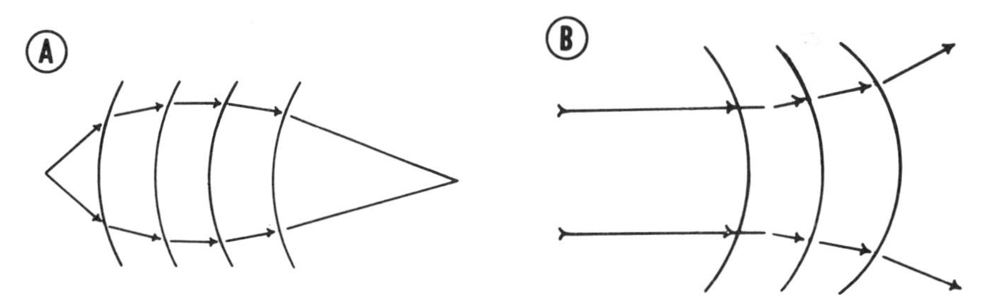

FIG. 37. Action of an electrostatic field.
(A) The converging action of convex equipotentials is illustrated.
(B) If the equipotentials are concave towards the electron source, the lens action is diverging.

the field. This situation can be visualized by considering an electrostatic field that is created by one convex charged plate and one planar plate that is at a different potential (Fig. 36B). Electrons moving from the surface of the charged plate can be considered to be "refracted" at the successive equipotential surfaces to produce a cumulative deviation of the electrons.* In this case, this results in bending the beam towards the central axis; thus such a field acts as a *converging lens* (Fig. 37A). A diverging lens action results from a field whose equipotentials are concave towards the electron source (Fig. 37B).

* By analogy, in light optics, one can simulate equipotentials by imagining the replacement of a region of continuous potential by slabs of successively increasing refractive indices. The total focusing effect of this system would then be due to the cumulative refraction of each of the successive surfaces and quite analogous to the simple electrostatic lens.

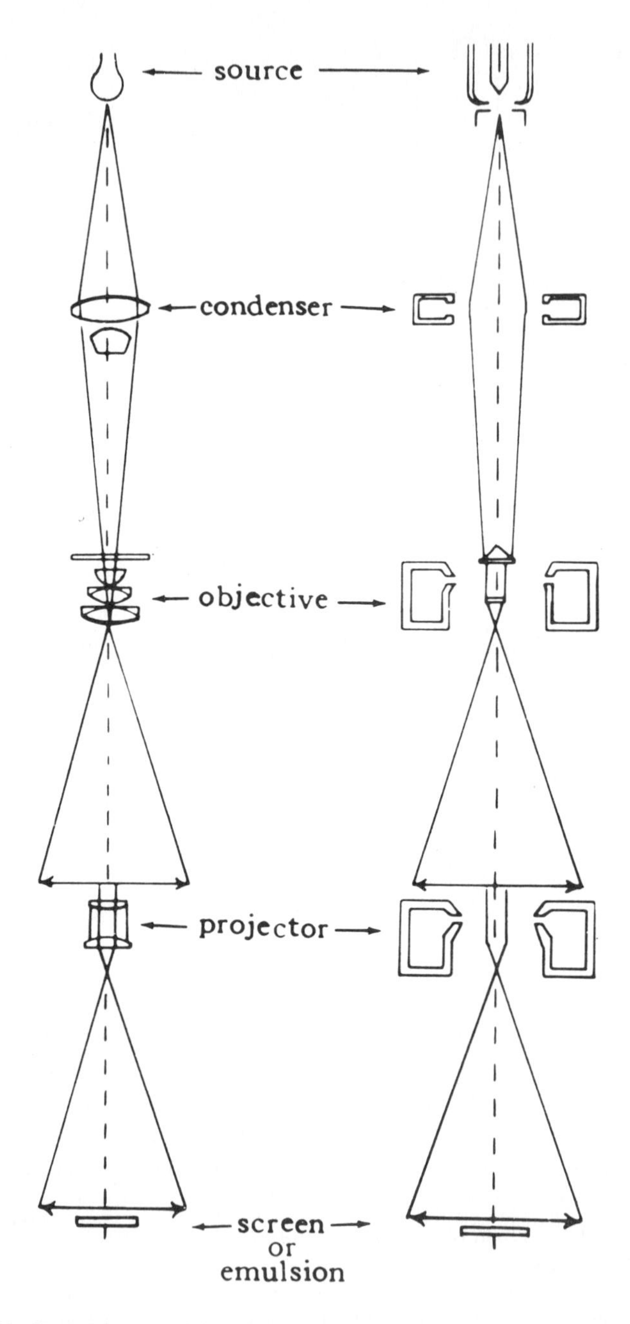

FIG. 38. Comparison of light and electron microscopes. The components of the light and electron microscopes are arranged so as to emphasize the analogy between these two systems. (Kohler illumination is used in the above schemes but the more recent systems make use of critical illumination.)

F. ANALOGY BETWEEN THE LIGHT AND ELECTRON MICROSCOPES

In a previous section, the similarities between the light and electron microscopes were reviewed. It should also be emphasized that both instruments are analogous so far as the arrangement and function of their components are concerned (Fig. 38). Thus both microscopes when used for photographic purposes can be conveniently divided into three systems, each of which contains the same basic components.

I. ILLUMINATING SYSTEM—serves to produce the required radiation and to direct it onto the specimen.
   (a) *source*—for the emission of radiation which is used to form an image.
   (b) *condenser lens*—regulates the convergence (and thus the intensity) of the illuminating beam on the specimen.

(Interposed between the illuminating and imaging systems is the specimen stage.)

II. IMAGING SYSTEM—the lenses which together produce the final magnified image of the specimen.
   (c) *objective lens*—focuses the beam which has passed through the specimen so as to form a magnified intermediate image.
   (d) *projection lens* (or ocular)—magnifies a portion of the intermediate image to form the final image.

III. IMAGE RECORDING SYSTEM—converts the radiation into a permanent image that can be directly viewed.
   (e) *photographic emulsions*—usually carry out this function.

Chapter 4

THE ELECTRON MICROSCOPE*

A. ILLUMINATION SYSTEM

This system contains two units: the electron gun which is the source of the electrons, and the condenser which regulates the intensity of the beam and directs it onto the specimen.

1. Electron Gun**

The parts of the electron gun, as noted before (see p. 34) and as will be evident from the discussion below, have the properties of and act like an electrostatic lens.

The gun consists of three components: the filament, shield, and anode (Fig. 39).

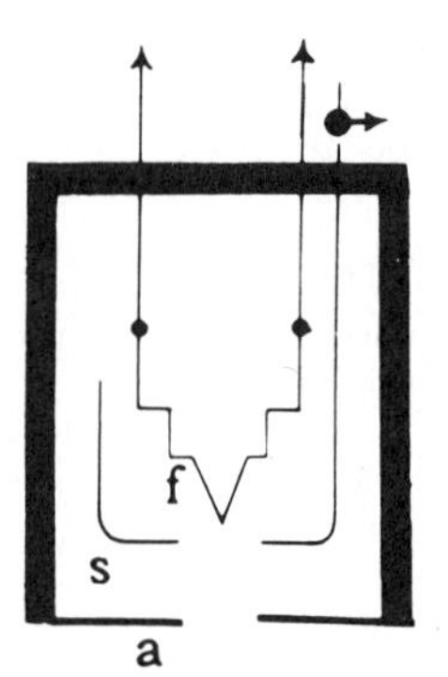

FIG. 39. Electron gun. The three components of the electron gun, the filament, *f*, shield, *s*, and anode, *a*, are schematically illustrated.

---

* While a number of different types of electron microscopes have been developed for specialized applications, the most generally useful and successful instrument is the transmission electron microscope. This type, as its name indicates, is intended for the study of very thin samples by transmitted electrons. The generalized model that will be described will be one containing magnetic lenses only. and thus can be called a *transmission magnetic* electron microscope.

** A review of the fundamentals of electronics can be found in Appendix D. It serves as the basis for the discussion in this section.

a. FILAMENT (cathode, emitter)

The filament is a thermionic cathode. It is usually a V-shaped piece of pure tungsten wire, about 0.1 mm in diameter (.004), which can be electrically heated to incandescence (see Fig. 43A). When a current is passed through the filament, the apex of the V becomes the hottest part and, therefore, the effective electron source. By the use of apertures and lens systems, these electrons are "channeled" so that they can be utilized in illuminating the specimen.

The prerequisites for the electron source are that it be small, symmetrical and intense. It should be small because (1) electrons from only a relatively small area of the source can be delivered to a small area of the specimen; (2) electrical power is consumed by the source and, therefore, the smaller the source the smaller the expenditure of power; and (3) the total beam current delivered by the gun is a function of the total emission of the cathode. A large source produces a large beam current and places severe demands on the high voltage supply. Since only a small current needs to be delivered to the specimen and the final image, the minimum size source permits economy in high voltage source design. Symmetry mainly provides convenience during alignment. Operation at high intensity is required because sufficient numbers of electrons must pass through the smallest resolvable area of the specimen during, for example, a photographic exposure, so that the square root of that number (the signal-to-noise ratio) will be high enough to provide a reasonable representation of that area (see p. 70).

There are two characteristics that a good thermionic emitting material possesses. The first is a sufficiently low work function (see p. 17) so that a copious supply of electrons is available at reasonable operating temperatures, while the second is a long useful operating life. The latter is maintained by preventing oxidation and minimizing evaporation of the metal (by providing that negligible amounts of oxygen are available during emission and that the operating temperature is just high enough to provide satisfactory electron emission).

Tungsten is one of the most suitable thermionic emission materials (see Table III). The reason for this is that although tungsten has a relatively high work function, it has the highest melting point of all metals and a copious supply of electrons can be obtained below its melting point. However, the operating temperature is usually near enough to the melting point so that some of the metal is continuously evaporated *in vacuo* during use. Filament life typically averages 15–40 hours. An additional advantage is that the work function of tungsten is not seriously affected by the presence of residual traces of air as are the oxide-coated type of cathodes with initially lower work functions.

TABLE III

THERMIONIC EMISSION MATERIALS

Material	Work Function (volts)	Melting Point (°K)	Operating Temperature (°K)	Suitability for E.M.
sodium	1.9	371	—	alkalies and alkaline earths while having a low work function vaporize at temperatures too low to permit adequate thermionic emission.
calcium	2.5	1123	—	
barium	2.0	990	—	
thoriated tungsten	2.6	—	1900	can be used but is deactivated by residual air.
oxide-coated nickel	0.5–1.5	—	1000	can be used but is deactivated by residual air.
tantalum	4.1	3250	—	can be used.
tungsten	4.5	3653	2650	most suitable.

The filament is attached to the negative terminal of the high voltage supply (usually through a resistance network; see Fig. 41B).

b. SHIELD (focusing electrode, grid cap, Wehnelt cylinder)

The shield, the equivalent of the grid of a triode (see p. 299), is an apertured cylinder. It lies immediately in front of the filament and its aperture, which is usually 1–3 mm in diameter, is centered over the filament tip. A relatively small potential difference is applied between the filament and the shield.

c. ANODE (positive electrode)

The anode is an apertured disc, which like the remainder of the microscope is grounded and is attached to the positive terminal of the high voltage supply. The cathode is, therefore, held at a large negative potential with respect to the anode. The anode is coaxially aligned with the filament. The emitted electrons are accelerated in the space between the filament and anode by the potential difference across this gap. After leaving the electron gun, the electrons pass down the field-free space in the column at a constant velocity.

d. NON-BIASED AND BIASED EMITTERS

In electron guns the filament-heater current supply usually delivers a few amperes and is often derived from an ac source. There are two basically different methods of connecting the accelerating potential to the electron gun. In the non-biased system (Fig. 40A), which was standard in early commercial electron microscopes, the filament and shield are connected through a pair of balancing resistors directly to the negative terminal of the high voltage supply. By these means the filament tip and the shield are maintained at the same potential at all times during the heater current cycle; they may be considered, in a sense, as one electrode. The presence of the shield aperture near the filament shapes its equipotential surfaces (see p. 34) to provide a focusing or converging lens action. Focusing is controlled by adjusting the distance between the filament and the shield aperture along the optical axis.

A number of different biased systems have been developed but the self-biased system, which is discussed here, is the one that has been found to be most satisfactory.

In the self-biased system (Fig. 40B), the filament is connected to the high voltage supply through both a bias resistor and a pair of balancing

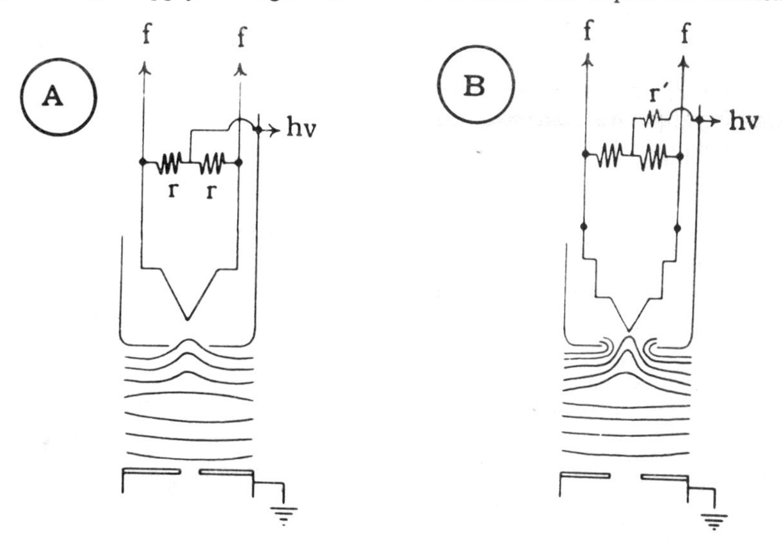

FIG. 40. (A) Non-biased emitter. A schematic diagram of the electrical circuit of a non-biased gun. The filament and shield are connected through a pair of balancing resistors, *r*, directly to the negative terminal of the high voltage supply, *hv*. Leads, *f*, go to the filament heater current supply. The equipotentials of the field when the circuit is activated serve as a converging lens.
(B) Biased emitter. A schematic diagram of the electrical circuit of a biased gun. The filament is connected to the high voltage supply through a bias resistor, *r′*, and the shield is attached directly to the *hv*. The equipotentials of the field when the circuit is activated serve as a strongly converging lens.

resistors while the shield is attached directly to the high voltage supply. The position of both electrodes when the microscope is operating is fixed. The filament, therefore, has a potential differing from that of the shield by an amount equal to the voltage drop (bias potential) across the resistor, which is on the order of −200 V. When beam current flows through the bias resistor this results in the filament being less negative than the shield. The negative field of the latter may be pictured as a throttle, controlling the extent to which the positive field of the anode can penetrate toward the filament to extract electrons from it. Thus the action of the shield is comparable to that of a grid in a self-biased triode. The bias resistor hardly affects the number of electrons *given off* by the filament, but rather it influences the number of emitted electrons which *pass through* the shield and anode apertures to the condenser lens. It is only these electrons that are the effective source of the illumination which participates in image formation.

The operational condition for the non-biased gun involves a state of "saturated emission." To reach this condition, the temperature must be raised until the emission is so high that it becomes *space charge limited* (i.e., the filament develops such a large positive charge from electron loss that the number of electrons entering the *space charge* is equal to those returning to the filament). This severely shortens filament life.

In summary, the principal difference between the non-biased and biased gun is that, in the latter case, the negative field of the shield results in the formation of a potential "well" in front of the filament into which electrons emitted from the filament flow and from which they can be drawn by the anode. The electron cloud or space charge which is so established acts as the effective electron source *instead* of the surface of the filament tip. The negative bias potential substitutes for part of the space charge, thus reducing the amount of current that must pass through the filament to produce a condition where a homogeneous space charge, rather than the inhomogeneous surface of the filament, acts as the electron source.

e. SELF-BIASED GUN

In practice, almost all microscopes currently manufactured have incorporated the biased system in their electron source and the type usually used is called the *self-biased* gun. In such an arrangement, the shield aperture acts as a *strongly* converging lens (Fig. 40B).

In all guns, an increase in filament heater current will result in a rapid rise in beam current (i.e., total electron emission). The self-biased gun arrangement has the desirable property that the shield throttles the trans-

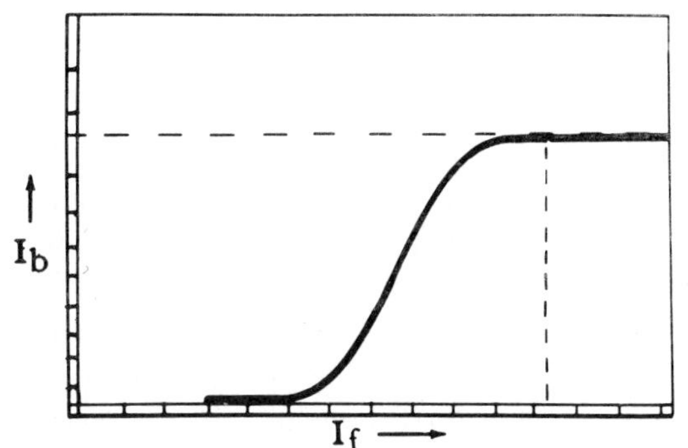

FIG. 41. Operational level of a self-biased electron gun. As the filament current, I_f, increases there is a rapid rise in beam current, I_b, to a flat maximum where I_b is independent of I_f. This is the optimal operational level of the electron gun.

mitted beam intensity to a desirable operational level (Fig. 41). Here the two factors controlling beam current, filament temperature and bias voltage, are in balance in such a way that small changes in either are automatically compensated by the other. Under these circumstances, the biased gun is said to be *saturated*, in allusion to the similar condition that occurs at higher temperatures when unbiased guns become space charge limited. Increasing the filament temperature further does not serve any useful purpose but rather decreases filament life because the beam current is thereafter essentially independent of any increase in filament temperature. The latter, if increased, results indirectly in a concomitant increase in the negative field of the shield which immediately throttles emission.

The main advantages of the biased gun are its:

(1) Greater intensity of illumination for a given filament temperature.
(2) Smaller cross-section of the beam at cross-over.*
(3) Constant illuminating intensity and constant current load on the high voltage supply.
(4) And, incidentally, less critical importance of filament position on gun, height and centration.

The intensity of illumination on the specimen at cross-over with the usual saturated self-biased gun is fixed (i.e., not subject to operation control). Since the electron bombardment may raise the specimen temperature to very high levels (hundreds of degrees centigrade), some types of specimens cannot be studied without defocusing the condenser. This is overcome in some microscope models by the use of a *variable self-bias gun* in which either the effective value of the bias resistor or the spacing of the filament and shield are controllable. By these means, even though the level of the beam current at any particular setting is fixed, as in a self-biased gun, the intensity of the beam can be varied by selecting a new setting.

* Cross-over is that condition where the filament or space charge image is focused on the plane of the specimen.

f. OPERATION OF THE SELF-BIASED GUN

When a microscope is operated with a self-biased gun at constant accelerating potential and fixed condenser current (with the lenses set to focus the image of the filament on the viewing screen), and the filament current is gradually increased to saturation, the following changes in the image of the source can be observed at low magnifications on the final screen (Fig. 42).

With a rise in filament current (i.e., filament temperature), a small faint spot becomes visible. A slight increase in filament current will intensify and expand the spot, and gradually also produce around it a bright ring broken at two opposite points. With a small additional increase in filament current, the central spot merges with the other components, the breaks close and the ring contracts towards the center. Finally, at saturation the image is a single bright spot which is larger, brighter and more uniform than the initial central one. Any further increase in the filament heater current produces no appreciable change in the intensity or appearance of this spot.

These changes in the emission pattern are interpreted as being due to the rapid buildup of space-charge-limited conditions (see p. 42) as the emission is increased. Thus the initial spot seen (Fig. 42A) is due to emission from the tip of the filament and represents an image of the filament itself, whose surface structure is even sometimes visible. The incomplete ring seen in the second stage (Fig. 42B) is produced by the electron emission from parts of the filament away from the tip (which are observed out-of-focus and astigmatically distorted). Gradually during this stage the space charge is formed around the axis. At this time, the anode field is still powerful enough to extract electrons *directly* from the filament through a "canal" in the negative field formed by the electron cloud. With a further slight increase in electron emission, the accumulation of electrons in the region and on the shield is sufficient to cancel the force of the extracting anode field and thus further removal of electrons directly from the filament itself is prevented (Fig. 42C). Under these conditions, electrons appear to be removed from the space charge rather than the filament. When "saturation" is reached, the space charge becomes so intense and the negative shield field, acting with the positive anode field, has become sufficient so that the only electrons that are extracted are those from the axial portion of the cloud. These produce the final intense spot seen (Fig. 42D). With any further increase in the filament current there is no additional increase in electron extraction; i.e., the beam current remains constant.

From what has just been said, it follows that when an electron mi-

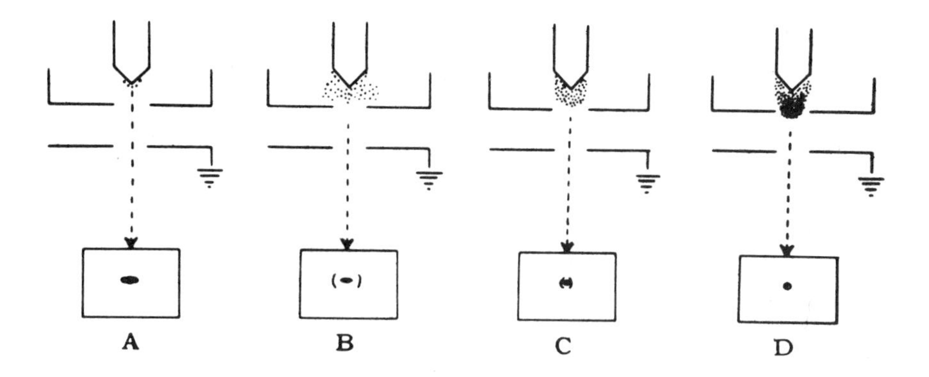

FIG. 42. Emission patterns with increasing filament current.
(A) The initial spot represents emission directly from the filament tip.
(B) In the second stage a ring of satellites is seen and is produced by electron emission from parts of the filament away from the tip.
(C) A further increase in electron emission results in an accumulation of electrons (i.e., formation of a space charge from which electrons are extracted).
(D) At saturation the space charge has become intense and only electrons from its axial portion are extracted to produce the intense spot.

croscope with a self-biased gun is put into operation, the filament current should be increased up to the point where the beam current meter reading levels off. Since this condition is reached quite rapidly, little further control of illumination through manipulation of the filament temperature can be attained and in most microscopes it is, therefore, the main function of the condenser to provide such control. (The variable bias gun, as was discussed above, does however permit additional independent control over intensity of illumination).

A self-biased gun need have only positioning controls for adjustment in a plane perpendicular to the axis of the instrument and/or a tilting arrangement to achieve centration (see p. 84).

g. BRIGHTNESS OF ELECTRON SOURCE

An electron microscope can use only a very small beam angle because of the spherical and chromatic aberrations of the objective lens (see pp. 53, 54), therefore, only equally small illuminating beam angles are useful. As a result very high beam *brightness* is required. The brightness of a portion of an electron beam is defined by the transmitted current divided by the area and the solid angle of spread of the transmitted electrons, i.e., current per unit area per unit solid angle, or intensity per unit solid angle.

To increase brightness above the value attainable with tungsten, it is necessary to use a cathode with lower work function at a comparable

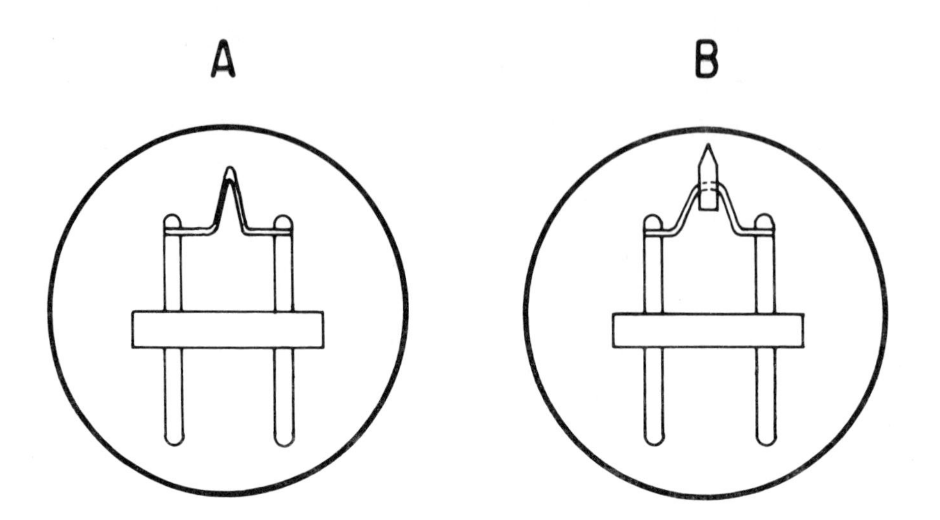

FIG. 43. Electron gun filaments.
(A) A standard tungsten filament.
(B) A pointed filament.

temperature. From this point of view, oxide-coated or thoriated filaments would be superior to pure tungsten ones. But, as noted earlier, their lifetimes at high temperatures are inconveniently short.

Brightness may be improved by the use of *pointed filaments*. Pointed filaments (Fig. 43B) are short pieces of tungsten wire etched to a fine point (*circa* 0.1μ in diameter) or a single tungsten crystal attached to a supporting electrically heated tungsten filament. The tip is away from the asymmetrical fields which surround the supporting piece and, as a result, a very small symmetrical source is obtained. More importantly, the accelerating potential gradients are very high near the tip, resulting in field emission (electrons "tunnel" through the potential barrier of the work function) which gives higher brightness even at low beam current and relatively low temperature. Pointed filaments are not yet in wide use.

2. Condenser Lens

The basic function of the condenser lens is to focus the electron beam emerging from the gun onto the specimen to permit optimal illuminating conditions for visualizing and recording the image. This depends upon the condenser's focal length (i.e., the distance from the center of the lens or lens aperture to the point of convergence of the rays at cross-over, for a source at infinity).

The condenser lens of the electron microscope is a relatively weak lens having a focal length of the order of a few centimeters which can be adjusted

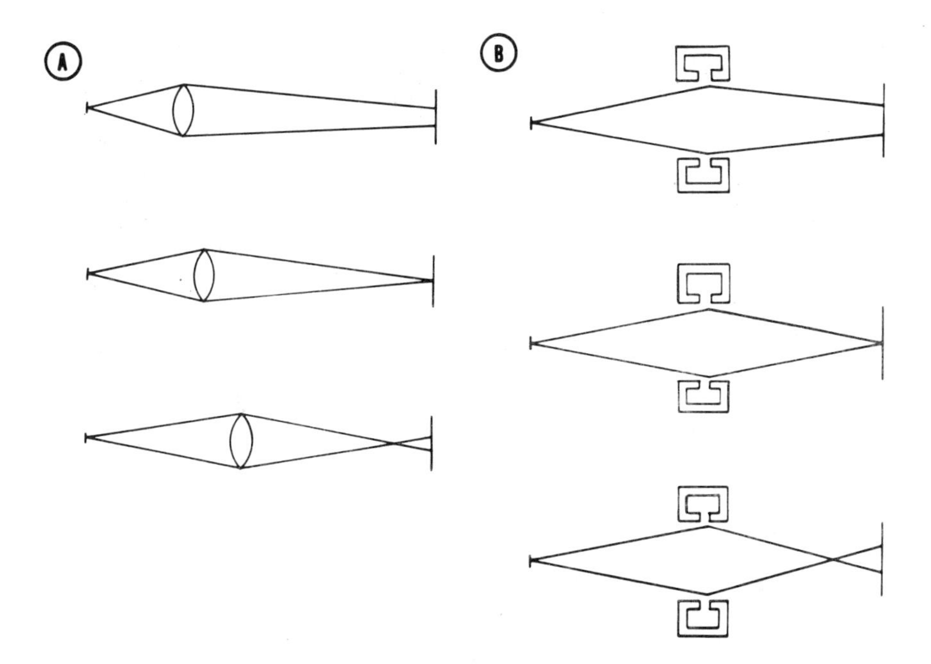

FIG. 44. Focusing of the condenser.
(A) The lenses of a substage condenser are represented by a single bi-convex lens. When the position of the lens is changed by moving it along the axis, focusing of the illumination source can be obtained.
(B) For a magnetic lens condenser whose position is fixed, focusing takes place by varying the focal length which is accomplished by changing the intensity of the coil current.

over a considerable range of values. As is the case in vertical movement of the substage condenser (with fixed focal length) of the light microscope (Fig. 44A), changing the *focal length* of the electron condenser lens shifts the position of the image of the electron source along the optical axis of the instrument (Fig. 44B). By these means two interrelated properties are simultaneously controlled. They are:

(a) Aperture angle (convergence) of the illuminating pencils accepted by the objective*.
(b) Intensity of the illumination at the specimen level.

Each of these properties will now be briefly discussed.

---

* A *pencil* of rays is made up of the cone of rays which participate in the formation of the image of a single point of the specimen.

a. APERTURE ANGLE OF ILLUMINATION

The aperture angle of illumination, α_c, is the half-angle of the illuminating pencils at cross-over. It is determined by the radius of the condenser aperture, r, and the image distance, s' (see Fig. E–1, p. 304). Mathematically, the dependence on focal length of the aperture angle (for small angles), as derived from Newton's equation for thin lenses (see Appendix E), is given by the equation:

$$\alpha_c = \frac{r}{f_c\left(\frac{s}{s - f_c}\right)} \tag{8}$$

where:

α_c = condenser aperture angle.
r = radius of the condenser aperture.
f_c = focal length of the condenser.
s = distance from the specimen to the lens.

The relationship between focal length and aperture angle is graphically illustrated in Figure 45.

To vary the focal length in a magnetic lens it is only necessary to change the *coil current* (see p. 30). The mathematical relation between these two factors can be expressed as:

$$f_c = \frac{k \cdot V}{i_c^2} \tag{9}$$

where:

f_c = focal length of the condenser.
k = constant depending on the geometry of the pole pieces and number of turns in the coil.
V = accelerating potential.
i_c = lens current.

Graphically, the relationship between focal length and the square of the coil current is similar to the relation between aperture angle and focal length so long as $s \geq f_c$.

b. INTENSITY OF ILLUMINATION

As in the case of the light microscope, the intensity of illumination is proportional to the square of the aperture angle, that is:

$$I = k \cdot \alpha_c^2 \tag{10}$$

where:

I = intensity of illumination in the plane of the specimen.
k = an instrument constant.
α_c = condenser aperture angle.

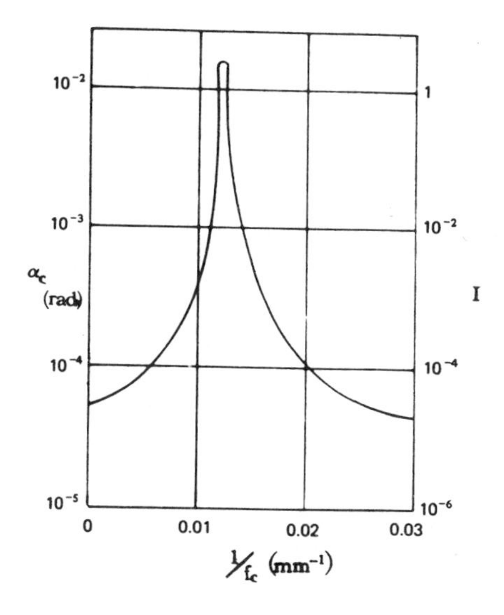

FIG. 45. Relationship of focal length (as $1/f_c$) to aperture angle, α_c, and to intensity of illumination, I (in relative units), for a typical magnetic condenser lens (angle measured at the objective aperture and transmitted intensity measured in the image plane).

If we were to substitute for α_c from equation (8), we would see that the intensity is, therefore, dependent on the focal length. A graphic representation of the relation between intensity and focal length (Fig. 45), as would be expected, is similar to that for aperture angle because of the direct proportionality of I to α_c.

It is obvious that, in practice, the maximum intensity of illumination is obtained by adjusting the focal length (i.e., the lens current) so that the cross-over of the pencils is on the plane of the specimen. By either increasing or decreasing lens strength, the cross-over takes place above or below the specimen plane, and the intensity is reduced because the illuminating aperture, α_c, is maximal only at cross-over and the beam is also spread out over a larger part of the specimen area.

As indicated above, the aperture angle is also determined by the size of the aperture opening. The larger the aperture, the more electrons pass through the condenser to strike the specimen, and thus the higher is the intensity of illumination.

c. CONDENSER LENS OPERATION

Until a few years ago, it was common practice for the condenser setting for viewing and focusing to be set differently from that for photography. However, it was found that by changing the "in-focus" position of the condenser just prior to photography a number of possible errors could be introduced. These include specimen drift (see p. 94), and changes in focus due to changes in the weak lens effects of charged apertures. To avoid these errors the same condenser setting is now usually maintained throughout the visual and photographic procedures for high resolution. The *ideal* operating conditions are obtained by:

(1) Providing the condenser with a physical aperture of such a dimension that the illuminating aperture angle at cross-over is equal to the optimal objective aperture as defined on page 58.

(2) Adjusting the bias potential of the gun or the axial position of the shield (see p. 43) to produce an illumination intensity which is just adequate for visualization and photography at a given magnification.

For electron microscopes where the flexibility for making the above adjustments is not present, the attempt is made to approximate these ideal conditions.

Since the focal length of the condenser can be relatively long, the magnetic field for this lens can be relatively weak. For this reason, some condenser lenses are not equipped with pole pieces and usually they have a smaller coil than the other lenses. The condenser is located between the illuminating source and the specimen so that it forms an image of the source on the specimen. Provision is usually made to allow a physical aperture to be inserted into this lens. The condenser is also usually provided with means for centering in a plane perpendicular to the axis of the instrument.

B. IMAGING SYSTEM

The imaging system consists of an objective lens and one or more projector lenses.

1. Objective Lens

The objective lens forms the initial enlarged image of the illuminated portion of the specimen in a plane that is suitable for further enlargement by the projector lens (see Fig. 38). The objective magnetic lens is the most critical component of the microscope. In its construction one must consider the following criteria, most of which will be discussed in detail below:

(a) That in order to obtain a high magnification at the given image distance, which is fixed by the length of the column connecting the objective lens with the projector lens, the specimen must be

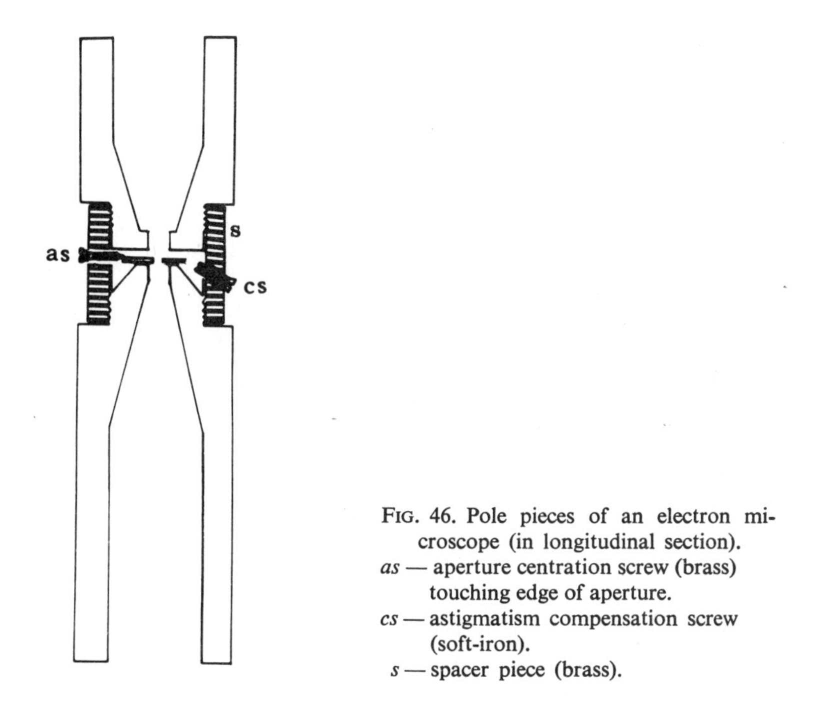

FIG. 46. Pole pieces of an electron microscope (in longitudinal section).
as — aperture centration screw (brass) touching edge of aperture.
cs — astigmatism compensation screw (soft-iron).
s — spacer piece (brass).

situated close to the focal plane of the objective. This is made possible by constructing the lens so that the specimen can be positioned right down into it (see frontispiece). In most cases, the initial magnification provided by the objective lens is approximately 100 X.

(b) That the focal length should be as short as practicable (1–5 mm) to insure minimum chromatic and spherical aberration be attained.

(c) That, at a very minimum, clearance must be provided for inserting the specimen and a physical aperture in or close to the objective lens.

(d) That provision must usually be made for inserting electrical or magnetic devices in, or close to the objective for correcting minute asymmetries in the lens field.

a. POLE PIECES

A typical modern objective lens is illustrated in Figure 34D. Like the other lenses it consists of a soft-iron casing around a coil. The objective is usually larger than the condenser and usually provides for more windings and, therefore, a potentially higher field strength. A non-magnetic (brass) spacing piece is often located on the inner wall of the casing, occupying the magnetic gap between two conical, hollow, soft-iron pole pieces (Fig. 46)

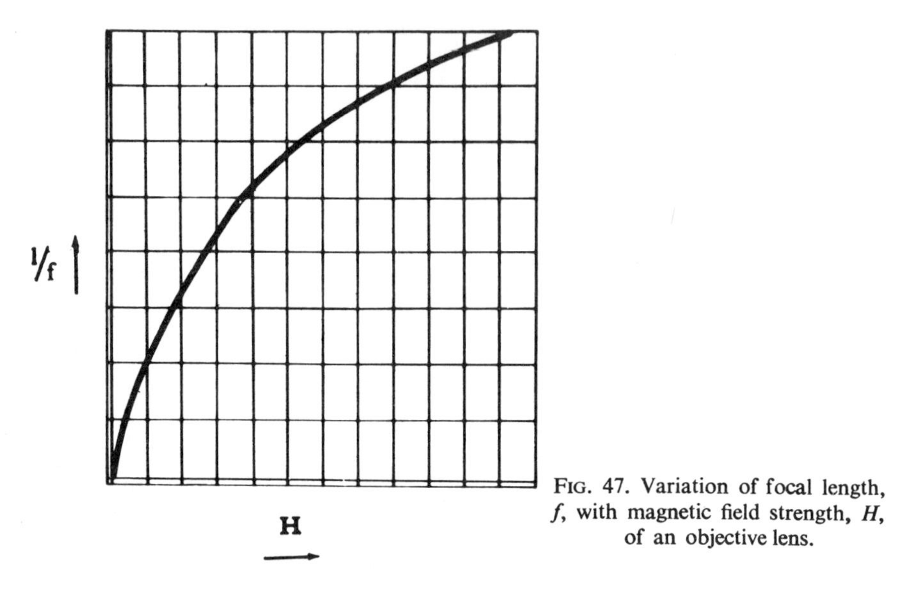

FIG. 47. Variation of focal length, *f*, with magnetic field strength, *H*, of an objective lens.

so that they are precisely axially aligned and separated by a very small free gap. The intense magnetic field set up occupies only a few cubic millimeters along the axis, producing a short focal length. The relationship between focal length and magnetic field strength is shown in Figure 47.

Pole pieces are present in all objective lenses and are frequently found in the other lenses. When present, the focal length is roughly proportional to the diameter of the bore of the pole piece. Since space is needed to permit the introduction of a movable specimen holder in the field of the lens, the spacing of the pole pieces at the gap cannot be reduced beyond a certain limit to attain a shorter focal length lens. At the gap, the surfaces of both pole pieces usually have the shape of truncated cones; or one can have this shape while the other may be untapered with a flat face. In any case, the planes of the facing surfaces should be perfectly parallel.

It is essential that there be good contact between the pole pieces and the soft-iron casing of the solenoid. This is attained by making the outside diameter of the pole pieces just slightly less than the bore of the iron casing of the lens core.

An important consideration that must be emphasized is that pole pieces usually cannot be produced completely free from mechanical and magnetic imperfections. These may be due to imperfect machining or to inhomogeneities of the iron. Such irregularities will induce an asymmetry in the magnetic field, which must be eliminated in order to attain the maximum performance of the lens. The detection of objective lens asymmetry and its compensation are discussed on page 89.

b. LENS ABERRATIONS

As with the light microscope, electron microscope objectives possess a number of inherent defects. In electron optics, however, some of the important aberrations cannot be corrected but their effects can be minimized. The three principal intrinsic lens defects which will be discussed here are spherical and chromatic aberration and distortion (see p. 64). Aside from these, there are others which collectively can be considered to be extrinsic defects that may contribute to producing deviations from perfect imaging. Included in this group are defects induced by misalignment of lenses (see p. 81), asymmetry of the objective lens (see p. 89), and distortions due to "stray" electric and magnetic fields (see pp. 93,94).

Spherical Aberration

This defect belongs to the class called *geometric aberrations* because it is produced by the geometry of the lens field. Other defects in this group, which are absent on the optic axis but increase rapidly with distance from the axis, are the *field aberrations*, coma, curvature of field and distortion. Spherical aberration is the most important of the geometric defects in electron microscopy. It cannot be corrected; it occurs on axis and thus is one of the principal factors limiting resolution.

In an earlier discussion of the properties of glass lenses, it was convenient to consider them as optically perfect in that parallel rays of light are refracted so that they all converge at an identical focal point (see Fig. 30). We could, therefore, assign for each lens a single focal length. In actuality, however, the focal length of a simple lens with spherical surfaces is not unique but varies over different portions of the lens aperture. Thus, in the case of a biconvex type of lens, focal lengths are shorter in the periphery and so rays passing through this region have a focal point closer to the optic axis than those passing through the more central portions of the lens (Fig. 48A).

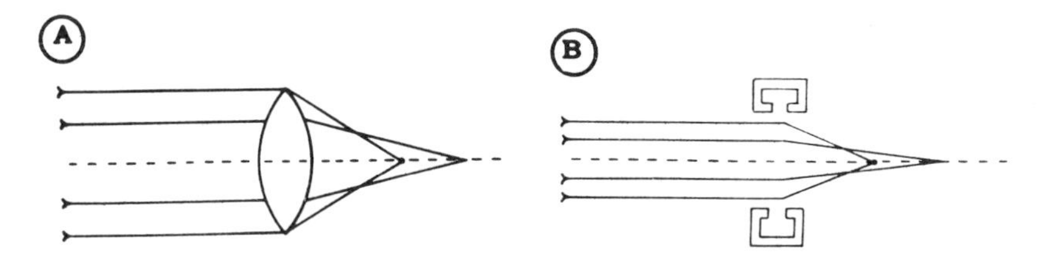

FIG. 48. Spherical aberration.
(A) For a glass lens.
(B) For an electron lens.

Similarly, in the objective lens of the electron microscope, spherical aberration is also present because the zones of the lens farther from the axis have a different refractive power than those around the axial region. Electron lenses, like biconvex optical lenses, are always characterized by "positive" spherical aberration* in that their rays which pass through the marginal zones are focused nearer to the lens than paraxial rays (Fig. 48B).

In the case of a magnetic objective lens, the spread of focal points will extend over a measurable portion of the optical axis. The limit of resolution in the presence of spherical aberration is given by:

$$d_s = k_s \cdot f \cdot \alpha_0^3 \qquad (12)$$

where:

d_s = separation of two object points which are just resolved.
k_s = dimensionless proportionality constant characteristic of the objective lens (which in general is reduced as the bore is reduced and the gap is increased).
f = focal length.
α_0 = objective aperture angle.

A discussion of this equation can be found in Appendix G. Equation (12) indicates that the aperture angle and focal length are the critical factors which determine the magnitude of spherical aberration.

Chromatic Aberration

In light microscopy, chromatic aberration is present only if the light contains rays of different colors (i.e., different wavelengths). These rays when passing through a *simple* lens are refracted to a different extent and come into focus at different points along the optical axis. Thus, for example, blue light comes into focus closer to the lens (i.e., has a shorter focal length) than red light (Fig. 49A). Achromatic compound lenses are used to correct this aberration.

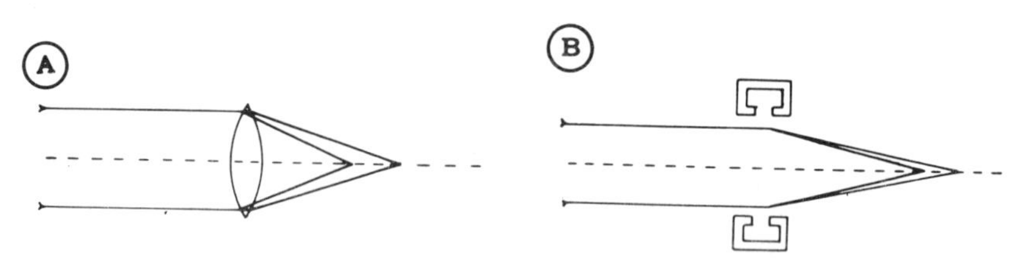

FIG. 49. Chromatic aberration.
(A) For a glass lens. (B) For an electron lens.

* Because all magnetic electron lenses are convergent, spherical aberration cannot be corrected by the addition of negative lenses, nor by mechanical changes (such as the production of aspheric surfaces) as is the case for light optical lenses.

In electron microscopy, the situation is analogous to chromatic aberration in the light microscope when electrons of different velocities, therefore, of different wavelengths*, pass through a simple electron lens. Those of greater velocity will be acted upon by the field for a shorter time and will be deflected less by the field, and thus will be focused at a point further along the axis than electrons of slower velocity (Fig. 49B). It is not possible to design achromatic compound electron lenses.

The equations for the limit of resolution in the presence of chromatic aberration are:

$$d_{c_v} = k_c \cdot f \cdot \alpha_0 \cdot \frac{\Delta V}{V} \tag{13}$$

$$d_{c_i} = 2k_c \cdot f \cdot \alpha_0 \cdot \frac{\Delta I}{I} \tag{14}$$

where:

d_{c_v} & d_{c_i} = separation of two object points which are just resolved, considering voltage and current respectively.
k_c = dimensionless constant (with values between 0.75 and 1.0).
f = focal length.
α_0 = objective aperture angle.
V = accelerating potential.
ΔV = maximum departure from V of electrons contributing to the image.
I = current.
ΔI = maximum departure from I.

The equations for chromatic aberration are discussed in Appendix H. As in the case of spherical aberration, aperture angle and focal length are critical factors.

There are three factors which may produce electrons having different velocities (i.e., showing finite values of ΔV). These are variations in emission velocity, variations in accelerating potential, and energy losses resulting from interaction with the atoms of the specimen.

(1) The initial velocity of the electrons emitted from the thermionic electrode may be different. This source of chromatic aberration is negligible since typical accelerating potentials are in the range from 50–100 KV and the initial energy of the emitted electrons is usually less than one electron-volt (i.e., $\Delta V/V \leqq 10^{-5}$). This is below the level where voltage variations are a disturbing factor as is apparent if one substitutes the known values in equation (13).

(2) Variations in the high voltage supply produce fluctuations in the accelerating potential and, therefore, in the velocity of the electrons which pass down the column. To assure that this source of error

* Since according to de Broglie (see eq. 2, p. 14) the wavelength of the electron is dependent on the velocity.

is minimized, the high voltage supply must be designed and constructed to severely restrict fluctuations by negative feed-back. The need for a highly stabilized accelerating potential is especially important in instruments where a stability near 1/100,000 of the voltage is required for resolution capability in the neighborhood of 2 Angstroms.*

(3) Although some electrons suffer changes in direction without concomitant changes in velocity (e.g., are elastically scattered), there must always be others which undergo changes in both velocity and direction as a result of their interaction with the atoms of the specimen. The induced changes in direction are clearly desirable attributes for it is one of the main mechanisms whereby contrast in the final image can be attained. For this reason some chromatic aberration must be present in electron microscopes. Parenthetically, in light microscopy a monochromatic beam is not altered in wavelength in transmission through the specimen and thus this source of chromatic aberration is unique in electron microscopes.

Since the focal length of objective lenses is dependent on the magnitude of the lens current, stability of the latter is also essential in order to secure sharp images. A fractional fluctuation in lens current, $\Delta I/I$ causes a change in focus exactly equivalent to two times the same fractional change in accelerating voltage $\Delta V/V$ and thus, lens current fluctuations are another source of chromatic aberration. Because of the factor of two difference, lens

---

* The maximal permissible variation in the accelerating potential can be determined from the equation for the limit of resolution in the presence of chromatic aberration (eq. 13):

$$d_{c_v} = k_c \cdot f \cdot \alpha_0 \cdot \frac{\Delta V}{V} \qquad (1)$$

which can also be written as:

$$\frac{\Delta V}{V} = \frac{d_{c_v}}{k_c \cdot f \cdot \alpha_0} \qquad (2)$$

The typical values for the various factors are:

d_{c_v} = 10Å; the limit of resolution as set by diffraction and spherical aberration.
k_c = 0.75; a dimensionless constant.
α_0 = 4.5×10^{-3} radians; which is about the optimal objective aperture angle (see p. 58).
f = 3 mm; which is a typical focal length (see p. 54).

which, when substituted in eq. (2), give:

$$\frac{\Delta V}{V} = \frac{10\text{Å}}{0.75 \cdot 4.5 \times 10^{-3}{}_{radian} \cdot 3\,mm} = 1.0 \times 10^{-4} = 0.0001 \text{ of the voltage.}$$

current stability near 1/200,000 is required for resolution capability in the neighborhood of 2 Angstroms.*

Variation in velocities of electrons manifests itself in two ways. It appears as:

(1) A change in the focal length of the objective which translates an object point into an enlarged disk of confusion in the image plane (see Fig. H–1, p. 313).

(2) A change in magnification (for both electrostatic and magnetic lenses) and a change in rotation about the axis (only for magnetic lenses) result in image movement off axis. The magnitude of the movement increases with the radial distance of the point from the axis, being zero on the axis.

In other words, for axial points a change in sharpness of the image point takes place, while for non-axial points changes in sharpness, magnification and rotation occur. These manifestations are also due to the fact that both focal length and image rotation are dependent on beam voltage (and magnetic field strength).

c. LIMITS SET BY OBJECTIVE LENS ABERRATIONS

From the equation expressing the limit of resolution in the presence of spherical aberration:

$$d_s = k_s \cdot f \cdot \alpha_0^{\,3} \qquad (12)$$

it is apparent that this geometrical defect can be reduced by reducing focal length, f, and the aperture angle, α_0, as much as possible. So far as the first factor is concerned, in practice, the objective lenses most suitable for high resolution work are those which have the shortest possible focal lengths. With a compensated magnetic objective lens working at a focal length of about 1.8 mm, a resolving power of 2 Å has been achieved. The need for such short focal length points up the need for an extremely intense magnetic field (i.e., the stronger the field, the shorter the focal length) and this can only be *easily* established by the use of rather small-bore pole pieces.

* The maximal permissible variation in the intensity of the lens current can also be determined from the equation for the limit of resolution in the presence of chromatic aberration (eq. 14):

$$d_{c_i} = 2k_c \cdot f \cdot \alpha_0 \frac{\Delta I}{I}$$

Since eq. (14) differs from eq. (13) by a factor of two, then, substituting the same values as in the footnote on p. 56,

$$\frac{\Delta I}{I} = 0.0005 \text{ of the current.}$$

Designing objective lenses with focal lengths under 1 mm poses a number of severe problems. The small pole piece gap and bore required make positioning of the specimen holder and objective aperture extremely difficult and leave virtually no room for other accessories (such as the Heide anti-contamination device; see p. 100) and astigmatic compensators (see p. 91). However, the use of superconducting solenoids *without pole pieces* may eventually provide a means of achieving usable lenses with very short focal lengths.[20]

On the other hand, while a reduction of the aperture angle does succeed in reducing the effect of spherical aberration, if α_0 is drastically cut down it would simultaneously decrease resolution. For, from Abbe's law governing maximum resolution (i.e., minimal diffraction effect; p. 11), the aperture angle, α_0, should be as large as possible. It is apparent, therefore, that an attempt to reduce spherical aberration by reduction of aperture angle is in conflict with the conditions for minimal diffraction effect. As a result, the choice of aperture angle must represent a compromise between these two factors.

In light microscopy, no such compromise is required since the spherical aberration of light microscope objectives can be essentially completely eliminated and nearly the largest possible aperture angle can be utilized.

From the curve showing the relationship of spherical aberration and diffraction to resolution (Fig. 50), it is apparent that the *optimal aperture angle* for a 3 mm focal length lens at 50 KV is 4×10^{-3} radians, for which the corresponding limit of resolution is 10 Å. At this angle the two effects are approximately equal; diffraction making it inadvisable to decrease the angle beyond the optimal level and loss in resolution due to spherical aberration forming the barrier to an increase in α_0.

It is obvious from the above that it would be desirable to keep the limiting effects of chromatic aberration smaller than that which sets the resolution limit by diffraction and spherical aberration. Thus, for example, the maximum permissible variation in accelerating potential, which is a major factor responsible for chromatic aberration, would, for 10 Angstroms resolution, have to be one ten-thousandth of the voltage used (see footnote on p. 56). Similarly, it is also necessary to limit the permissible variations of the current (i.e., focusing field) in the lenses (see footnote on p. 57).

A most critical factor which determines the extent of chromatic aberration is the thickness of the section used. This is because it determines the magnitude of the change in voltage (ΔV) of the electrons which emerge from the specimen. If ΔV is excessive due to very thick specimens, the resultant increase in chromatic aberration will seriously limit resolution.

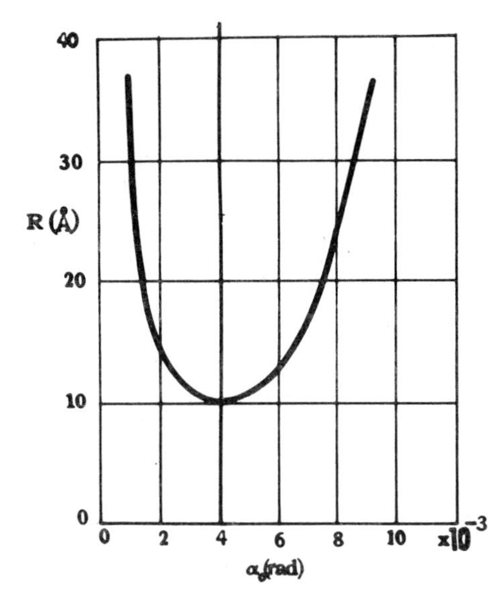

FIG. 50. Limit of resolution. The curve shows the relation between resolving power, R, and aperture angle, α_0, of a 3 mm focal length objective lens at 50 KV. This curve is the sum of the resolving power limits imposed by diffraction and spherical aberration.

It should, however, be noted that for microscopes equipped with short focal length lenses and having accelerating voltages of 100 KV, high resolution is attainable even with sections as thick as 0.3 μ.[21]

d. CONTRAST AND IMAGE FORMATION

The similarity between image formation in the light and electron microscopes has already been emphasized. It is also evident that the details of the images produced by both systems are visible only as a result of variations in specimen contrast. The analogy between the two microscopic systems is, however, not complete because the mechanism by which the intensity differences in the image are produced are different for each system.

In the light microscope, differential absorption of light, which depends mainly on staining the specimen, results in the visible differences in various parts of the image. In the electron microscope, the portion of the beam absorbed* with the thickest useful specimen is infinitesimal.

To explain the mechanism of image formation as it takes place in the electron microscope, it is best if each point of the specimen be considered as being situated at the apex of a narrow cone of electrons of angle equal to or less than the aperture angle (Fig. 51). Thus such a cone, or "pencil" of electrons, contains all those particles which participate in the imaging of an

* Those electrons which give up all their energy on collision with the specimen are *absorbed.* Those which interact with the specimen by giving up part of their energy are *inelastically scattered.* Those which give up none of their energy but change their direction are *elastically scattered.*

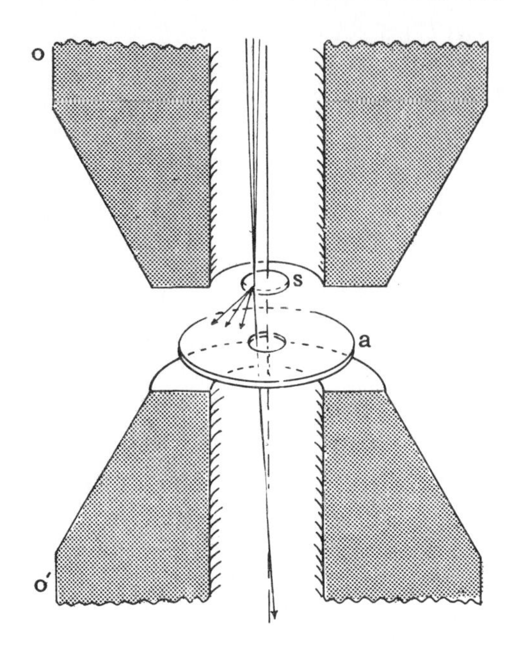

FIG. 51. Image formation. The mechanism of image formation is diagrammatically illustrated. Shown is a lengthwise section through objective lens pole pieces, between which are seen the specimen (*s*), and physical aperture (*a*). The "subtractive" action on a pencil of electrons which images an individual specimen point is demonstrated.

individual specimen point. If such a pencil were to traverse the specimen without any change, it would pass through the lens aperture in its entirety. In reality the shape of the incident pencil on traversing the specimen is altered due to the "collisions" of some of its electrons with electrons and atomic nuclei of the specimen. This results in electron scattering and the narrow pencil upon emergence from the specimen is widened. Those electrons of the pencil from the object point that are strongly scattered by more than the angle defined by the objective aperture are prevented from reaching the corresponding image points. This loss of electrons will result in a diminution of intensity at these points. The contrast of the image, therefore, is due to the cumulative effect of the described image-forming mechanism which results in the variations of intensity at the individual points. The objective lens can, therefore, be described as producing the image by the "subtractive" action of the aperture.

The amount of scattering which occurs at any particular specimen point is dependent upon its physical density and thickness (and is essentially independent of the atomic number, chemical composition or other specimen properties[22]). Thus the scattering power of a particular area of a specimen is directly proportional to its *mass density* (mass per unit area) which is the product of the density and thickness. A number of different methods have been used to increase the mass density and thus the contrast of parts of a biological specimen. These methods belong to the field of electron staining (see p. 169) and cyto- and histochemistry (see p. 177).

The "subtractive" action results from the use of a physical aperture. The size required will depend on the focal length of the objective used. Thus a lens of 4 mm focal length with an objective aperture of 80 μ, 4 mm from the specimen, will subtend a half-angle of 10^{-2} radians. Electrons which are scattered through larger angles will be eliminated. In practice, when working with an externally centerable physical aperture, the size that is used varies from 10 to 200 μ. The contrast is considerably enhanced with small apertures but, of course, resolution will be decreased according to Abbe's equation.

e. DEPTH OF FIELD

When a lens is set to give the sharpest focus on a plane, the eye can observe and the photographic plate can record particles that are located somewhat above or below the focal plane. The range of object distances over which structures can still appear reasonably sharp is called the *depth of field.* In light microscopy, the most significant factor in determining this quantity is the numerical aperture (N.A.) of the system.

As indicated in Table IV, the greater the N.A. the smaller the depth of field. In the case of an oil-immersion objective this quantity, for all practical purposes, is of the order of 0.2 μ and only structures which lie within that axial distance from one another will be simultaneously in sharp focus. The only way to increase the depth of field is to reduce the N.A., an improvement which is obtained at the expense of resolution (see p. 58). In practice this can be done by using an objective of lower N.A. or reducing the substage N.A. by stopping down the condenser iris diaphragm which, in principle, involves a reduction in the aperture angle of the cone of light passing through the system.

TABLE IV

RELATIONSHIP OF N.A. TO MICROSCOPE PERFORMANCE

Achromatic Obj. N.A.	Depth of Field	Magnification	Focal Length
0.2 -0.3	10 μ	10X	16 mm
0.65–0.5	1–2 μ	40X	4 mm
1.2 -1.3	0.5 μ	100X	2 mm

For the electron microscope, the situation is basically similar. For there exists a range within which the fine focusing control produces no visible change in an image which has been critically focused (see Fig. F–1,

p. 308). This, in reality, means that the focal length of the lens could be varied so that the focal point moves over a short axial distance above and below the true position of the object without producing a noticeable change. In other words, several planes can be "in focus" at the same time. This is due to the fact that the depth of the field is dependent on the aperture angle and is mathematically expressed as:

$$D_{f_i} = \frac{2r}{\alpha_0} \tag{11}$$

where:

D_{f_i} = depth of field.
r = radius of disc of confusion (see Fig. F-1).
α_0 = objective aperture angle.

The derivation of equation (11) is given in Appendix F.

f. OBJECTIVE LENS OPERATION

In the electron microscope, *focusing is accomplished by varying the focal length of the objective lens.* As already discussed, this usually involves changing the current which passes through the coil. Since the adjustments in focal length may vary over a relatively wide range, while at the same time extreme precision in focusing is also required, several controls of progressively increasing sensitivity are usually provided.

As with the components of the electron microscope that already have been discussed, so too must the objective lens be aligned or centered (see p. 81). Adjustments to move the lens (or its pole pieces) in a plane perpendicular to the optical axis are provided in most instruments.

The scattered electrons which strike the lens elements may result in the formation of deposits on their surfaces. These deposits tend to accumulate an electrostatic charge during operation which can modify the lens field significantly, producing astigmatism, distortion, and/or image shifts (see p. 94). To minimize this problem, lenses must be periodically cleaned quite meticulously.

2. Projector Lens

The projector lens, as its name implies, serves to project the final magnified image on the screen or photographic emulsion. This lens has a construction similar to the objective lens except that it has no provision to accommodate the specimen holder.

The properties of the projector lens contribute to the final formation of the electron image in the following important respects:

(a) A broad range of focal lengths.

(b) The great depth of focus of the resulting high magnification system.

(c) A difference in the manifestation and a less serious effect of its aberrations on resolution.

The projector lens of a *two-stage unit* (i.e., an imaging system made up of one objective and one projector lens) is usually provided with a focal length with a greater range than the objective. The fact that pole pieces which are used with this lens can be made symmetrical, since the specimen does not have to be accommodated, permits use of a shorter focal length. The full axial length of the magnetic field takes part in the imaging process rather than only the upper part of the field, as is the case for the objective lens.

a. DEPTH OF FOCUS

The depth of focus (see Fig. F–2, p. 308), which corresponds to the depth of field in *image* space, can be determined by an equation which is similar to that for the depth of field (see Appendix F):

$$D_{f_0} = \frac{2r_0}{\alpha_0} \cdot M^2 \qquad (15)$$

where:

D_{f_0} = depth of focus.
r_0 = radius of the smallest resolvable area in the specimen plane.
α_0 = objective aperture angle.
M_t = total magnification of the lens system (e.g., $M_0 \times M_p = M_t$; see p. 66).

This equation shows that the depth of focus is determined by the objective aperture angle and the square of the product of the magnification of the individual lenses (the total magnification). The resultant depth of focus in typical electron microscopes is extremely large (several meters). This means that the position of the viewing screen relative to photographic film is relatively non-critical with respect to focus. There is no need to compensate for the difference between the two since the image is in focus over a large distance along the axis of the microscope. (Of course, the *magnification* on the screen and film will be different if they are at different distances along the axis.)

Similarly, the depth of field of the projector lens (eq. 11) exceeds that of the objective due to the fact that the aperture angle of the pencils entering the projector lens is far smaller, being of the order of 10^{-4} radians or less.

This means that when the focal length of the projector lens is varied in order to alter the magnification and focus of the final image, this operation can be carried out without being concerned about the exact object distance of the projector lens (i.e., the distance from the intermediate image). In other words, *changing the magnification will not produce detectable defocusing of the image* (in the high magnification range, i.e., 10,000 X or higher).

The problem of lens aberrations is also different in the case of the projector lens. Since the aperture angle of the imaging pencils entering the projector lens is very small, normal spherical and chromatic aberrations which have been shown to be proportional to the aperture angle (eq. 12, 13 and 14), do not, in practice, affect the realizable resolving power (i.e., the sharpness of the final image) because the magnification required of the projector is only of the order of 300 X.

While the presence of spherical aberration in the projector lens does not cause a significant spread of each image point, another geometrical error, *distortion*, produces a variation in magnification across the image as a whole. Since the outer parts of the magnetic field of the lens are relatively too strong, pencils of rays which pass this region are refracted more sharply than paraxial pencils. This results in the off-axial pencils crossing the axis earlier and thus striking the image plane at a greater distance from the axis than expected of an ideal lens (Fig. 52A). As a result, the peripheral region of the image is more highly magnified than the paraxial region. When, as in this case, the intermediate image is a real image, then the resultant form that the distortion takes is known as *pincushion distortion* which is observable especially when the power, that is the magnification, of the projector is

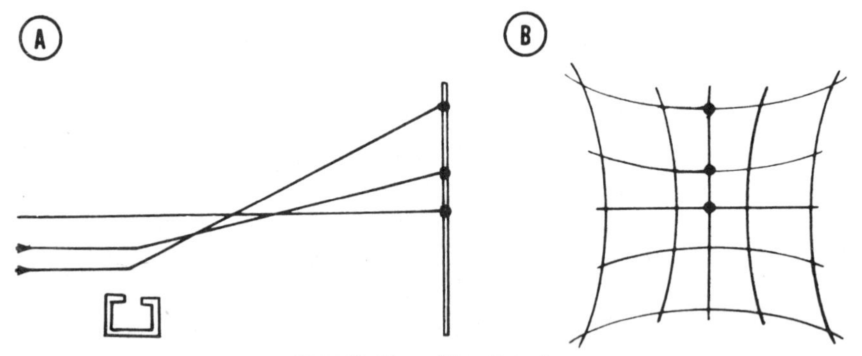

FIG. 52. Pincushion distortion.

(A) Two parallel rays are shown, one nearer to the axis than the other. The latter is refracted by the magnetic field *more* sharply than the former, resulting in a distorted image.

(B) The distortion effect as reflected in the electron image of a metal grid of uniform squares.

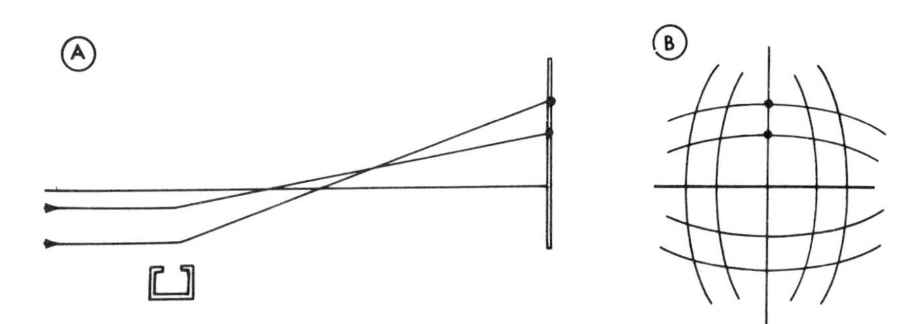

FIG. 53. Barrel distortion.
(A) Two parallel rays are shown, one nearer to the axis than the other. The latter is refracted less sharply than the former resulting in a distorted image. (B) The distortion effect as reflected in the electron image of a metal grid of uniform squares.

low (Fig. 52B). It is this phenomenon that sets the lower limit to the useful magnification that can be achieved in an electron microscope with a given set of pole pieces.

If the same magnetic lens is used with a long object distance and short image distance (used as a demagnifier), a type of distortion known as *barrel distortion* results (Fig. 53A,B). For a given magnification it is possible to decrease net distortion by the proper choice of pole pieces. It can be shown by using a series of lenses (e.g., a three-stage unit; see p. 66), some of which magnify the image and others which demagnify, that it is possible to arrange the pincushion and barrel distortion to compensate for each other and still produce a highly magnified image.[23]

b. MAGNIFICATION AND FINAL IMAGE FORMATION

Since the projector lens of the electron microscope, like the ocular of the light microscope, is the principal component responsible for magnification beyond that of the objective, it is appropriate that these subjects be considered here.

In the light microscope, the objective forms the initial image of the object in the image plane of the objective lens. The ocular usually uses this image as its object and forms a final image 10 inches beyond the ocular (see Fig. 3). When the eye is placed directly behind the ocular, the combination of the ocular and the eye may be considered as a new short focal length lens (where the image distance is equal to the distance from the crystalline lens to the retina). The total light microscope magnification of the final image is the *product of the magnifications of the two individual optical components: the objective and the ocular*. (In computing visual magnification through optical systems, the magnification contributed by the eye is usually considered to be constant and is, therefore, not included.)

In a simple electron microscope (see Fig. 38), the objective produces the initial (intermediate) image in its image plane which coincides with an object plane of the projector lens. Electrons from a small area of the intermediate image, concentric with the optical axis, pass into the projector where they are refocused in its image plane to form a more highly magnified *final electron image* which coincides with the screen or photographic emulsion.

c. RANGE OF MAGNIFICATION

With a two-stage unit, the first intermediate image formed has an essentially fixed magnification (determined by the positions of the specimen and the objective and projector lenses) which is on the order of 100 diameters. The single projector lens provides a variable magnification in the range of 50 X–300 X. For such a system the total magnification is:

$$M_t = M_0 \cdot M_p \tag{16}$$

where:

M_t = total magnification.
M_0 = magnification of objective lens.
M_p = magnification of projector lens.

Where it is desirable to work at a broader range of magnifications, two arrangements to attain this goal are possible. The first involves the selection from an interchangeable set of projector pole pieces and, by these means, the intensity of the magnetic field (i.e., the focal length) is varied and thereby the desired range of magnification can be selected. This method is cumbersome. As a result, the alternate method currently incorporated in most instruments involves permanently inserting a second (variable) projector (intermediate) lens into the system. By using magnetic projector lenses of the *three-stage unit*, which consists of the objective, intermediate and projector lenses, a greater range of magnification can be attained. The magnification of such a system is determined by the equation:

$$M_i = M_0 \cdot M_t \cdot M_p \tag{17}$$

M_t can be varied over ranges such as 1000 X to 100,000 X by varying lens currents. Another significant advantage of this type of arrangement, in addition to flexibility, is that the microscope itself can be of more compact construction. The reason for this is that since the values of the individual magnifications are significantly reduced, this results in shorter image dis-

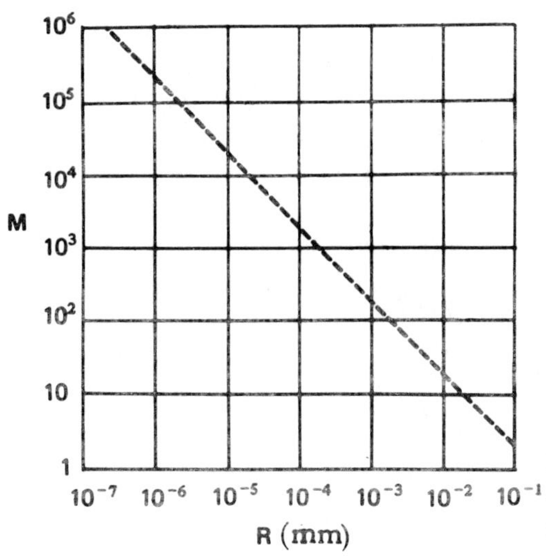

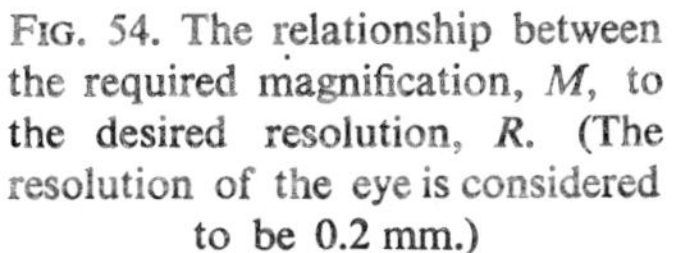

FIG. 54. The relationship between the required magnification, M, to the desired resolution, R. (The resolution of the eye is considered to be 0.2 mm.)

tances and a reduction in the total column length. A third advantage of this system is that it permits designs essentially free of image distortion.

So far as the upper limit of the final magnification is concerned it may be possible to magnify any intermediate image without limit. This in itself may be valueless if further details of the object are not finally discernible to the eye. The maximum useful magnification is determined by the ratio of the limit of resolution of the eye to that of the instrument (d), which is mathematically expressed as:

$$M = \frac{d_e}{d} \tag{18}$$

where:

M = magnification.
d_e = limit of resolution of the unaided eye in mm.
d = limit of resolution of the instrument in mm.

Since d_e (conservatively speaking) is equal to 0.2 mm, at 12 inches viewing distance at low levels of illumination, equation (18) can be written as:

$$M = \frac{0.2}{d} \tag{19}$$

The relationship between resolving power and required magnification is seen on Figure 54. From this graph it can be seen that where 10 Å is to be resolved, a final magnification greater than 200,000 X would usually be useless for viewing distances at 12 inches (i.e., would provide "empty magnification").

A fluorescent screen (see p. 69) emits light almost uniformly through 180°. The unaided eye at a distance of 12 inches from the screen accepts light from a solid angle ranging between one and two degrees. Therefore, the bulk of the light output of the screen is lost.

The apparent size of an image on the screen is determined by the size of the image formed on the retina. In the unaided eye, the size of the retinal image is determined by the angle subtended by the object as viewed by the eye. Greater detail may be examined by permitting the eye to approach closer to the screen. Since the eye cannot focus on an object closer than its near point (about ten inches), a converging lens is placed between the eye and the screen. The virtual image thus formed subtends a greater angle as viewed by the eye and thus provides greater detail for examination (at a new, closer near point).

A 10 X magnifier of N.A. = 0.09 will provide (on the retina) an image that is 10 X the height of the image of the object viewed at the nearpoint of the unaided eye. This represents an area increase of 100 X. Under these conditions the lens is proportionally closer to the screen and accepts light from an angle of about ten degrees. Since this solid angle accepts 100 X the amount of light viewed by the unaided eye, the same amount of light energy strikes the retina per unit area and the enlarged image appears as bright as the image formed by the unaided eye.

When an optical magnifier (of adequate aperture) is used to observe a fluorescent image, the total magnification demanded of the electron system is reduced. This permits lower electron current densities at the specimen level or higher screen brightness than when the unaided eye is used and the added enlargement (which reduces brightness in proportion to the square of the added magnification) is provided electronically.

A practical limit to this optical magnification is usually set by the graininess of the fluorescent screen. The ultimate practical limit would be set by the spread of the high voltage electrons in even a "grainless screen."

These considerations, together, decide the practical division of electronic and optical magnification in any particular visual study.

C. IMAGE TRANSLATING SYSTEM

The information which the electron microscope provides is contained in the final image in the form of variations of electron intensity over its area. This electron image must be converted into a visible light image in order to be seen by the eye. This is done both by means of a fluorescent screen and photography.

1. Fluorescent Observation Screen

In the electron microscope, the instantaneous and continuous transformation of an electron image into a visible one is accomplished by means of a "fluorescent"* material which has the property, when bombarded by radiation of one wavelength, of emitting light of a longer wavelength. For electron microscopy, sulfides containing a trace of some metal, when irradiated with electrons act by emitting light in the visible range. A yellow mixture of zinc and cadmium sulfides (plus a trace of an activator like silver or copper) has been found to be one of the most suitable materials for use in electron microscopy. This mixture, in the form of finely divided particles, is coated on a plate with the aid of a dilute binder such as gum arabic or collodion. The visible image can be viewed directly or observed through a low power optical magnifier.

The screen provides the opportunity for:

(a) Orientation and location of the desired field of view.

(b) Study of the general characteristics of the image of the specimen.

(c) Accurate focusing of the lenses.

(d) Alignment of the entire instrument.

The resolution obtainable in the image seen on the screen depends primarily on the grain size of the fluorescent material and secondly on the spreading of the beam as it penetrates into it. It has been determined that a coating of 50–100 μ in thickness, which is desirable for work in the 50 to 100 KV range, can, with an image of good contrast, provide a resolution of 35–50 μ. Since this is better than the limit of the resolving power of the unaided eye, a 3 X–10 X viewing telescope or magnifier can be profitably used. In any case, the contrast of the image observable on the screen is limited and thus the finer details of the image may not be visible. The photographic method of image translation overcomes this latter limitation.

2. Photographic Recording

A photographic recording material consists of a plastic or glass base coated with an emulsion that is made up of a layer of gelatin in which is embedded a photosensitive silver halide. The electron beam acts by the liberation of free silver from the silver halide grains and produces, after conventional development, a photographic negative of the final electron

* Fluorescence and phosphorescence are distinguished by the time between the excitation (collision with an electron) and the emission of light. The so-called "fluorescence" of the screen material continues for periods longer than 10^{-8} seconds after excitation (usually for periods up to 10^{-1} seconds) and, therefore, is more properly phosphorescence than "fluorescence." Zinc and cadmium sulfide are typical electron phosphors.

image. A print of the image on the film is known as an *electron micrograph* and serves as a permanent record. The fine-grained negative contains a more detailed and higher contrast image than that produced on the fluorescent screen. By photographically enlarging such a negative, a micrograph can be obtained which makes the detailed information of the final image visible to the eye. Since such a record is superior, both qualitatively and quantitatively, to the image visible on the viewing screen, photographic image translation has become the routine means used in the critical study of material that is being investigated with the electron microscope.

3. General Considerations in Image Translation

a. BEAM INTENSITY LEVEL

In viewing the specimen image, or in its photography, there are three general considerations which together determine the working level of beam intensity during each of these two activities.

(1) The beam intensity should be adjusted so as to avoid excessive electron bombardment of the specimen. Excessive irradiation can cause the effective temperature of the specimen to rise to hundreds of degrees centigrade and result in considerable specimen damage. Therefore, to minimize specimen "aberrations" both the maximum intensity of the beam and the length of exposure during visual examination and photography need to be carefully controlled.

(2) The level of beam intensity must, however, be such that an image of sufficient brightness will be formed on the retina to permit location of the desired field of view, precise focusing and examination of the electron image.

(3) Short exposure of the specimen during photography is desirable because it reduces the possibility of obtaining a blurred picture. Exposures longer than about three seconds may result in loss of resolution due to specimen drift and circuit instability (see p. 94).

b. CHOICE OF PHOTOGRAPHIC EMULSION

With any given photographic emulsion, a longer exposure (to a given negative density) requires a lower beam intensity and, therefore, produces less net specimen damage but increases the risk of blurring due to specimen movement or instrument instability.

The smallest point discernible on a photographic plate is determined by the size of granules of the photographic emulsion. This means that a photographic emulsion of smaller grain will provide greater resolution.

By permitting greater photographic enlargement, fine grain emulsions permit photography at *lower electronic magnifications*. This ought to permit the use of even lower beam intensities for the same exposure time. Fine grain emulsions, however, are photographically slower (i.e., require more exposure time and/or intensity) than coarser grain emulsions, and there is usually little net gain with respect to specimen damage when using fine grain emulsions.

At low electronic magnification, the image on the screen or film is *brighter* than at high magnification for a given illumination. This is a simple consequence of spreading the total energy over a smaller area and thereby obtaining a greater energy per unit area. The greater image illumination at low electronic magnification permits the use of the slower higher resolution photographic emulsion for a fixed amount of specimen damage. Fine grain emulsion has an additional advantage. It usually permits *greater contrast* than does a coarse grain emulsion (fast photographic emulsion) and, therefore, permits increased resolution of low contrast specimens.

To a good approximation, contrast for electron exposed emulsions is equal to about 2.3 times the developed density for *all* emulsions.[24] Useful contrast requires the recording of a large range of densities *with precision.* Most of the noise or imprecision in the recording of density in an electron photomicrograph is due to the statistical fluctuations in the number of electrons falling on an elemental area of the emulsion. This noise increases as the square root of the number of electrons. The information bearing signal increases in proportion to the number of electrons and, therefore, the net signal to noise ratio or precision of recording of density increases as the square root of exposure. For this reason slow, thin, fine grain emulsions, which can record very high densities and collect the information from larger numbers of electrons to produce a given negative density, can provide more *useful* contrast.

The parameters involved in film choice are the electronic magnification, the optical magnification, the grain of the screen and the grain of the film. The amount of optical magnification depends on the necessary increase in retinal image size (without loss of brightness) to permit accurate focusing (see p. 67). It is usually convenient to choose a photographic emulsion with fine enough grain so that at minimum electronic magnification the required optical magnification approximately matches the resolution of the screen. Under these conditions, the brightest possible focusable image will be formed on the retina for a fixed level of specimen damage if the level of illumination during both focusing and photography is held constant (see p. 49).

D. OTHER ELECTRON MICROSCOPE COMPONENTS

1. Specimen Chamber and Holder

The specimen chamber, located either between the condenser and objective lenses or within the objective lens, must be constructed so as to permit easy and rapid exchange of the specimen holders. In addition, incorporated into this component must be a specimen stage which makes it possible to:

(a) Move the specimen over a distance of about one-fourth of a centimeter in two coordinate directions at right angles to the optic axis, to allow bringing the desired field into view.
Position the specimen with a precision of about 0.1 μ, thus permitting accurate relocation of a desired area.
Control the movements of the two external stage controls, free from backlash greater than about 0.1 μ, thus maintaining the desired area in view.

(b) Maintain the positioned specimen stationary within about 1 Å for up to three seconds, thus permitting photographic recording of the image.

(c) Dissipate heat from the specimen and objective lenses, thus avoiding overheating the specimen.

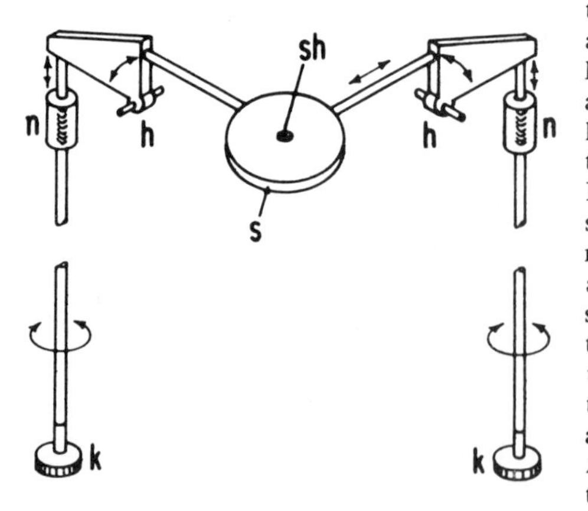

FIG. 55. Schematic diagram of a stage control unit. Rotation of the control knobs (*k*) results in movement of a fine screw located in a stationary nut (*n*) on the microscope column. This, in turn, produces a motion along the axis of the screw. As a result, the lever hinged to the column at *h* rotates around *h*. Since the two arms of the lever are at right angles to each other, the direction of the motion is changed. In so far as the second lever arm is shorter than the first, additional reduction of mechanical motion over and above that provided by the fine screw is also achieved. The motion of the short lever arm is transmitted to the stage (*s*) through a rod connected to the stage. The stage is backloaded against the lever arms by a spring. A specimen holder (*sh*) is located in the opening in the center of the stage.

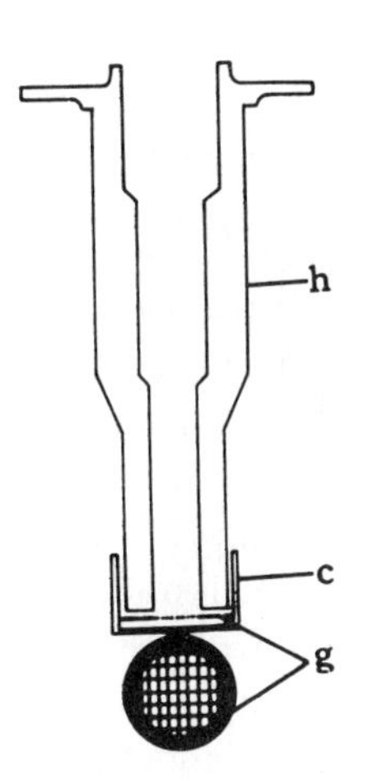

FIG. 56. One type of specimen holder and a specimen grid. (*h*) specimen holder; (*c*) grid cap; (*g*) grid.

The specimen stage (Fig. 55), which is made of a good thermal conductor, is designed to slide directly above the lens or between its pole pieces. The stage is held firmly in contact with the lens by some type of spring or gasket. The drive mechanisms for the stage movement usually act at 90° to each other.

The specimen holder (Fig. 56) is often cartridge-shaped and fits firmly into a conical hole in the center of the specimen stage block and its apical, specimen-containing end thus is recessed into the upper pole piece of the objective lens.

For specimen support, a *grid*, a circular disc of metallic mesh about 3 mm in diameter, is used (Fig. 56). The grid is usually made of copper, since this metal has the necessary high thermal conductivity and is non-magnetic, but for special purposes other material can be used. The grid is often placed at the bottom of a small cap which in turn can be slipped onto the tubular end of the specimen holder.

2. Objective Aperture

The usual objective lens is provided with a physical aperture to permit the use of an optimal aperture angle but, even more significantly, to provide usable contrast. While small apertures can improve contrast, if they are not kept clean they can degrade the image quality. If the edge of the aperture becomes covered with a thin, electrically insulating layer of contamination, then when subject to bombardment by the wide angle scattered electrons (see p. 60), it becomes charged. The aperture then acts as a weak electrostatic lens which, however, is capable of significant effects because of its proximity to the beam. Since the contamination process is not usually completely symmetrical, the "lens" ultimately formed will be asymmetric and thus introduce astigmatism.

With the aforementioned considerations in mind, the ease of cleaning was one of the three principal requirements for material selected for use for aperture discs; the other two being that they have a high electrical conductivity and be readily machinable to give openings that are round, smooth and very small. These requirements were to a large extent met by platinum and molybdenum. Nevertheless, such apertures had to be frequently cleaned (e.g., after 6-hour exposure) because of the buildup of contaminants. This usually meant "down time" and interruption of the work schedule.

To overcome the need for frequent cleaning, multiple aperture holders were developed for most microscopes. This permitted not only quick change of several apertures without loss of the column vacuum, but also the interchange of apertures of different sizes. Another, but not as yet widely adopted approach, is the use of *heated apertures*.[25,26]

A more successful approach to overcome the problem of frequent cleaning has been the use of *ultrathin metal apertures*. It has been found that such apertures have a negligible astigmatic effect. This may be explained as follows: The critical edge of the hole of the aperture has an area orders of magnitude smaller than the typical aperture. The lens effects due to a charged layer of contamination on such an edge are, therefore, very much reduced. Because of its relatively low heat conductivity, the electrons impinging on such an aperture can be used to raise its temperature sufficiently to remove contaminating layers while it is still in the microscope. Such apertures are now commercially available. They can also be readily made by evaporating a metal, such as silver, over spheres of dextran resting on a collodion or formvar membrane. Removal of the film support and spheres with solvent produces a thin (*circa* 0.5 μ) metal sheet with apertures corresponding in diameter to the dextran spheres.[27] Such apertures are serviceable for long periods (e.g., months). They need only be replaced if they close as a result of dust or excessive contamination buildup.

The diameter of the aperture opening used can be varied; most commonly used sizes being 15 to 50 μ. Above 75 μ, the presence of the aperture is of little consequence since relatively few electrons are scattered at angles sufficiently large to be removed by such an aperture, while below 25 μ the instrument's resolution may begin to be affected. This is due to the fact that at small aperture sizes we may be in the diameter range where the diffraction effect (rather than spherical aberration) limits the resolving power (e.g., at 50 KV, with 3.0 mm focal length objective lens, and a 6 μ aperture in a rear focal plane, α is 10^{-3} and the resolution is 20 Å, while at 24 μ, that is, at α_{opt}, the resolution is 10 Å; see Fig. 50). Thus, theoretically, the optimal aperture size should be determined from a resolution curve which will show the point where the effects of diffraction and spherical aberration are equal. In practice, where very thin, low-contrast specimens are being examined, use of an aperture substantially smaller than the theoretical optimum may be helpful since the theoretical limit may, in any case, not be achievable in such specimens because of contrast limitations. Such a choice will, however, have another drawback in practice since the smaller the aperture the more rapidly it becomes contaminated and thus produces astigmatism.

The position of the aperture within the objective lens (see Fig. 51) in relation to the specimen is a critical factor in determining image contrast, resolution and rate of aperture contamination. Thus for a particular desired α_0, if placed very close to the specimen, the aperture must be very small, the contamination rate will be high and the field of view will itself be limited by the aperture. If placed close to the rear pole piece, the aperture can be larger and the effects of contamination and field obstruction will be greatly reduced. Ideally, the aperture should be placed in the rear focal plane of the objective lens but this often does not significantly improve its effectiveness over that when it is positioned just in front of the pole piece.

3. Photographic Apparatus

While some instruments are equipped with a 35 mm camera, most are designed for use with plates or sheet film (frequently of more than one size). A typical camera complex (Fig. 57) may consist of the following elements:

(a) Camera chamber. This compartment usually is located in back of and at almost the same level as the fluorescent screen. In some instruments it can be evacuated independently by means of a connection with the main column vacuum pump circuit (see p. 78).

(b) Cassette magazine. This metal box is located in the camera chamber and holds a number of stacked cassettes each containing a photographic plate.

(c) Rack. This carrier mechanism serves to move a cassette to a suitable position under the screen and, after exposure, to a receiver.

(d) Cassette receiver. This compartment is often located in front and on a level beneath that of the fluorescent screen. In some instruments it can be closed off from the column by means of an airlock and can be evacuated by means of a separate vacuum connection to a fore-pump.

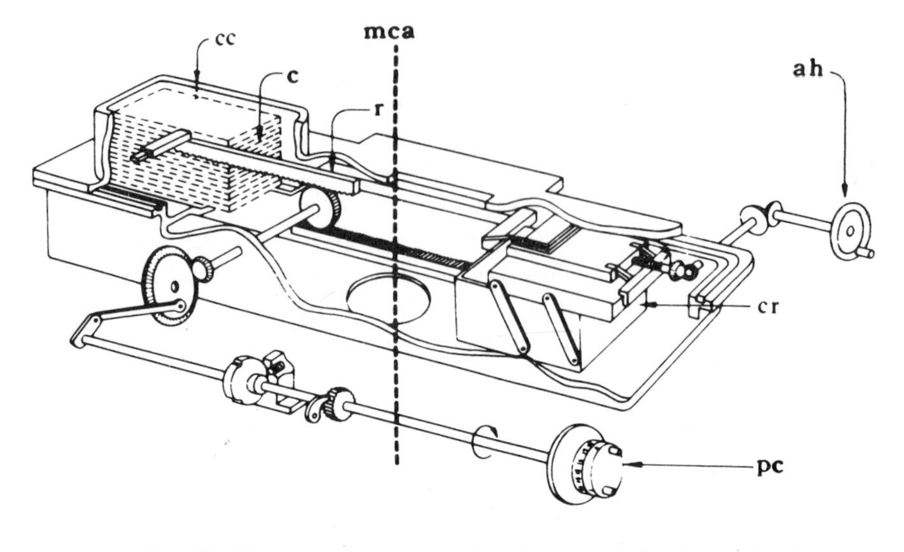

FIG. 57. Diagrammatic representation of a camera chamber mechanism. (Not shown is the cassette magazine.)

cc — camera chamber.
cr — cassette receiver.
c — cassettes.
mca — microscope column axis.
r — rack.
ah — air-lock chamber handle.
pc — plate counter.

4. Vacuum System*

It is necessary that the column of the electron microscope be evacuated for the following three reasons:

(a) To avoid collisions between electrons of the beam and gas molecules.

At or near atmospheric pressure, this would completely stop an electron beam over a path of the order of a millimeter (at 100 KV). Even at pressures of 3×10^{-3} torr (1 torr = 1 mm of Hg), this would result in a spread of electron velocities thus increasing chromatic aberration (decreasing resolution), as well as reducing the contrast of the image.

(b) To avoid high voltage breakdown (gas discharge) between anode and cathode.

(c) To avoid substantially decreasing the life of the tungsten filament which takes place when excessive amounts of gas, especially oxygen, are present.

It has been found that the above requirements are satisfied if the gas pressure is reduced to 10^{-4} torr because under these conditions the mean free path of electrons is about 2.5 meters which is, therefore, adequate for modern instruments whose column length is of the order of one meter.

---

* For a detailed discussion of vacuum pumps used with electron microscopes, see Appendix N.

To attain and maintain the desired vacuum level, a very efficient pumping system must be incorporated into the microscope system. The reason for this is that the vacuum is frequently altered by such operations as specimen, film, or filament replacement. When pole piece cleaning is necessitated, the entire vacuum needs to be broken.

At room temperature and at pressures below 10^{-3} torr, adsorbed gas from surfaces inside the vacuum apparatus is only slowly released and this gas source behaves as a virtual leak in the system. It has, therefore, been found advisable to use a pumping system of sufficient capacity to reduce the vacuum to below 10^{-4} torr in the presence of *virtual* and/or *real* leaks of the order of one liter of gas at 10^{-4} torr per second into the system and to run the pumps continuously during the operation of the microscope.

The basic components of the vacuum pumping system of an electron microscope consist of a rotary mechanical *fore-pump* which pumps at relatively low speeds (40–200 L/minute at 10^{-2}mm) and which is used to reduce the pressure to a level (*circa* 10^{-2} mm) at which pressure activation of a high speed (100 L/second at 10^{-6}mm or better), electrically heated, water cooled, high vacuum, oil *diffusion pump* is initiated (Fig. 58). Both types of pumps can be operated so as to be essentially vibration-free and thus can be positioned within the physical housing of the microscope.

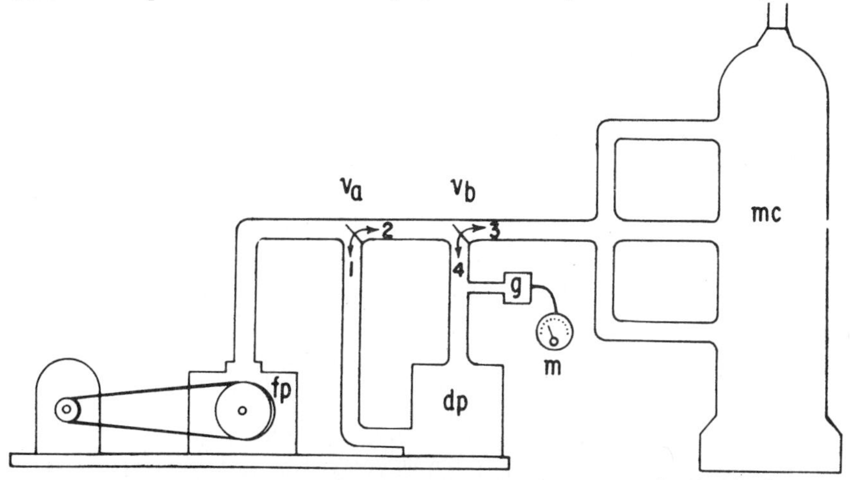

FIG. 58. A schematic diagram of the elementary components of a simple electron microscope vacuum system. Fore-pump (*f.p.*); diffusion pump (*d.p.*); vacuum gauge (*g*); vacuum meter (*m*); flap-valves (v_a, v_b). The pumping schedule has three phases. Phase I: With v_a in position 1 and v_b in position 3, the fore-pump rough pumps the main microscope column, *mc*. Phase II: With v_a in position 2, the fore-pump rough pumps the diffusion pump. Phase III: With v_b in position 4, the fore-pump pumps the diffusion pump and the latter pumps down the column to operational levels.

When the fore-pump has reduced the pressure in the oil diffusion pump to the order of 10^{-2} torr, the diffusion pump heater is turned on (Fig. 59). When the oil begins to boil, the streams of oil vapor are directed downward by the "jets" of the diffusion pump. The rather high momentum of the oil vapor is transmitted by collision to the gas molecules in the path of the oil vapor jets and propels them to the bottom of the pump, thus setting up a pressure gradient, increasing from top to bottom. When the oil molecules collide with the cool wall of the pump, they condense and return to the boiler. The fore-pump removes the gas at the bottom (high-pressure end) of the diffusion pump.

The specimen and camera chambers on many microscopes can be closed off from the main column by vacuum valves. Thus the vacuum in these chambers can be broken without disturbing the vacuum in the rest of the microscope. These two chambers can then, for example, be evacuated by a special fore-pump and the vacuum in the chambers is quickly restored.

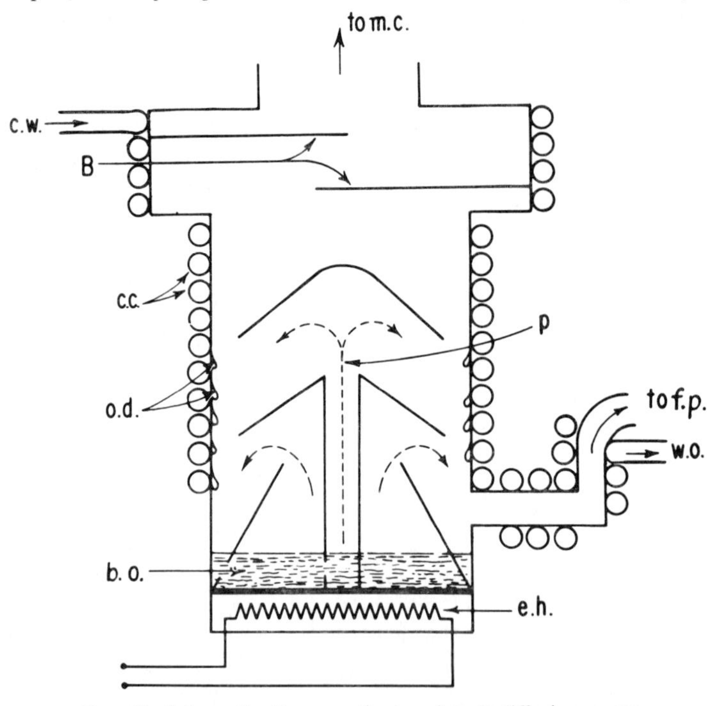

FIG. 59. Schematic diagram of a two-jet oil diffusion pump.

cw — cold water.
cc — coil carrying water.
d — condensing oil droplets.
wo — water out.
mc — microscope column.
b — baffle.
p — path of oil droplets.
bo — boiling oil heater.
eh — electrical heater.
fp — fore-pump.

This pump can also be linked up to a vacuum desiccator in which the photographic plates (or films) can be initially prepumped before being loaded into the camera. This is of considerable advantage because the adsorbed water in a gelatin photographic emulsion constitutes a very large virtual leak. On some instruments an air-lock mechanism is also used to isolate the gun housing, thus also permitting filament exchange without breaking the main column vacuum.

Newer model electron microscopes are now equipped with automatic vacuum systems. These frequently consist of push-button activated, programmed, motorized, piston-type valves, which provide for greater operational convenience and reduce the chance of operating valves in the wrong order. This is important because hot, diffusion pump, hydrocarbon oil may decompose when exposed to air, lowering the efficiency of the pumps, and a gust of air can carry clouds of oil vapors backwards, contaminating the entire vacuum system. (For this reason, the diffusion pump-cooling water is kept running for an interval after the pump heater is shut off to quickly reduce the temperature of the oil.) A water-failure switch is frequently incorporated into the system to turn off both the diffusion pump heater and high voltage systems. (If the pump heater were to remain active when condensation of oil vapor in the pump is stopped, the vacuum apparatus would be filled with oil vapor and, in addition, the pressure would rise and the high voltage supply would discharge at high current).

Vacuum gauges of high sensitivity are usually incorporated into modern microscopes and the vacuum system is also frequently equipped with a vacuum-failure switch. For normal operational needs it is not necessary to have an accurate measure of the vacuum pressure. A rough estimate can be obtained from the appearance of a gas-discharge tube connected to the column to which is applied an alternating potential of several KV. With such a device, rough changes in vacuum between 0.1 and 0.001 torr can be recognized. At lower pressures there is no visible discharge in such a tube and this corresponds closely to the desired operating range.

Molded rubber, or preferably Viton "O" rings, are usually used to seal the demountable vacuum joints of the electron microscope. These rings are placed in grooves and undergo compression in assembly of the seal. Excessive greasing of "O" rings to insure vacuum tightness is to be avoided, for the grease releases vapors which can contribute to specimen contamination (see p. 95). With good-fitting, non-moving gaskets whose surfaces as well as those of the metal joints are smooth and clean, greasing can usually be avoided.

5. Power Supply

The electrical power supply of an electron microscope must provide:

(a) Current to heat the tungsten filament and thus generate the image-forming electrons.

(b) High voltage to accelerate the emitted electron beam so that its particles have a sufficiently short and precise wavelength.

(c) Current to each magnetic lens to provide the necessary magnetic fields.

The regulation for the second and third items mentioned is very critical and should be maintained constant to the degree required to provide the desired resolution (see pp. 56,57) for a time-interval sufficient to focus the final electron image and permit it to be photographically recorded.

a. FILAMENT CURRENT SUPPLY

Modern electron microscopes usually use an ac heating current for their tungsten filaments, a more convenient source than batteries which would have to be insulated for 50–100 KV.

The beam current from all but the central portion of such an ac heated filament [which alone has a fixed bias with respect to the grid cap (see Fig. 40)] will fluctuate at twice the heating frequency. Such fluctuations in beam current may easily cause changes in the charging and discharging rates of critical contaminated surfaces in the column and, therefore, generate slight blurring effects. For this reason, some newer microscopes use rectified and filtered ac (i.e., dc) filament heater supplies.

b. HIGH VOLTAGE SUPPLY

The selected magnitude of the accelerating voltage of an electron microscope is dependent, in part, upon the required penetrating power of the beam. Since image contrast is mainly produced by differential electron scattering (see p. 60), the velocity of the beam selected should be considerably greater than that adequate to penetrate a suitable specimen.

Suitable accelerating potentials fall between 15–1000 KV (standard range, 25–100 KV). Some electron microscope power supplies are so designed as to permit selection of the desired accelerating potential at regular steps. Other instruments operate at a fixed potential.

The beam current which the high voltage source must supply is on the order of magnitude of 0.1 milliampere or less. Thus the accelerating potential supply will be of the high voltage, low current type, delivering from 2 to 10 watts of power.

With a limiting resolution of 10 Å, the maximum high voltage fluctuation permissible is of the order of 0.01 % (see p. 56). To attain such a degree of stability, stable high-gain amplifiers with negative feedback from a resistive voltage divider between the cathode and ground are used. For considerably greater stability, in some instruments the voltage divider is isolated in its own insulated oil tank and may even be thermostated.

c. LENS CURRENT SUPPLY

The supply of current necessary to energize each of the magnetic lenses is now also supplied by high stability feed-back amplifiers. Temperature-compensated resistive dividers in series with the coil windings are used. The lens coils are sometimes cooled with water from a constant temperature source.

E. OPERATIONAL REQUIREMENTS

In order that one may be able to obtain electron micrographs of high resolution, a number of requirements have to be fulfilled. These include (1) aligning of the lenses, (2) detection and elimination of lens asymmetry, i.e., minimizing residual astigmatism, and (3) elimination or control of potentially disturbing conditions, such as specimen mechanical and thermal drift, stray magnetic fields, instrument vibrations, etc.

1. Alignment

Alignment of the components of the column is necessary to minimize the effects of circuit instability which especially degrade off-axis resolution (through field aberrations), as an increasing function of distance from the optic axis. As already discussed, the extent of image magnification and rotation is dependent on the accelerating voltage and lens current. Thus slight electrical disturbances will induce image movement, the magnitude of which will increase directly with the distance from the optical axis (Fig. 60). This movement is relatively insignificant near an optical axis but can be the major source of loss of resolution at large distances from the axis.

The usual method of alignment is based on *voltage centration*. This involves determining the point in a focused image which remains stationary (namely, around which rotation occurs) when the accelerating potential is varied with fixed lens currents, i.e., the point at which the image is least sensitive to field aberrations as a result of variations in the high tension. This voltage center defines the position of *one of the optical axes of the magnetic lens system*. The *magnetic center*, which defines the other optical

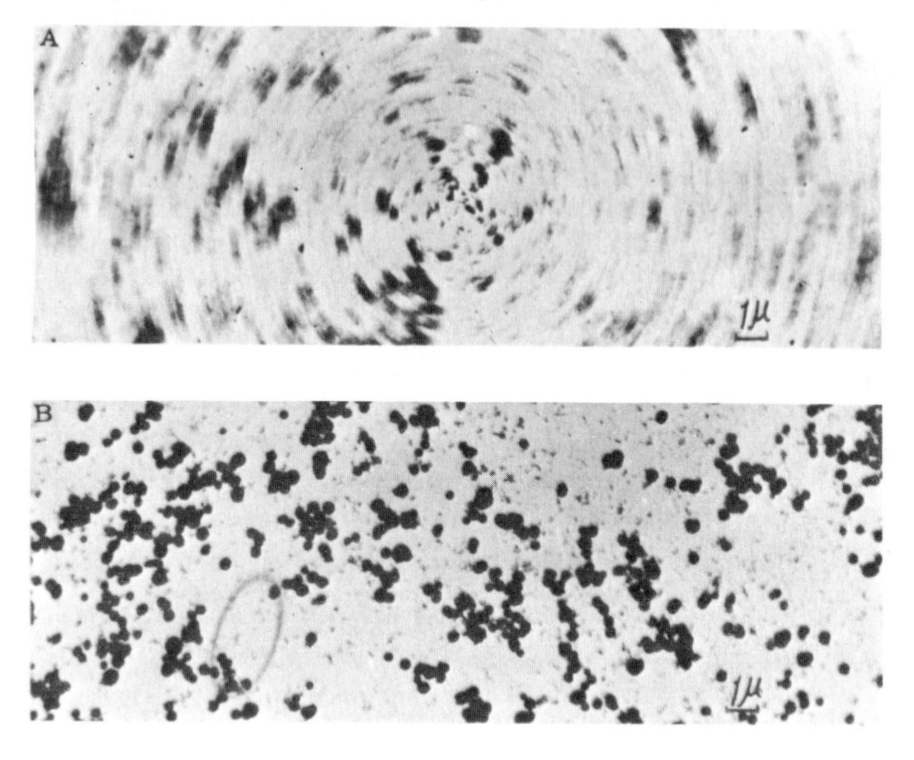

FIG. 60. Effect of voltage fluctuations on off-axis resolution. (Courtesy of Hitachi, Ltd.)
(A) An image in the presence of large, artificially superimposed fluctuations of 5 percent of the accelerating voltage. Note how rapidly resolution is degraded as one proceeds peripherally from the voltage center. Near the optic axis, chromatic field aberration is smaller than axial chromatic aberration, whereas off-axis it is many times larger.
(B) A similar field in the absence of such fluctuations.

axis, is determined by locating the stationary point in a focused image when the lens current is varied, the accelerating potential being fixed. The two axes are not necessarily coincident.

Voltage centration is the usual method of choice because the accelerating potential is both less well-regulated and more susceptible to erratic fluctuation and, therefore, the image movements due to voltage fluctuations will be minimum when voltage centration is employed. The microscopist is most fortunate when the mechanical axis of the magnetic lenses of his microscope closely coincides with both optical axes. Under such circumstances, the rotational effects of fluctuations in both voltage and current on resolution are minimized. In practice, however, the optical axes usually do not coincide with the mechanical axis. The voltage and magnetic centers represent the points of the image in the field of view for which the vector sums of all displacements resulting from fluctuations of the accelerating potential and lens current, respectively, are zero. That the two centers do not coincide is due to the fact that machine tolerances to attain *preset alignment* are not easily achieved and/or inhomogeneities in the iron or stray magnetic fields induce curvature of the optical axes. Thus commercial electron microscopes come equipped with controls that make alignment possible.

Early electron microscopes permitted wide translation of the components. Unfortunately, the extent of initial misalignment could be considerable and correction could be a very tedious process. Current microscopes are machined so that only that amount of translation of the components can be made as to compensate for the maximum divergence between the mechanical and voltage centers which may result during manufacture. Total "loss" of the beam at low magnifications due to misalignment of objective, intermediate or projector lenses is, therefore, now much less likely.

The following seven alignment conditions must be met (usually in the following order) for an electron microscope having a single condenser lens.*

(a) The beam emerging from the electron gun must be centered relative to the condenser lens aperture.
(b) The gun and condenser as a unit must be centered relative to the objective axis (and specimen).
(c) The tilt of the illumination system must be adjusted to be made parallel with the objective lens axis.
(d) The projector lens must be aligned relative to the objective lens axis.
(e) The intermediate lens must be aligned relative to the objective lens axis.
(f) The objective lens must be adjusted relative to the center of the screen (to provide voltage alignment).**
(g) The objective aperture must be aligned relative to the objective lens axis.

* For most electron microscopes the double condenser system (see p. 102), where present, is made as an integral unit. The manufacturing process is usually such that internal alignment tolerances can be met and thus no special additional alignment controls are needed. Thus alignment is straightforward and parallels that for a single condenser. This is true when the first condenser lens is regarded as a preset adjustment of spot size, and that variations of lens current are confined to the second condenser during the alignment procedure.

** If substantial movement of the objective lens is required for this step, steps (a) to (e) may have to be repeated, then followed by a *very slight* readjustment of the objective lens alignment.

a. ELECTRON GUN—CONDENSER ALIGNMENT

The electron source must be positioned on the condenser lens axis, otherwise uniform illumination will be obtained only at one condenser setting (i.e., when the beam is a narrow pencil). The effect of a misaligned electron gun is illustrated in Figure 61. It can be seen that with an increase in condenser lens strength, the illumination, as indicated by its region of maximal intensity (cross-over), will move laterally. Moreover, as the condenser current is varied over a wide range on either side of cross-over, the illumination sweeps across and off the field rather than spreading uniformly about the center. Alignment is attained by manipulating screws of the electron gun and condenser which translate these in planes at right angles to the optic axis until symmetrical expansion of the illumination spot on either side of cross-over is attained.

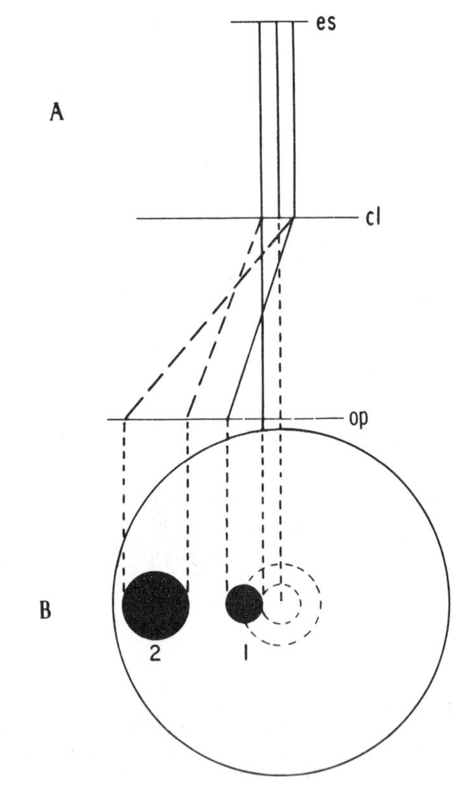

FIG. 61. Electron gun–condenser alignment.
(A) Ray diagram.
es — electron source. *cl* — condenser lens. *op* — object plane.
(B) View on final image screen.

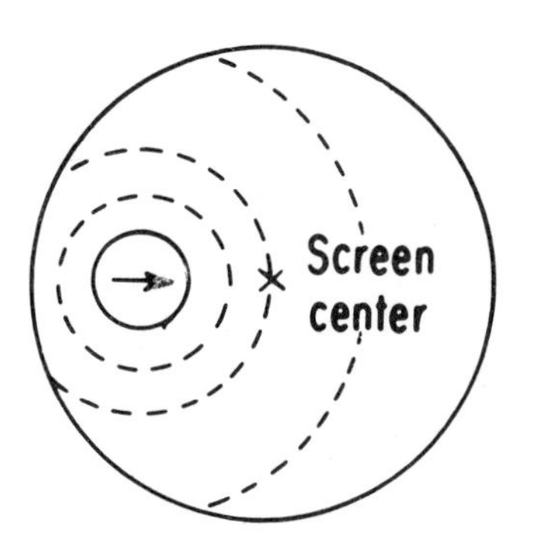

FIG. 62. Illumination — objective lens centration showing progressively focused illumination on final viewing screen.

Gun adjustment will normally need only infrequent checking since it will remain fairly steady during the life of a filament (with the possible exception of its last hour). When filament replacement takes place, it must be adjusted by mechanical centration of the tip of the filament (e.g., using a centering jig), and by setting correct height of the filament tip relative to the grid cap (e.g., using a depth gauge), since this adjustment fixes the magnitude of the beam current and brightness at saturation in fixed-bias guns. When these adjustments are made and the gun is reassembled, gun-condenser alignment may not be necessary if the previous filament had been carefully centered.

If a condenser aperture is introduced it may cause illumination sweep unless properly centered by lateral translation.

b. ILLUMINATION—OBJECTIVE LENS ALIGNMENT

A specimen is brought to focus at low magnification and the condenser is adjusted to cross-over (the brightest image of the source in the plane of the specimen). The condenser-gun assembly is translated to bring the brightest spot to the center of the field (Fig. 62).

c. ADJUSTMENT OF ILLUMINATION TILT

If the axis of illumination is inclined at an angle to the axis of the objective lens, movement of the image will take place with a change in the objective lens current. This can be seen in simplified form in Fig. 63, where, when a large tilt error is present, the image points appear to move from both sides towards central magnification point. This slight movement will be magnified by the other lenses so that the final image of the central point may come to lie so far from the screen center as to be entirely off the screen. As a result, a sweep of the image across the screen will take place as the objective lens current is altered. In addition to sweep, a rotation of the image will be seen at the final screen.

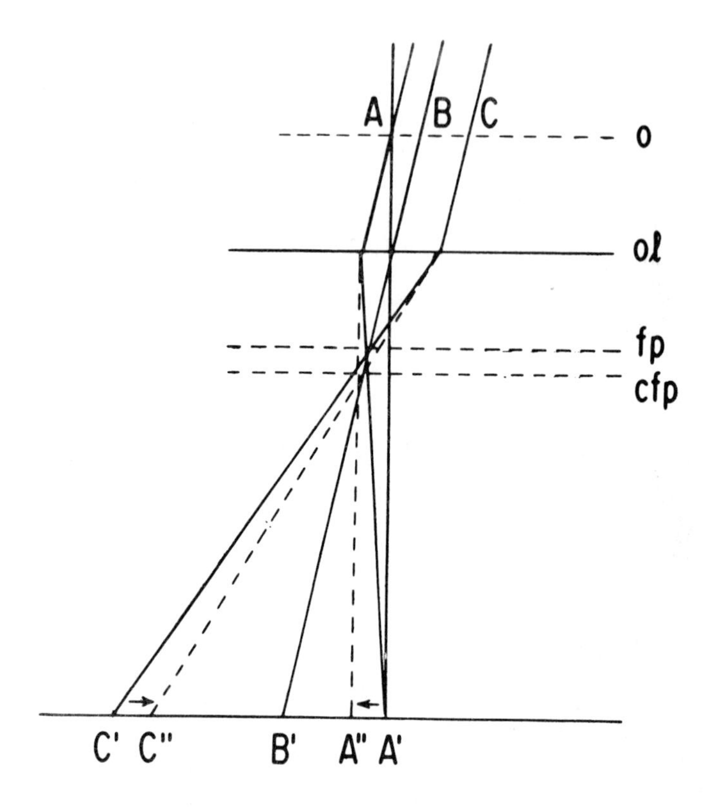

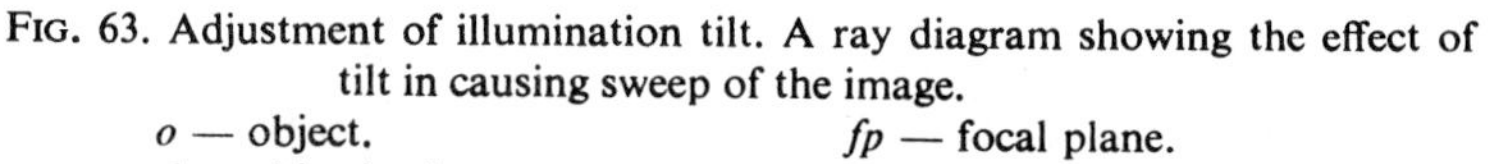
FIG. 63. Adjustment of illumination tilt. A ray diagram showing the effect of tilt in causing sweep of the image.
o — object. *fp* — focal plane.
ol — objective lens. *cfp* — changed focal plane.

Adjustment is made by placing some focused feature of the image in the center of the screen and then defocusing the objective lens. Any image movement away from the screen center is adjusted with the tilt controls, which swing the illumination axis, so that the image is restored to its initial position. When properly adjusted, it should be found that variation of the objective lens current will only result in rotation of the image around the screen center.

Tilt adjustment should again be corrected after (d) projector and (e) intermediate lens adjustment are made. This is due to the fact that since image sweep due to tilt is observed on the final fluorescent screen (rather than objective image plane), any misalignment of the above lenses will also cause *A'*, in Figure 63, to be displaced from the screen center and the final resultant image motion on the screen will not readily be related to the misalignment of the particular contributing elements.

d. OBJECTIVE LENS—PROJECTOR LENS ALIGNMENT

The tolerances for alignment of these two lenses are such that normally they are adequately satisfied by original factory alignment.

e. OBJECTIVE LENS—INTERMEDIATE LENS ALIGNMENT

This can be achieved by use of translation controls on the intermediate lens or on the entire upper half of the column with respect to the objective lenses.

f. OBJECTIVE LENS—SCREEN ALIGNMENT

By varying the voltage or by varying the current of the objective lens the point which does not move, i.e., around which rotation occurs, can be found. This point is then translated to the center of the field.

When this step is completed it is then necessary to return and "touch up" the alignment of the gun-condenser assembly to the objective axis (step b).

g. OBJECTIVE APERTURE ALIGNMENT

Most instruments are equipped with a removable objective aperture holder which permits cleaning or change of the aperture. When the aperture must be removed, realignment of the aperture with respect to the optical axis will be necessary. The general principles in achieving this are as follows:

(1) The specimen or thin support film is brought into focus so as to provide necessary scattered electrons.
(2) The beam is brought to cross-over.
(3) The current going to the filament transformer is reduced to the point where emission of electrons takes place only from the center of the filament.
(4) The lens system is then set so that the plane of the objective aperture is imaged on the screen. (This is usually near zero magnification).

Under these conditions, an image of the objective aperture formed by the electrons scattered by the specimen, as well as a bright and smaller image of the condenser aperture formed by the undeflected electrons, will be seen on the screen (Fig. 64). The objective aperture is manipulated until the images of the two apertures are concentric. The objective aperture is now aligned.

The end result of alignment should be to make the beam parallel and then coincident with the optical axes of the lenses (which may differ from their geometric, i.e., mechanical axis).

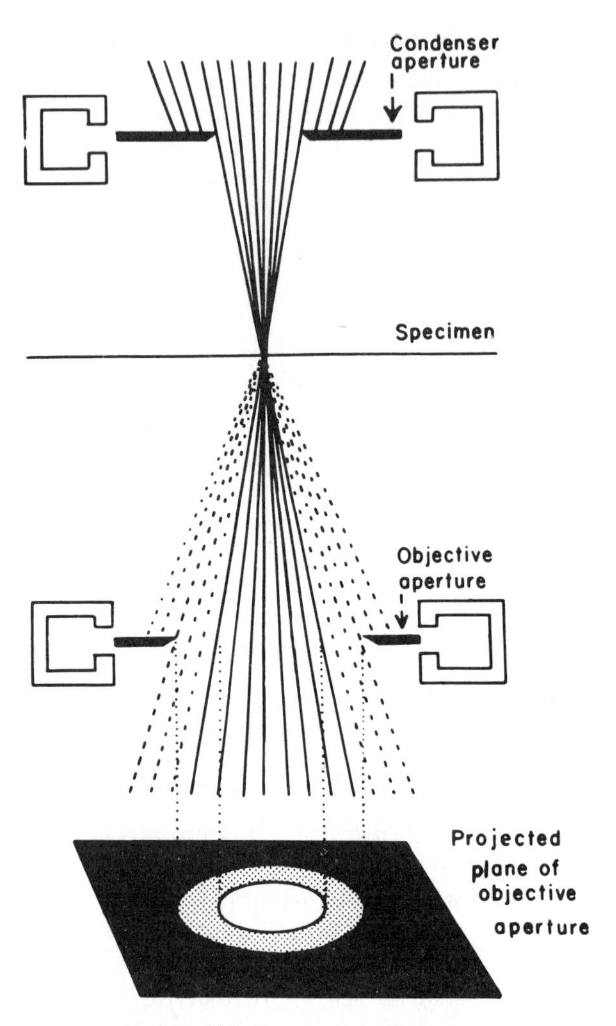

FIG. 64. Objective aperture alignment.

When all the alignment adjustments are made then:

(1) The illumination should be sufficiently bright to enable objects to be viewed and focused at the highest magnification for which the instrument was designed.
(2) The illumination should remain centered as its intensity is varied by changing condenser focus.
(3) The image should remain centered as the focus is varied, or as the voltage is varied.
(4) The image should remain centered as the magnification is varied.

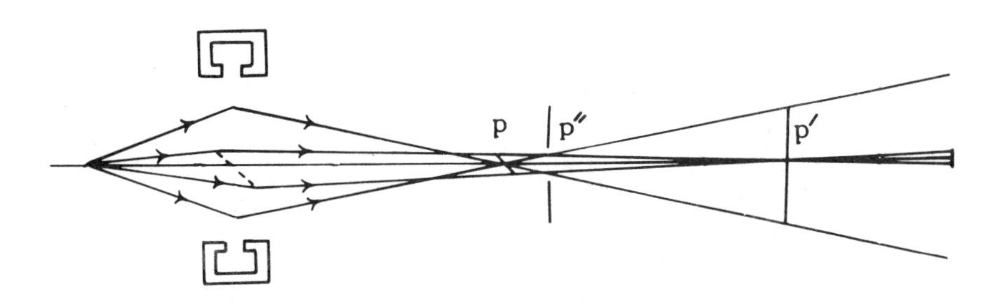

FIG. 65. Astigmatism. Rays parallel to the axis in two mutually perpendicular planes which pass through an astigmatic lens are brought into focus at two different points, P and P'. The disc of least confusion would be located between P and P' and is designated as the "in-focus" position, P''.

2. Detection and Correction of Lens Asymmetry

During the discussion of alignment, it was assumed that lenses and, therefore, their magnetic fields were perfectly axially symmetrical. In actuality, it is extremely difficult to machine lenses of short focal length. The tolerances required for the condenser and projector lenses are less critical and those that come with the instrument usually need no compensation. For the objective lens where the mechanical tolerances are most critical, asymmetry of the pole pieces, whether due to the departure from true circularity of the bore or inhomogeneities in the iron, always results in a slightly asymmetrical magnetic field being formed by this lens (see p. 52). A similar defect can also be induced by the presence of semi-conducting contaminants either on the lens surfaces or on its objective aperture.

An asymmetrical lens, in light optical terms, acts as though an additional weak cylindrical lens (see Appendix I) has been added to the objective of the existing convex, (i.e., "spherical") lens (Fig. 65). The result is that the latter now has different focal lengths for rays entering in planes parallel to the optical axis and at right angles to one another. Thus instead of forming a point image of a point object, as is the case for rays passing through an axial symmetrical lens, two mutually perpendicular line images occur at different distances along the optic axis in image space. When the entering rays are at right angles to each other, rays in intermediate planes are focused at intermediate points along the axis and nowhere is there a "point" image of a "point" object; therefore, this aberration is known as *astigmatism.*

In the electron microscope, this defect can be detected by taking a through-focus series of spherical particles, such as carbon, and examining the rings which are the diffraction images at their edges (Fig. 66). Small holes in fairly thick films are even more suitable objects for visualizing these

FIG. 66. Carbon particles (115,000X). The particles (from left to right) are over, under and in focus (compare with Fig. J-4).

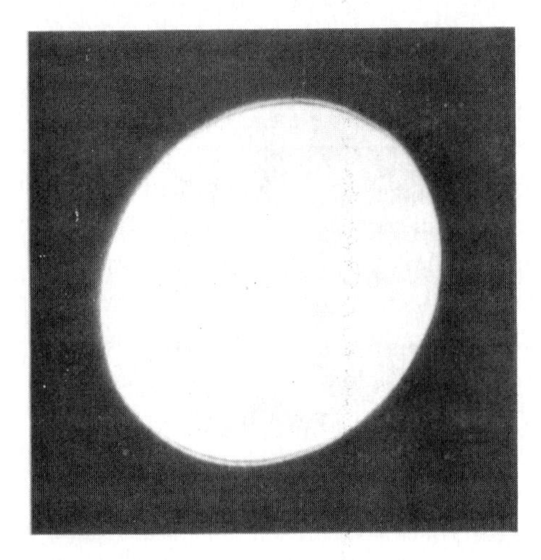

FIG. 67. Diffraction pattern of a small hole in a thick carbon film demonstrating the presence of astigmatism in the lens system. (For quantitative determination, see p. 236.)

Fresnel diffraction fringes. With a symmetrical lens in which the image is in perfect focus, a minimal diffraction fringe always exists. The presence of astigmatism, apparent in the fringes at opposite poles of the test object, only are visible in out-of-focus positions (Fig. 67). This is due to the fact that rays from one of the planes closer to focus give thinner diffraction fringes closer to the edge at one set of poles, while the rays from the other plane are further from focus and wider fringes further from the edge result. Astigmatism limits resolution since all image points appear as circles of confusion of a diameter related to the distance between the two focal planes for object detail in two mutually perpendicular directions.

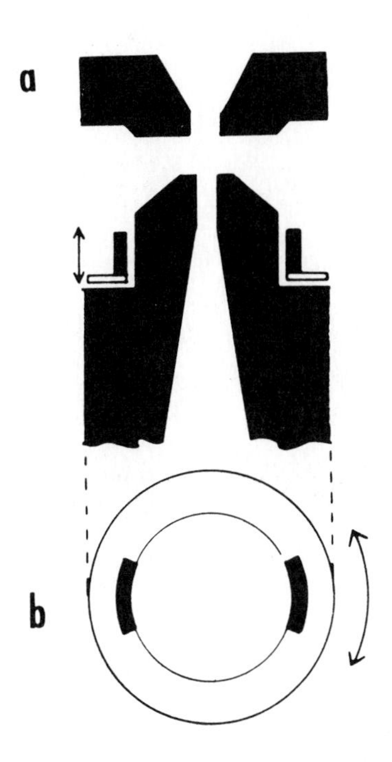

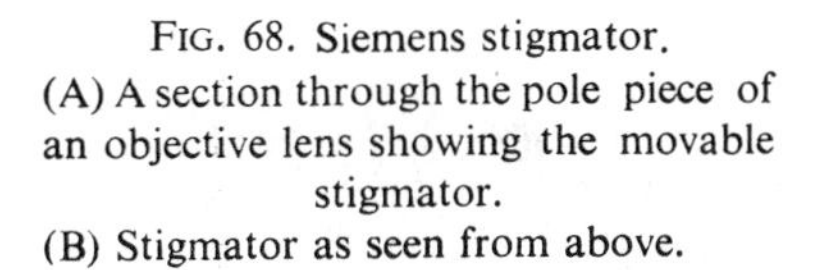
FIG. 68. Siemens stigmator.
(A) A section through the pole piece of an objective lens showing the movable stigmator.
(B) Stigmator as seen from above.

Astigmatism can be corrected by superimposing on the lens magnetic field another field of asymmetric distribution and of variable magnitude which can be positioned so as to oppose and cancel the existing lens asymmetry. The actual correction can be done most easily by means of a *stigmator*. Lenses equipped with stigmators are called *compensated objectives* and in current instruments it is possible to vary the strength and angle of the compensating device by external controls so that the astigmatism can be eliminated during operation of the instrument.

There are essentially three basic types of stigmators currently in use to achieve the correction. The first type is a magnetic stigmator, which acts as an additional weak cylindrical lens of strength just sufficient to correct the cylindrical component of the objective lens. Its axis of asymmetry can be oriented in a direction perpendicular to that of the objective lens' astigmatism. In practice, such systems are found in such electron microscopes as the Siemens Elmskop I and Philips 75 and 100B. In the first of these instruments (Fig. 68), it consists of two iron pieces on a ring of non-magnetic material which are mounted concentric with the objective axis. The ring can be rotated about the objective axis as well as shifted along the direction of the axis. By these means it is possible to independently compensate for the influence of both the magnitude and the direction of the inherent astigmatism of the objective lens.

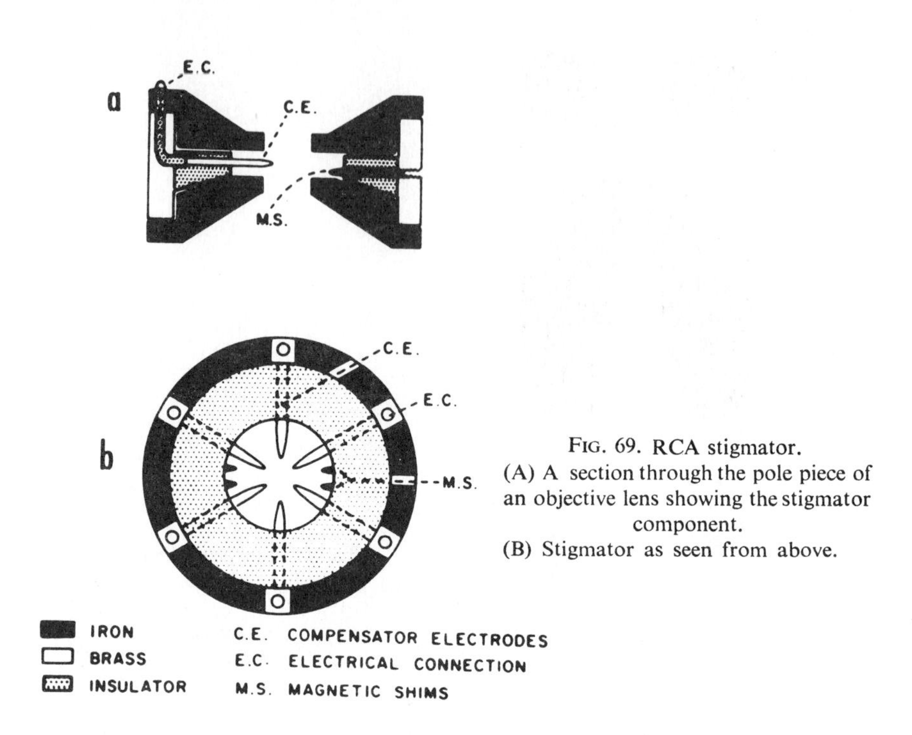

FIG. 69. RCA stigmator.
(A) A section through the pole piece of an objective lens showing the stigmator component.
(B) Stigmator as seen from above.

A second type of device consists of rod electrodes mounted symmetrically around the optical axis of the objective pole piece. By changing the distribution of potentials on the electrodes, one can vary both the direction of the axis and the strength of the weak electrostatic cylindrical lens which is produced. The RCA EMU–3 and –4 and the AEI–EM 6 are instruments making use of electrostatic compensation. As an example, the former has the following system (Fig. 69): six non-magnetic electrodes provide the electrostatic field and four adjustable iron shims vary the magnetic field. The latter serve as a coarse adjustment to reduce the astigmatism to the point where the continuously variable electrostatic compensator is capable of operating at high sensitivity.

The Philips EM 200 and 300 and Hitachi HU–11E make use of electromagnetic compensation systems to correct for astigmatism. In these instruments, both the condenser and objective lenses are provided with electromagnetic stigmators where the magnitude and direction of the electromagnetic shims are controlled by controlling the current in their coils.

Compensation for lens asymmetry should be carried out after the other "external" sources of image defects (see p. 93) have been eliminated. Correction for astigmatism is absolutely necessary where the desired level of resolution is below 50 Å.

3. Disturbances

Under ideal instrument conditions (see p. 190), resolution of about 2 Å can be attained. In actual practice, however, a number of disturbances have to be avoided in order to insure that potential limitations to the achievement of high resolution are not present. Disturbances (other than vacuum leaks, see p. 95) can be classified into two different types depending on the manner in which they are manifested on the screen. Thus there are those which produce a shift or "drift" of the image over a period of time that appears on the screen as image drift. Other disturbances produce a uniform or unidirectional blur in the image. Identification of the specific cause of the disturbance is possible by examining the electron image and correlating the change in appearance with a particular manipulation of the instrument.

The causes of image drift and image blur can be related to a number of different sources which induce one or both of these image defects.

a. STRAY MAGNETIC FIELDS

Electron microscopes are shielded against stray magnetic fields. Nevertheless, where such disturbance is intense the operation of the instrument may be seriously affected. Thus the presence of a slowly varying, strong, transverse magnetic field will make it difficult or impossible to align the column. When such alignment difficulty is found, a test should be made to detect the presence of such an external field.

b. MECHANICAL DISTURBANCES

To minimize possible erratic or systematic displacement of the specimen as a result of mechanical dislodgement of the specimen or its holder from its resting position by mechanical shocks or vibrations, the specimen chamber is usually designed so that the specimen holder is held rigidly when in its operating position. Nevertheless, since the specimen holder must be readily movable and replaceable, it cannot be permanently fixed in position and there may, therefore, be backlash or give in movable components of the stage mechanism. For this reason, external vibrations such as those coming from the passing of heavy trucks or trains can result in image drift. Thus the location of the electron microscope in a building should be where potential for disturbance is minimal. Where building vibration is known to exist, shock absorption mounts may be desirable. Also the rotary mechanical pump should be shock-mounted and even mechanical controls on the microscope, such as the shutter, should be operated without erratic motion to avoid vibration and thus minimize image shift or drift. If the mechanical vibrations are intensive and continuous, image blurring will result.

c. ELECTRICAL DISTURBANCES

If the lens has been centered on the voltage axis, small drifts in lens current will produce image shifts. Slight variations in the objective lens current at frequencies below five cps, though causing slight oscillation of focus, may produce larger rotation of an image about the objective lens magnetic axis at large distances from that axis. Small fluctuations of the high voltage supply or lens supplies result in small changes in the effective strength of each of the imaging lenses so that the magnification and/or focus of each lens is altered slightly and this blurs the image.

A more common variety of this type of disturbance is due to a higher frequency alternating (e.g., 60 cps) magnetic field in the vicinity of the microscope. This may result from the proximity of a large ac carrying cable, transformer, or voltage stabilizer. The most sensitive part of the beam path is just on the image side of the specimen. Such an ac disturbance manifests itself as a blurring of the image in one direction which is consistent from specimen to specimen. If, however, it is the region between the condenser and objective that is affected by the disturbance, then the presence of the transverse alternating magnetic field manifests itself by acting as a beam wobbler (see p. 101) and is only detectable in out-of-focus planes.

d. SPECIMEN INSTABILITY

Image shifts may also be due to specimen instabilities (specimen drift). These may be due to thermal drift, e.g., the absorption of heat by the specimen grid so that thermal expansion results. This disturbance is evident even under normal operating conditions when shortly after activation of the beam an image shift is apparent. Thermal equilibrium between the specimen grid and its holder is quickly attained at constant levels of illumination between the specimen grid and its holder and image stability is then improved. If, however, the absorbed heat of the grid cannot be quickly dissipated by transfer to its surroundings, thermal equilibrium will not occur quickly and a slow image drift, detectable with the magnifying viewer, may persist for a long period of time. To insure image stability, good thermal contact must be provided by fitting the specimen grid accurately into its cap which, in turn, should be securely clamped on to the holder. The latter should then be placed in its mount so that good contact is made.

Another cause of specimen drift is broken or loose support film of the grid square under study. Under these conditions, thermal effects can induce changes in tension in the film which cause movements relative to the position of the tear. Also the specimen and supporting membrane develop substantial positive charges (due to secondary electron emission) during exposure to the beam and variations in intensity and/or position of the illuminating beam will cause variations in charge on the specimen and specimen holder. The resulting changes in repulsive forces between these charged elements can cause appreciable movements of the torn supporting film (similar to those operating in a gold leaf electrometer).

e. CONTAMINATION

The slow irreversible deposition of organic matter from the residual gases or greased areas occurs on the beam-exposed surfaces of the column. Upon direct or indirect bombardment of such insulating layers by electrons they acquire an electrostatic charge and can act as lenses or beam deflectors. Depending on the magnitude, position and rate of change of the charge a varying degree of image shifting or blurring results. To minimize these disturbances, cleanliness of all the beam contacting surfaces, especially of the objective aperture, must be maintained.

4. Vacuum Leaks

The need for maintaining an adequate vacuum level in the column has already been emphasized (see p. 76). The mechanical and oil diffusion pumps which provide the vacuum need little routine care. The potential source of difficulty is an air leak of such magnitude that the capacity of the pumping system is exceeded. The presence of a serious vacuum leak will be indicated by the vacuum gauges or the activation of the vacuum failure switch.

A vacuum leak is located by the process of elimination. Since leaks may occur due to improper closure of the gun, specimen, or photographic chambers, these should be checked first to see if they are properly closed. More frequently, however, the leak is located in the vacuum system or microscope column. Thus, if the aforementioned chambers are secure and the leak persists, the main valve between the pumps and column should be closed and where the vacuum gauge is on the column, the pressure should be noted. If it rises rapidly, the leak is obviously in the column. A next step may be to determine if the leak is at the site of the external manipulative controls. Such leaks will become evident if, with the main vacuum valve open (i.e., normal operating condition), a pressure rise is noted when the specimen stage, fluorescent screen or plate cassettes are manipulated. If the leak is still not localized, then the reliability of the vacuum valves may be checked. This can be done by noting the vacuum at various stages of pump down from atmospheric pressure. If the valves seem to be in order, a further search may require stepwise dismantling of the column.

Prior to undertaking the rather laborious task of dismantling the column the search may be facilitated by using an organic solvent or gas as a "leak guide" and applying it to particular column junctions. Thus, where a large leak is present, wetting a junction may occasionally result in a drastic rise or drop in the pressure. A useful method, feasible where a gas discharge tube is present, depends upon noting the color of the discharge after applying the solvent. For example, if the suspect joint is wet with alcohol and if alcohol passes into the column at the defective site, a striped whitish glow will be produced. The most sensitive method involves the use of a commercially manufactured leak detector. This is a simplified, sensitive mass-spectrometer which is connected to the vacuum system and adjusted to detect the helium line. Thus, if helium gas is released near the site of the suspected leak and it passes into the column, an electric signal will be produced whose amplitude is proportional to the size of the leak and the closeness of the source of helium to the leak.

If dismantling of the column is unavoidable, it should be started at the top. The column is closed off at the level of the section removed (but left connected to the pumps). When the leak disappears, the joint at the level of the last section removed is implicated as the source.

5. Specimen Damage

A thin specimen placed in an electron microscope for examination is exposed to at least four factors which may alter it and thereby limit resolution and distort the data that can be secured from it. The four factors are: exposure to the vacuum, to the electron beam, to hydrocarbons, and to other residual gases.

a. EXPOSURE TO VACUUM

Almost all specimens prepared for electron microscopy must be in a dried state. They are not necessarily damaged by the absence of atmospheric pressure. (However, any damage which results from the initial need to dry the specimen may, in a sense, be attributed to the vacuum of the electron microscope.)

b. EXPOSURE TO ELECTRON BEAM

At the outset it should be noted that some beam effect is unavoidable because inelastic scattering (p. 59), which is associated with any beam-specimen interaction, transfers energy to the specimen. The type and severity of radiation damage is dependent on the following parameters: chemical and physical composition and constitution of the object, the geometry of the object and the geometry of the cross-section of the electron beam in the object plane, and the intensity of the beam and the composition of the residual gases in the vacuum.

Organic material shows radiation damage even with low electron current densities and, within wide limits, the damage is proportional to the product of beam current density and irradiation time. Under these conditions, the first step in these interactions of the specimen and electrons which produce changes occurs within a very short time (10^{-4} seconds) and results in either direct ionization or excitation of the molecule, which then sometimes is followed by bond rupture or formation. Polymers, in particular, undergo cross-linking or scission. With further exposure, extensive loss of hydrogen, oxygen and nitrogen atoms takes place and carbonization and/or sublimation results (see Chapter 10). Sublimation was a particularly critical problem when methacrylate was the principal embedding medium used. It, however, occurs in varying degrees with all other embedments. This artifact-inducing phenomenon can be minimized by slow irradiation of the sections with a low intensity electron beam before exposing them to high intensity, then by using the minimal intensity needed for operations being performed (e.g., scanning, focusing, photography) and by exposure for as short a period of time as needed. Under these conditions, carbonization and cross-linking (as opposed to sublimation) seem to be favored. After carbonization the specimen can usually withstand exposure to considerably higher current densities. It has been shown that the use of a cooling chamber does not prevent *direct* irradiation damage in organic specimens.

c. EXPOSURE TO HYDROCARBONS

The contamination of the specimen when exposed to the beam was noted more than two decades ago.[28] With the development of high resolution electron microscopes it has become evident that this factor can seriously limit their performance.

The major sources of hydrocarbon contaminants are the vacuum pumping oils, vacuum greases and material desorbed from the interior walls and surfaces of the microscope. The hydrocarbons condense on the specimen (and other) surfaces and are cross-linked and polymerized in place after exposure to the electron beam. Under unfavorable conditions, the thickness of the contamination layer can grow quickly. Limitation of the electron microscope observations of such contaminated specimens will naturally follow. This limitation consists of a loss in resolution due to loss of contrast and increased chromatic aberration. The production of erratic electrostatic charges on critical optical components can produce further blurring effects (see p. 34).

The most logical approach to the reduction of contamination is by inhibiting one or more of the factors that are responsible for its growth. Since growth depends largely on three factors there are, theoretically, three modes of influencing or reducing contamination.

(1) Reducing the partial pressure of hydrocarbons.
Drastic reduction in the amount of hydrocarbon vapors in the column can be obtained by removing them at their source. This can be done by (a) using special diffusion pump oils, which should be kept clean, (b) using special sealing greases and gaskets, (c) by using a liquid nitrogen trap between the pumps and the column and (d) cleaning all column surfaces with non-organic solvents. Nevertheless, desorption of hydrocarbons from the walls will be a continuing source of contamination. Thus it is preferable to reduce the partial pressure of the hydrocarbons in the immediate neighborhood of the specimen. This can be done by keeping the surfaces surrounding the specimen holder cool so that they act as a local cold trap for the residual gases (see p. 99).

(2) Control of electron beam intensity.
It has been found that beam current density (amperes per square centimeter) affects the rate of contamination. The contamination rate is proportional to the partial pressure of the hydrocarbons and to the number of electrons hitting the specimen per unit surface area per unit time. However, it is possible to obtain a substantial reduction of contamination (setting aside the effect of heating, which will be discussed below) at low current densities (i.e., below 10^{-3} amp. per sq. cm), although this only permits working at low magnification. Thus control of contamination by the reduction of beam intensity would result in severely limiting the range of ultrastructural studies.

(3) Increasing specimen temperature.
At increased specimen temperatures, the time spent on the surface

by a hydrocarbon molecule becomes shorter and, therefore, the probability that it will be polymerized there is diminished and the rate of contamination is decreased. At high enough temperatures (*circa* 250°C), condensation of the organic vapors is effectively prevented and contamination is largely avoided. Thus the contamination of the specimen is frequently especially heavy near the grid bars, since the grid bars are conductors of heat. This results in the periphery of a grid square being cooler and it is here that contamination is most rapidly laid down. The centers of the grid square are hotter and, therefore, the rate of contamination in this area is slower. By judicious control of specimen illumination, the regions away from the grid wire can be kept above about 250°C, but below temperatures which would cause melting or rapid thermal decomposition. Intensive heating can itself cause both thermal drift and specimen damage and so this approach is a poor but occasionally useful compromise. It can only be successfully employed after careful carbonization at lower beam currents (and temperatures).

d. EXPOSURE TO NON-HYDROCARBON RESIDUAL GASES

The non-hydrocarbon residual gases, the most numerous molecules of which are water, have under normal conditions little noticeable effect upon the specimen. However, when the partial pressure of hydrocarbons is reduced, as with a cooling chamber, the growth of carbonaceous contamination is reduced and no longer masks the effects of the non-hydrocarbon gas molecules. Under these conditions, removal of specimen material in the beam may become visible. Water molecules which are absorbed onto the surface are ionized by electron impact and this results in reactions with the carbon of the biological specimen so as to oxidize it and thereby cause its loss as CO or CO_2. This kind of etching of the specimen, while perhaps useful for certain special studies, is generally undesirable.

The most desirable method for reducing contamination is by means of a very drastic reduction in the partial pressure of the hydrocarbons in the vicinity of the specimen. This can be done by the use of an appropriate cold trap in proximity to the specimen. If such a trap is kept below −140°C, it has been found that it will also serve to simultaneously sufficiently reduce the partial pressure of the non-hydrocarbon residual gases to eliminate the deleterious effects of such residual gas molecules, as well as of the hydrocarbon vapors.

The design of a suitable cold trap that does not disturb the instrument's operating conditions (e.g., by producing thermal drift) and does not restrict the aperture angle or field of view presents severe problems. These have been discussed by Heide,[29] who designed an efficient prototype.[30] Commercial anticontamination devices, following Heide's design principles which are quite effective in reducing contamination, are now being offered as optional accessories by a number of manufacturers.

F. OPERATION OF THE ELECTRON MICROSCOPE

1. Operating Adjustments

Manipulation of the electron microscope is analogous to that of the light microscope. The specific differences are due to the fact that the lenses of the former have a variable focal length and have smaller aperture angles. The functional analogy between the two microscopes has been summarized on Table V.

TABLE V

FUNCTIONAL ANALOGY BETWEEN LIGHT AND ELECTRON MICROSCOPES

Lens	Function to Control	Usual L.M. Adjustment	Usual E.M. Adjustment
condenser	illuminating aperture angle & intensity of illumination	aperture diameter of condenser & its position along optic axis	aperture diameter of condenser & condenser current control
objective	change in focus of intermediate image	position of objective (or specimen) along optic axis	objective current control
projector	change in magnification of final image	replace objective and/or ocular lens	projector and/or intermediate current control and/or pole-piece interchange

2. Photography

The need for and the advantages of photographing the final electron image have already been discussed (see p. 69). The routine procedure involved in this operation can be summarized as follows:

(a) The field of view is selected at low magnification and at low beam intensity and preliminary centering and focusing is carried out.

(b) The magnification is raised so that the desired level of resolution of detail in the electron image (visually near the level of resolution on the photographic emulsion) may be attained.

(c) The intensity of the illumination must be raised but is kept to a level just adequate for critical focusing and/or reasonably short exposures.

(d) The image is finally brought into critical focus by varying the current (i.e., focal length) of the objective.

(e) The photograph is then taken without changing the condenser setting (i.e., intensity of illumination). To insure that it is in exact focus, especially at higher magnifications, a number of photographs are taken with the objective lens current varied slightly in both directions from the assumed critical focus. Such a group of photographs are known as a *through-focus series*.

Until a few years ago it was the standard procedure to decrease the intensity (i.e., aperture angle, α_c) of illumination by about one order of magnitude by reducing the condenser lens current just prior to photography. This was usually done so as to increase the depth of field. However, since a change in the "in-focus" position of the condenser may secondarily introduce a number of possible kinds of image degradation (see p. 50), current preferred procedure requires that the condenser setting remain essentially unchanged throughout final focusing and photography.

One method, incorporated in some microscopes to aid focusing at low intensity of illumination (i.e., small illuminating aperture) for both visualization and photography, makes use of a built-in deflector type of focusing mechanism commonly known as a *beam-wobbler*. This involves the use of magnetic coils fitted between condenser and specimen to deflect the electron beam from side to side over an angle of about 10^{-2} radians. This temporary increase in the aperture angle produces a marked decrease in the depth of field from the normal conditions. As a result, the in-focus (i.e., the range of variation of sharpness of the image with change in objective current) is reduced and departure from the exact focus is more readily recognized. The aperture angle can also be increased by inserting a larger physical aperture in the condenser lens. However, this results in a very large increase (often excessive) in the beam intensity on the specimen. The beam-wobbler, as an aid in focusing, is especially valuable at low magnifications when a specimen of poor contrast is used or when the intensity of illumination must be kept low.

3. Determination of Magnification

Upon installation of most microscopes, a magnification calibration chart is supplied by the manufacturer. Since the actual magnification will depend on such factors as the position of the specimen, voltage employed, image distortion or asymmetry, it is not uncommon that manufacturer's calibrations may have an error of the order of $\pm$ 10%. When greater accuracy is necessary, such as in studies of particle size, special procedures must be used. For routine work, determination of magnification is made with simplicity and reasonable accuracy by the use of standard reference objects of known size, such as polystyrene latex spheres. Another means of calibration of magnification is by using a diffraction grating replica. These are now available with, for example, 50,000 lines per inch, and are used at a low range of magnification.

In making measurements of particles using low magnification micrographs, measurements should be made in the central region. This is due to the fact that a particle of well-defined size can appear larger or smaller in the periphery than in the center, depending on the kind of geometric distortion (p. 64) that the microscope's projector lens may produce. Since at high magnification only a very small central region of the image appears in the electron micrograph, this source of error is usually negligible.

4. Test of Resolution

A number of methods have been proposed to determine the resolution and limit of resolution of an electron microscope. The three most commonly used involve:

(a) Measurement of the minimum distance between particles which are recognizable as such.

(b) Measurement of the minimum recognizable particle diameter. This method is much less reliable and meaningful than (a).

(c) Fresnel fringe width at focus (see p. 320).

G. ACCESSORIES FOR THE ELECTRON MICROSCOPE

1. Double Condenser

The shorter the focal length of the condenser, the smaller is the illuminated area at cross-over for a given electron gun. However, as pointed out previously (see p. 63), the focal points in short focal length lenses usually are located between the pole pieces of a lens. Since both a short focal length condenser and an objective have focal planes located within their pole pieces, it is necessary to add a second long focal length lens to the condenser to permit the imaging of such a small cross-over on the specimen plane within the objective (Fig. 70). Such a two-lens condenser is known as a *double condenser*.

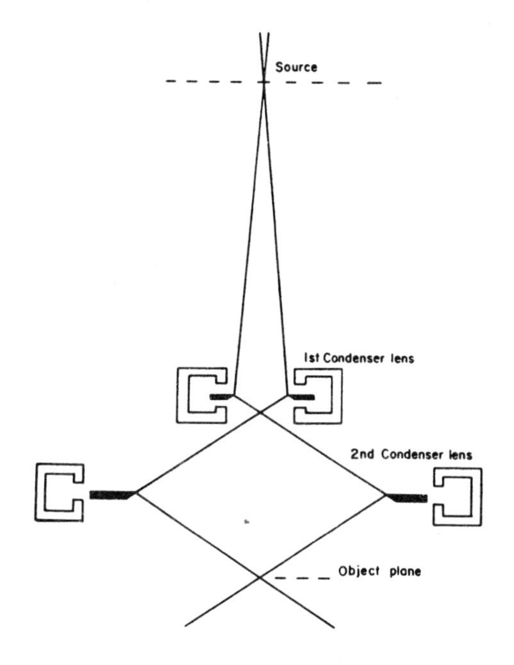

FIG. 70. Double condenser.

Through the use of a double condenser, controlled small-area illumination can be obtained. Each of the lenses can be individually excited. The first lens is strong and produces a (variable) demagnification of the image of the source (as does the ordinary condenser). The second condenser lens projects this reduced image onto the specimen. The advantage of the double condenser is that it provides a means of reducing the beam diameter at the plane of the specimen. However, such a system does not in and of itself increase the "brightness" of the beam since this limit is set by the aperture angle of the condenser which cannot usefully exceed that of the objective. The double condenser eliminates unnecessary irradiation of specimen areas outside of the field of view and, therefore, also cuts down background scattering from such areas.

2. Image Intensifiers

The last few years have seen the commercial introduction of image intensifiers as electron microscope accessories. These units are usually designed to:

(a) Permit direct high magnification image observation at comfortable brightness levels.

(b) Permit specimen observation at lower beam intensities (which thus would reduce the chances for specimen damage).
(c) Permit the image to be electronically scanned and processed and, therefore:
extend the range of image contrast beyond that provided by direct visualization on the screen,
permit simultaneous specimen observation by groups of viewers thus facilitating teaching and demonstration,
permit continuous recording (on magnetic tape) of transient phenomena.

There are a number of different electronic systems in current usage. As a group they are yet to fully meet the level of anticipated promise because of their limited resolution except at the highest electron microscope magnifications. Photographic recording still appears to be the best means of securing and recording most electron microscope information.

3. Stereomicroscopic Accessories

Because of the small objective aperture which provides for a large depth of focus (see p. 63) and because of the large depth of field (see p. 61), recording of stereoscopic views of objects is rather simple with the electron microscope. This is especially useful when viewing *replicas of deep structures* (which by conventional methods do not distinctly reveal the form of the surface) and *with thick* (0.1–0.2 μ) *biological sections* of tissue.

To secure stereoelectron micrographs, the specimen holder is tilted through an angle of 5°–10° between two successive exposures.* In practice the specimen is centered, the first photograph is exposed; the specimen is tilted, recentered, and refocused, if necessary, and then rephotographed. If the specimen plane is rotated excessively, axial dimensions are exaggerated. When the two micrographs are viewed through a stereo-viewer, a three-dimensional image of the object is seen. Prints for stereoscopic viewing need to be precisely mounted at correct orientation and separation to secure satisfactory results.

4. Electron Diffraction Accessories

The apparatus used for transmitted electron microscopy can usually be used for electron diffraction studies. The results of both types of studies

* Tilting is accomplished either by rotation of the specimen holder around an axis in the plane of focus, by a built-in stage control or alternatively, by remounting the specimen support with a shim under one edge sufficient to produce the 5°–10° tilt.

of an object in many cases can provide kinds of information that supplement each other. A discussion of electron diffraction can be found in Appendix K.

H. OTHER TYPES OF ELECTRON MICROSCOPES

Two other types of electron microscopes have been commercially available. They will be briefly considered in this section.

1. Electrostatic Electron Microscopes

While the optical principles discussed in the preceding sections also apply to the electrostatic electron microscope, the major features of instruments of this type need some elaboration. Such instruments, which use electrostatic lenses (p. 34) as their focusing elements, have their main advantage in the simplicity of their power supplies. This is due to the fact that no separate lens power supplies are needed and some instability in the accelerating potential can be tolerated with the employment of unipotential lenses with the negative center electrodes connected to the cathode. This is because the lens focal lengths depend only on the ratios of electrode potentials to electron energies. On the negative side, electrostatic microscopes (a) require even more precisely aligned lens elements, (b) have limited accelerating voltages (*circa* 50 KV) due to difficulties of arc-over between closely spaced lens electrodes for short focal lengths, and (c) have a larger chromatic aberration coefficient than in magnetic lenses (which in these microscopes is very significant for *specimen-induced* spread in the energy of the imaging beam of electrons), even though it is relatively insignificant for power supply fluctuations.

2. Scanning Electron Microscopes and Microprobes*

While the principle of scanning electron microscopes was put forward as early as 1935 by Knoll,[31] only recently have technological advances been adequate to develop commercial instruments capable of reasonably high resolutions (*circa* 250 Å).

Image formation in the scanning electron microscope differs from that of the conventional light or transmission electron microscope. In the scanning electron microscope, the image is formed on a cathode ray tube after first converting information from the specimen surface into a train of electrical signals (Fig. 71.). Specifically in this instrument, the electron

* For a detailed discussion of SEM and microprobes, see Section Two below.

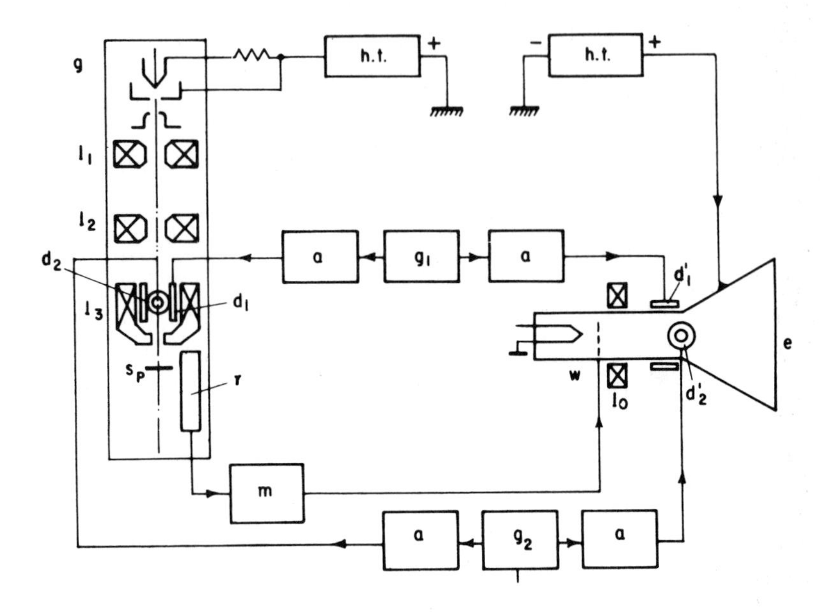

FIG. 71A. Diagram of a scanning electron microscope. On the left—the microscope itself. Gun (g); lenses (l_1,l_2,l_3); deflection coils (d;d_2); specimen (sp); receiver (r). On the right—television tube. Electrode (w) for modulating the brightness; CRT magnetic lens (l_0); deflection coils (d'_1;d'_2); fluorescent screen (e); generators (g_1,g_2) for the scanning, both specimen, and the image, which supply the pair d_1 and d_2 and the pair d'_2 and d_2, respectively, after amplification in a. The electron signal which is received at r is transmitted to w after amplification in m. High voltage supply (h.v.).

lenses are used only to demagnify the image of the source to produce a finely focused electron spot. This probing spot is deflected to scan the specimen under study in a manner comparable to a TV camera. A suitable detector placed near the specimen continuously collects the secondary electrons, back-scattered electrons and/or transmitted electrons (or in some cases, x-ray fluorescence, or even light fluorescence) emitted by or transmitted through the bombarded surface. The signal which is obtained is amplified and then used to modulate the brightness of a television or cathode ray tube, adjusted to scan synchronously with the probe. The variations in brightness recorded produce the highly magnified "image" of the object.

It may be of interest that scanning electron microscopy also has its analogue in light microscopy as the flying-spot microscope. A point source (for example, a compact arc lamp or laser) placed on the optical axis in the plane of the real image beyond the eyepiece is imaged "backwards" through the microscope onto the plane of the specimen as a demagnified point of light. If the optic axis between the eyepiece and the source is bent (typically at right angles) by a mirror, movement of this mirror will move the point image across the specimen. In one version of the flying-spot microscope, two such mirrors, usually of the galvanometer type, are driven at right angles to one another, in order to generate a raster scan of the object. A photodetector is placed below the specimen plane to collect the light transmitted by the specimen, and is used (as in scanning electron microscopy) to modulate the brightness on a cathode-ray tube screen. The cathode-ray is deflected synchronously with the deflections of the mirrors, producing a magnified transmission image of the scanned specimen on the cathode-ray tube screen.

Another version (Fig. 71B) dispenses with the moving mirrors and uses instead one cathode-ray tube as the light source in place of the arc lamp; the beam is deflected electrically and synchronously with a second (imaging) cathode-ray tube, again giving a television-type transmission image on the screen. Flying-spot microscopes have been used for ultraviolet imaging, for automated cytophotometric research, and in one automated differential white blood cell counting instrument. Variants of the flying-spot microscope are also being used in current designs of commercial optical videodisc readers and for read-out of computer trillion-bit optical memories.

3. High Voltage Electron Microscopes (Fig. 72)

a. GENERAL CONSIDERATIONS

One of the ultimate goals of molecular biology has been to describe the arrangement of the atoms of molecules or subcellular elements within live cells. Some have hoped that electron microscopy would significantly contribute to the attainment of this goal, perhaps bypassing laborious biochemical research paths to the same objectives.

Studies of biological macromolecules have shown that disordering can be induced by dehydration; therefore, there have been efforts to study intact hydrated specimens. However, the goal of atomic resolution of intact wet biological structures with the electron microscope presents several challenging problems: (1) how can a wet specimen be studied in an instrument which requires a vacuum; (2) how can the resolution and contrast be increased to resolve atoms; and (3) how can the disruption of the specimen's atoms by the probing electrons be minimized or bypassed?

As the accelerating voltage is increased, the ability of an electron to penetrate a specimen increases. In the region above 1,000 KV, it becomes feasible to penetrate a very thin, wet specimen (e.g., one on the order of 1 μm

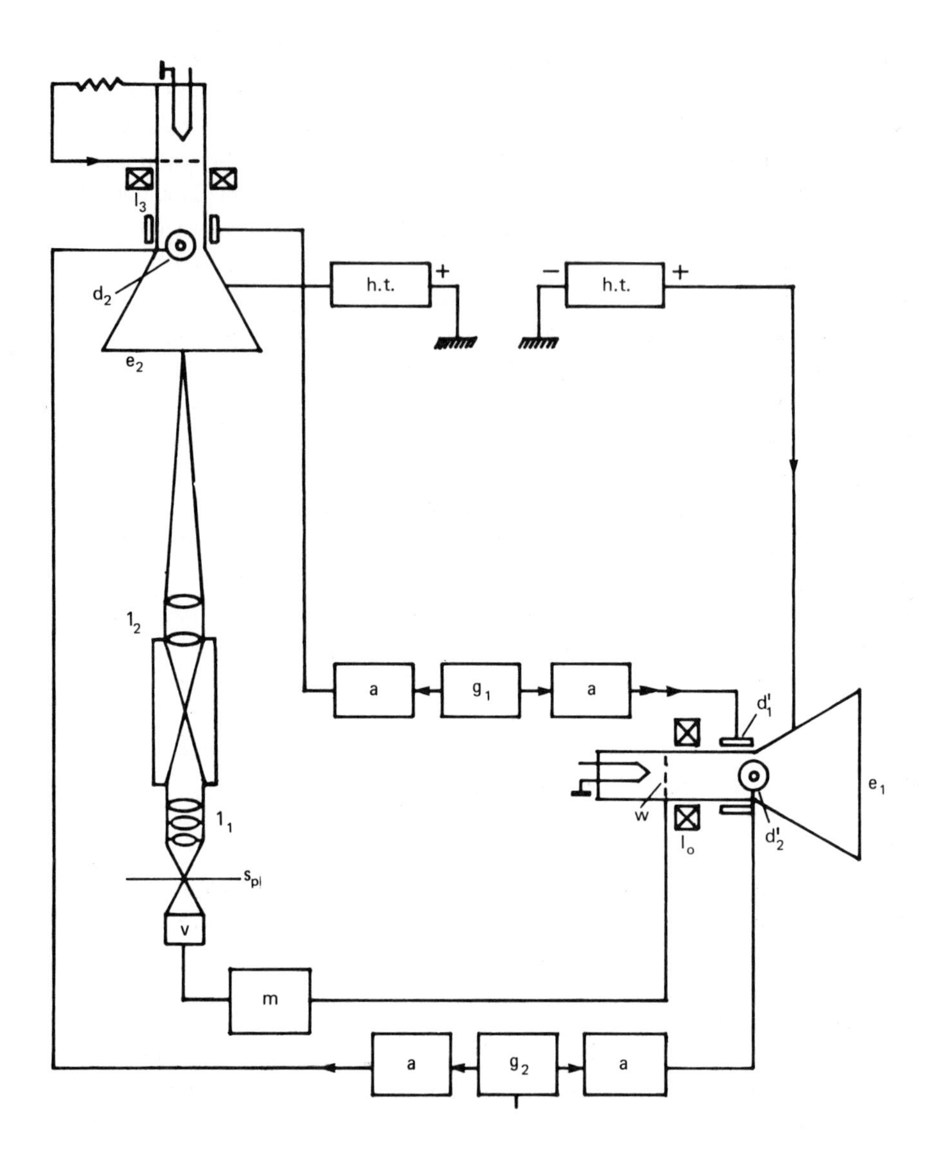

FIG. 71B. Diagram of a flying-spot microscope. On the left—the light microscope itself. Microscope objective lens (l_1); microscope ocular (l_2); specimen (*sp*); photodetector (*v*). Above—the cathode ray tube. To the right—cathode ray tube display. Electrode (*w*) for modulating display brightness, (brightness of source is held constant). CRT magnetic lenses (l_0, l_3); deflection coils (d_1, d_2, d'_1, d'_2); fluorescent screens (e, e_2); sweep generators (g, g_2), for scanning both the specimen and the image; sweep amplifiers (a). The electric signal generated at the photodetector, *v*, is transmitted to *w*, after amplification in *m*. High voltage supply (h.t.).

FIG. 72. A 1000 KV high voltage electron microscope (JEM-1000D). Note the conventional but somewhat more massive microscope beneath the enormous high voltage supply. (Courtesy of JEOL [U.S.A.] Inc.)

thickness) sandwiched between two even thinner vacuum-tight windows, and obtain a usable level of resolution in its image. This was one of the promises of high voltage electron microscopy.

Unfortunately, even the 1 μm path through the specimen and windows produces a $\Delta V/V$ that results in poorer resolution and lower contrast than the best that can be achieved in a thinner section in a vacuum at lower voltages. (Cf. the discussion of STEM, p. 253.)

Biological material in its natural hydrated state is extremely delicate. It readily deteriorates when exposed to electron bombardment (see p. 98 above). This is reflected by the fact that hydrated crystalline biological material whose diffraction pattern indicates ordered spacings of 4 Å or less disappears in seconds or may be entirely absent after short exposure to electron beams of even low current density. Consequently, only a few electron diffraction pictures of nonmineral, wet biological specimens have been obtained. High resolution electron microscopy of live, wet cells obviously requires that beam interaction be kept to a minimum.

One significant approach toward achieving low beam interaction is to reduce the fraction of the incident radiation that deposits energy into the specimen by inelastic scattering. This reduction occurs when high accelerating voltages — i.e., those in the 500 to 1,000 KV range — are used.

''Imaging'' of minute, wet biological crystals presents a special opportunity for the utilization of the high voltage electron microscope. An illuminating electron dose which would be sufficient to provide adequate signal-to-nose ratio in the image of a unit cell of a crystal would destroy that unit cell. One hundred times this dose can be spread out over tens of thousands of unit cells in the three-dimensional crystal to produce an electron diffraction picture (see p. 242 below) where the information from a large number of undamaged unit cells adds together to produce a usuable image.

It has been established that an improvement in the lifetime of diffraction patterns of organic crystals, on the order of 10 to 20 times, may be obtained by increasing the voltage from 50 KV to 1,000 KV. For constant beam current at the higher voltage, the resolution in the diffraction pattern is also increased and the desired atomic resolution of materials unavailable in other than microscopic crystalline form is thus achievable with the high voltage electron microscope.

Because of the higher penetration power of the higher energy electrons, thicker sections can be penetrated at a given level of resolution. When the contrast loss that goes with increasing voltage can be tolerated, this permits useful stereomicroscopic studies to be performed on thicker segments of a cell that can be accomplished at lower voltage, and has substantially aided in the interpretation of the three-dimensional structure of organelles such as the structure of the mitotic spindle.

b. OPERATIONAL CONSIDERATIONS

Selecting the optimum operating high voltage acceleration may not be a straightforward decision as it is in conventional TEM. As noted earlier (see p. 195), Heidenreich considers that for the out-of-focus phase contrast mode of imaging, the optimum voltage is 200 KV. The use of an image intensifier TV system, or on-line computer processing of the image, or a combination of both, may provide an alternate means of restoring contrast and thereby making use of voltages higher than 200 KV more desirable. Testing a variety of voltages may be the most advantageous approach, but 1000 KV appears at present to be the most convenient upper operating limit. This is because of the excessively high cost of voltage generators and lenses beyond this level, and serious problems in dealing with hard X-rays that are produced.

High voltage microscopes must be provided with adequate shielding in the objective aperture zone, as apertures are a source of hard X-rays. The use of a holder carrying multiple apertures is also very desirable since the aperture diameter must be decreased as the voltage is increased to maintain constant resolution with minimum loss of contrast (see pp. 191–193). For high voltage instruments, stability of the accelerating potential and objective lens current is especially critical. While significant progress has been made in developing superconducting objective lenses for use at 100 KV, it has been rather limited in constructing such lenses for use at higher high voltages.

I. DIFFERENCES BETWEEN THE LIGHT AND ELECTRON MICROSCOPES

At the conclusion of the theoretical section of this book, the *analogy* between the light and electron microscopes was emphasized. At this point it may be profitable to summarize the *differences* between the two microscopes even though they have already been discussed in detail.

(1) Magnetic electron lenses do not have fixed focal lengths like optical lenses. Their focal length, which is dependent upon the strength of the field, can, therefore, be varied by changes in the current which passes through the coil.

(2) It is not necessary to change the objective or projector (ocular) lenses when different magnifications are desired. This is attained

by varying the focal length of the projector lens. The magnification of the objective is fixed. "Zoom" oculars now provide analogous control of magnification in the light microscope.

(3) In the light microscope, because of the small depth of field, different focal levels of the specimen can be directly seen, relatively independently, by manipulation of the fine focus control. These cannot be separately appreciated in electron microscopy by visual observations (but information about different specimen levels can sometimes be recovered in stereoscopic photographs and in the scanning electron microscope).

(4) The mode of image formation is different in both microscope systems. In the light microscope, this is due to differential absorption of light, while in the electron microscope, loss of electrons outside of the objective aperture angle induced by scattering by the individual parts of the specimen results in variations of intensity in the image (i.e., image contrast).

Part Two

METHODOLOGY

Chapter 5

INTRODUCTION

THE LAST DECADE has seen significant advances in the techniques used in the preparation of biological specimens for electron microscope study. Thus, at the present time, the microscopist has at his disposal a variety of useful fixation and embedding media, different types of reliable microtomes, several forms of cutting edges and a multiplicity of staining techniques.

This part of the book will be essentially devoted to a discussion of the *rationale* of the various steps used in the preparation of material for both routine and specialized electron microscope investigation. A number of books are available which discuss, in detail, the *practical aspects* of specimen preparation for electron microscope investigation (see p. 363).

Technical methodology has reached the point where reproducibility of specimen preparation is possible. This reproducibility has strengthened the belief that photographic records of such electron microscope images are relatively faithful reflections of the morphology of the native living protoplasm. Electron micrographs of well-preserved specimens, revealing ultrastructural patterns of great detail, fine texture and high degree of orderliness (as well as exquisite beauty) are becoming commonplace.

The use of varied fixation, embedding and staining techniques, introduced during the last decade, has served to reinforce and consolidate the bulk of the information that was secured prior to 1960 by the almost exclusive use in tissue preparation of osmium tetroxide fixation and methacrylate embedding. However, in some areas, especially at high resolution levels, the newer methods have necessitated reevaluation of earlier observations.

An additional advantage that has accrued from the multiplication of the specimen preparation techniques has been the lowering of the risk of misinterpretation that could result from specific technique-induced artifact, since observations need no longer be restricted to a single kind of preparation. It should be noted, however, that while electron microscope methodology has now become somewhat more routine, nevertheless, it remains a highly sophisticated art and is still time-consuming, technically complex and not always fully predictable.

Chapter 6

HISTORICAL REVIEW

DURING the decade from 1940 to 1950, as was noted earlier (see p. 3), the promise of applying the electron microscope as a biomedical research instrument became a reality. It became obvious that crude fragmentation of structures for direct analysis or for the study of surface replicas was not, by itself, adequate. Moreover, it began to be generally appreciated quite early in the 1950's that determination of ultrastructure by examination of thin sections was a more promising approach since it could provide information about internal organization and structural relationships within the specimen. In attempting to secure thin sections that were suitable for examination with an electron beam, it was natural to begin with the basic technological preparatory approach of classical histological studies. It became apparent that, while the traditional sequence of steps in tissue processing might be applicable, special media and devices would have to be developed that would be suitable for the distinct needs of electron microscopy. The basic needs were for: (a) a fixative that would preserve the natural state of the specimen and render it resistant to the effects of dehydration; (b) an embedding medium to fill the volume originally occupied by the water of the specimen, the consistency of which was firmer than paraffin and which could yet be sectioned; (c) a cutting edge that was both sharp and durable enough to cut thin sections; and (d) an ultramicrotome that had both precision and reliability to cut sections thinner than 1000 Å (0.1 μ). In the past 15 years, the problems presented by the above requirements have been essentially overcome and, as a result, the techniques involved in preparing thin sections are now both predictable and controllable. The images of the sections seen under the electron microscope provide a close representation of the organization of the structural elements in the section and this, in turn, has been demonstrated to match the structure of living cells as seen at the limits of resolution of the light microscope and, in some cases, at the level of resolution of x-ray diffraction.[32,33]

The highlights of the methodological advances that made preparation and study of biological material with the electron microscope profitable are summarized below and will serve as a brief historical review.

1947 Claude[34] introduced osmium fixation and naphthalene embedding, the first successful slow-speed, single-sections pass ultramicrotome, and use of a trough for floating-off the edge of the knife.

1948 Pease and Baker[35] proposed double embedding for electron microscopy and introduced the first modification of a conventional microtome for thin sectioning for electron microscopy (Fig. 73).

1949 Newman, Borysko and Swerdlow[36] introduced methacrylate as an embedding medium for biological specimens.

1950 Gettner and Hillier[37] discussed the advantage of leaving the embedding medium in the section for viewing in the electron microscope (rather than removing it; the practice of most previous workers).

—— Latta and Hartmann[38] introduced the glass knife as a cutting edge for ultrathin sections.

1951 Sjöstrand[39] published the first study of frozen dried tissue with the electron microscope.

1952 Palade[40] demonstrated that buffered osmium tetroxide at a near physiological *p*H, provided excellent fixation of biological material at electron microscopic levels of resolution.

1953 Porter-Blum[41] introduced the first *widely adopted* ultramicrotome.

—— Fernández-Morán[42] introduced the application of diamond knives to thin sections.

1955 Watson[43] emphasized the advantages of carbon films instead of Formvar or collodion to support tissue sections.

—— Sheldon, Zetterquist and Brandes[44] made the first successful efforts of combining enzyme histochemistry (alkaline phosphatase) and electron microscopy, followed up extensively for acid phosphatase and other phosphatases by Essner and Novikoff[45,46] and others from 1958 on.

—— Hall[47] introduced the "negative" staining technique, followed up by improved high-resolution variants in 1959, by Brenner and Horne[48] (see Fig. 88).

—— Hall[47] introduced quantitative densitometry to the electron microscopy of biological specimens, followed up later by Zeitler and Bahr[49] and others.

1956 Morgan, Moore and Rose[50] pointed up the importance of differential sublimation of tissue components and embedding medium in the electron beam, in interpretation of electron images.

—— Borysko[51] pointed up the significance of polymerization damage when using methacrylate.

—— Maaløe and Birch-Anderson[52] introduced an epoxy resin as an embedding medium which was of only limited use. Glauert, Rogers and Glauert[53] subsequently developed the use of "Araldite" which was a more effective epoxy resin.

—— Luft[54] introduced potassium permanganate as a fixative for electron microscopy (see Fig. 76,77).

—— Gibbons and Bradfield[55] introduced the first simple method for (positive) staining of thin sections.

—— Palade and Siekevitz[56] and Novikoff[57] showed the feasibility of electron microscope analysis of cell fractions (see Fig. 100).

—— Kellenberger, Schwab and Ryter[58] introduced the polyester Vestopal W, as an embedding medium for electron microscopy.

1957 Steere[59] introduced the freeze-etching technique which was later perfected by Moor and Mühlethaler[60] (see Fig. 102).

1958 Watson[61] introduced lead staining of tissue sections. This technique was later improved upon by Millonig[62], Karnovsky[63] and Reynolds.[64]

—— Swift and Rasch[65] introduced the use of specific nuclease digestion followed by staining with heavy metals for electron microscope nucleic acid cytochemistry.

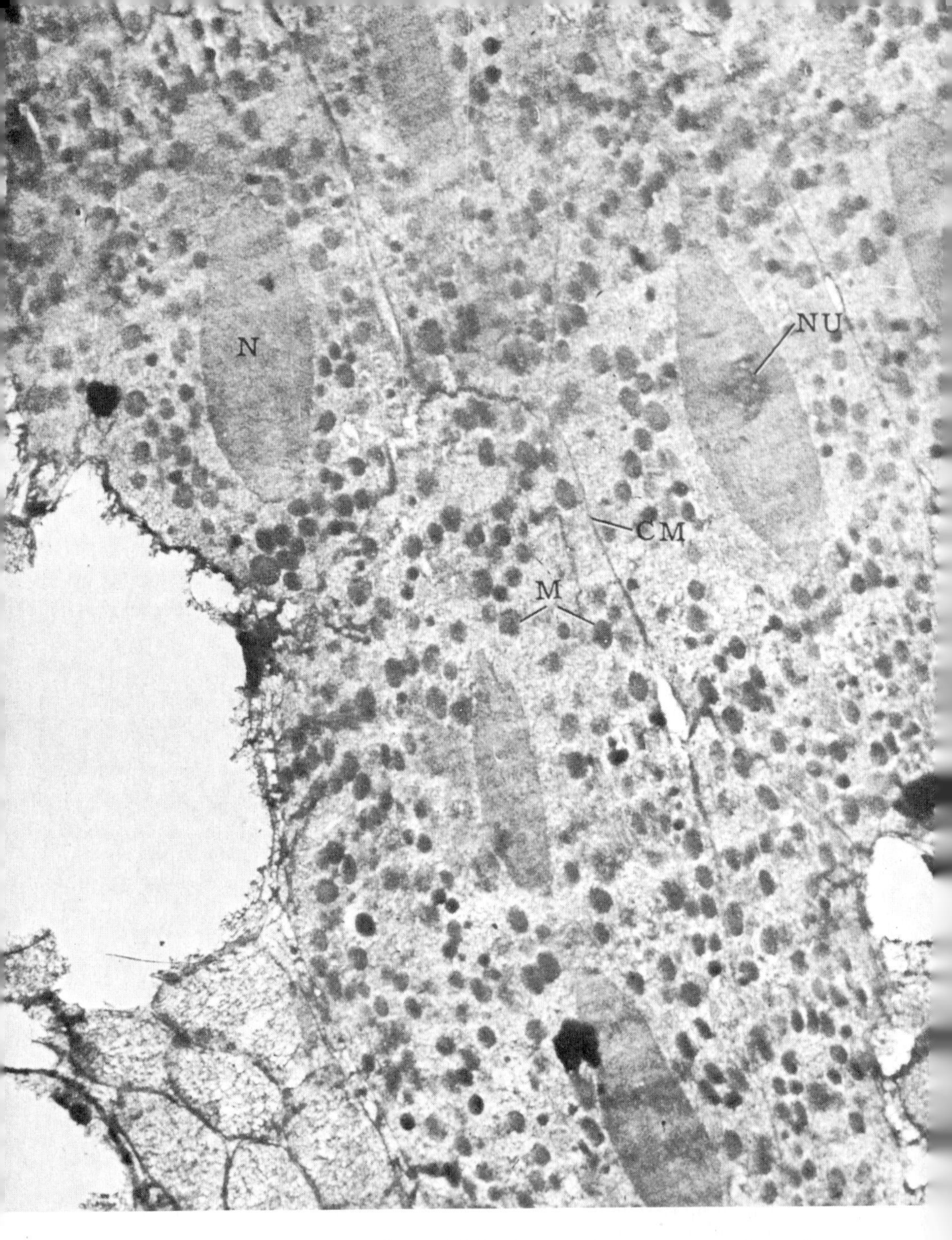

FIG. 73. Demonstration of early (1948) electron microscope preparation technique. Rat liver prepared by perfusion with 2% osmium tetroxide and double embedding, celloidin-paraffin, and then removing the paraffin with benzene and sectioning at 0.2μ. Among the structures visible are cell membranes (*CM*), nuclei (*N*), nucleoli (*NU*) and mitochondria (*M*). 5000X.
(Courtesy of Drs. D. C. Pease and R. F. Baker.)

——	Wissig[66] introduced the use of gold, thorotrast and ferritin as specific ultramicroscopic particulate tracers (see Fig. 92).
1959	Lehrer and Ornstein[67] introduced enzyme localization using organic enzyme histochemical "stains".
——	Wachtel, Lehrer, Mautner, Davis and Ornstein[68] introduced formalin fixation, followed by enzyme histochemical tests, followed by osmium fixation. Significant advances in this technique were reported in 1962, by Holt and Hicks.[69]
——	Singer[70] introduced a technique permitting ultrastructural immuno-electron microscopy with ferritin labelled antibody (see Fig. 95).
——	Fernández-Morán[71] and Rebhun[72] (in 1961) both introduced freeze-substitution techniques for electron microscopy. This was followed in 1965 by the highly promising technique of Van Harreveld, Crowell and Malhotra.[73]
——	Kleinschmidt and Zahn[74] introduced the surface-spread film technique for the preparation of nucleoproteins and nucleic acids (see Fig. 105).
1960	Rosenberg, Bartl and Lesko[75] introduced the water-soluble resin hydroxy ethyl methacrylate as an embedding medium for electron microscopy, followed up successfully by Leduc and Bernhard[76] and others, especially for enzyme histochemical studies (see Fig. 80).
1961	Luft[77] published his report on the use of Epon 812 which he had introduced several years earlier, as an embedding medium for electron microscopy.
——	Watson and Aldridge[78] introduced the first *specific* stoichiometric heavy metal stain for electron microscopy of nucleic acids (see Fig. 89).
——	Caro[79] demonstrated the potentialities of electron microscope autoradiography for identifying the site of metabolic events within animal cells, followed up by Salpeter and Bachmann[80] and others (see Fig. 97).
1962	Wolfe, Axelrod, Potter and Richardson[81] localized norepinephrin by electron microscopy.
——	Palay[82] introduced a successful method for fixation perfusion of brain tissue. The basic procedure was subsequently applied to other tissues (see Fig. 74).
1963	Sabatini, Bensch and Barrnett[83] introduced gluteraldehyde (usually followed by OsO_4) as a fixative for electron microscopy. This was especially significant to subsequent studies of microtubules which are best preserved by this fixation procedure (see Fig. 75).
1964	Karnovsky[84] improved the localization of cholinesterase in the electron microscope using thiocholine esters by a new precipitating reaction.
1965	Karnovsky[85] introduced the use of horseradish peroxidase as an ultramicroscopic tracer (see Fig. 93).
——	Pease[86] introduced inert glycol-type dehydrating agents to preserve tissues for electron microscopy (see Fig. 79).
1966	Karnovsky and Revel[87] introduced the use of lanthanum hydroxide as an ultramicroscopic tracer (see Fig. 94).
——	Nakane[88] reported the successful use of enzyme tagged antibody for the electron microscopic localization of antigen.

References to more recent innovations and modifications in methodology can be found in the Bibliography (see p. 361) as well as in the more current issues of such publications as the *Journal of Cell Biology, Journal of Ultrastructure Research, Journal of Histochemistry and Cytochemistry,* and other journals (see p. 362).

Chapter 7

OBTAINING THE SPECIMEN

ONE OF THE critical factors in obtaining good experimental results is the isolation of the tissue in as close as possible to *in vivo* condition, prior to placing it into the fixation medium. Since all chemical fixatives have a finite rate of penetration, the size of the specimen has been found to be a critical factor in insuring successful fixation.

A. PREPARATION OF EXCISED SPECIMENS

In most cases, the tissue to be studied is excised from a living animal. With practice, one can remove tissue from any site in the animal and cut it gently into small pieces within one or two minutes. Thus minimum trauma and a minimal interval of time occurs between depriving the tissue of its circulation and placing it in the fixative.

The slicing up of the excised tissue is best carried out on a non-absorbent surface (e.g., on a thick polyethylene slab) in one or two drops of the fixative. To minimize the mechanical disruption of the tissue near the plane of the cut, it is advisable to use only new razor or scalpel blades. The actual slicing should proceed as rapidly as possible to insure quick contact of the specimen with full strength fixative (i.e., to minimize dilution by tissue fluids and plasma). Rapid slicing also minimizes the possibility of tissue damage resulting from chemical interactions of some fixatives with the razor blade.

The final size of each tissue mass to be processed should be such that it usually will be less than one mm thick in its smallest dimension in order to insure rapid fixation of the center of the tissue. With fixatives such as the aldehydes, which have a more rapid penetration rate than osmium tetroxide, somewhat larger pieces can be successfully used. After slicing, the tissue should be quickly transferred to a suitable small vessel that contains a quantity of the fixative equal to six or more times the volume of the tissue, in a manner that avoids mechanical damage to the specimen. Finally, to insure good exposure to the medium, periodic agitation of the vial is desirable, especially during the first 10 minutes of fixation.

B. *IN SITU* FIXATION

Tissue can also be fixed *in situ* in the living anesthetized animal. This can greatly minimize the effects of both cutting on unfixed tissues and of anoxia resulting from cutting off circulation. Where the organ surface is to be sampled, it is essential to remove any existing capsule which might serve as a barrier to the penetration of the fixative. The fixative is applied by directly flooding the organ surface with substantial volumes of fixative for a short time. The preserved surface is then undercut and a thin slice of tissue is removed and cut into blocks. These are transferred to a small vessel for further prolonged fixation.

When deeper parts of an organ are to be studied, *in situ* fixation can be facilitated by incising the organ immediately following the application of the fixative so as to expose its deeper surface. Another approach, applicable to organs containing a lumen, is the injection of the fixation medium directly into this cavity.

C. PERFUSION

Because some tissues, for example, those of the central nervous system, are too delicate to withstand much manipulation or even short periods of anoxia, it has been found that, in principle, *in situ* vascular perfusion of the tissue would be the ideal approach to secure optimal fixation conditions. A restriction to the use of this approach lies in the fact that fixatives are often vasoconstrictors. When a fixative is used alone, many of the arteries become closed very shortly after contact with it which prevents adequate amounts of the fixative from reaching the capillary bed in the tissue. To counteract this difficulty, several approaches have been used in an attempt to induce vasodilation immediately prior to perfusion. These include (1) precooling the animals under anesthesia, (2) increasing carbon dioxide levels, and (3) preperfusion with isotonic liquid or (4) induced breathing of gaseous vasodilators. Unfortunately, these approaches have so far met with little reproducible success. An important, if not critical, factor in perfusion that apparently has not been taken into account in some of these early studies, is the need to provide a "drainage route" for the fixation medium and to control vascular bed pressures. Provision of a "drainage route" makes it possible to quickly dilute out the body fluids.

A few years ago, Palay *et al.*,[82] perfected a perfusion technique that requires considerable operative skill as well as some specialized equipment

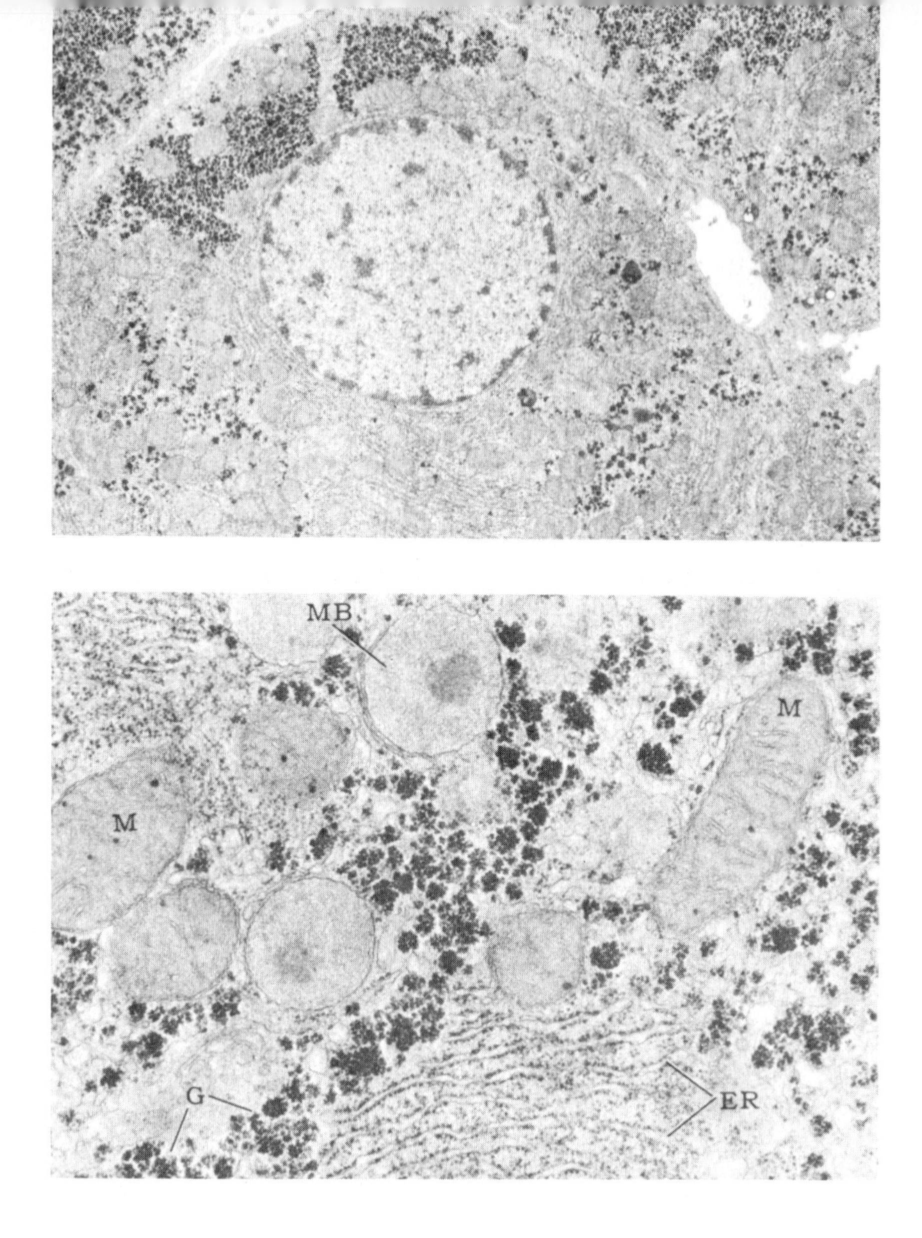

FIG. 74. Demonstration of perfusion fixation. Rat liver perfused with buffered glutaraldehyde and postfixed in OsO_4. (Courtesy of Dr. H. D. Fahimi.) (A) A low power survey electron micrograph showing one entire and portions of five liver cells. The hepatocytes have a dense cytoplasmic matrix and they all contain large aggregates of particulate glycogen. The chromatin is aggregated in the form of small masses along the periphery of the nucleus. 3500X.
(B) A higher power electron micrograph showing a portion of a liver cell. Mitochondria (*M*) are seen in both cross and longitudinal section. They contain both cristae and small dense granules. These help to differentiate this organelle from the microbodies (*MB*). An array of rough endoplasmic recticulum cisternae (*ER*) are very evident as are the particulate aggregates of glycogen (*G*). 25,000X.

to carry out. This approach, which was subsequently modified by a number of investigators, often yields superbly fixed brain tissue. An example of the method follows: After opening the abdominal cavity and following cannulation of the descending aorta, a vasodilator (e.g., sodium nitrite) is slowly injected through this cannula followed by a small volume of warm balanced salt solution and then followed by fixative. Adequate drainage is provided during perfusion by opening the posterior vena cava (and occasionally jugulars). Following perfusion, the tissue is excised and sliced into small blocks which are transferred into vials containing fresh fixative to allow completion of fixation.

It is obvious that perfusion is possible for tissues other than those of the central nervous system (Fig. 74). However, it is questionable whether it is desirable to routinely make use of such an elaborate approach (i.e., one which is time consuming and requires considerable technical skill) with tissues that are not as sensitive to anoxia or mechanical damage. Moreover, with care, the immersion method can be successfully used, even for studies of tissues excised from the central nervous system, especially if gluteraldehyde is used as the primary fixing agent and is followed by postfixation with osmium tetroxide.

D. HANDLING OF SMALL SPECIMENS

The previously described approaches are not applicable for use with small specimens such as protozoa, cell suspensions, algae, and bacteria. Such microscopic specimens can be prepared by being placed directly into the fixative. If the specimens are very small, they can be centrifuged to form a pellet immediately after fixation. The speed of centrifugation necessary to insure that a stable pellet is formed is dependent upon the nature of the specimen. The speed should usually be sufficient so that the centrifuged mass is compact enough to be cut or teased into small aggregates which can be treated as though they were solid tissue blocks. The pellet is cut up either after fixation or during or after dehydration (or, if very small, is used *in toto*).

For very small particles and where high speed centrifugation may be harmful, low speed may be used and the pellet then embedded in soft agar for further processing. This eliminates the need for centrifugation at every step and minimizes the possible loss of material during preparation.

E. HANDLING OF HUMAN MATERIAL

It is a well-recognized fact that results obtained with human material have often been disappointing. It has been found that the needle biopsy method of obtaining specimens has not proven reproducibly useful and that, at present, surgical biopsies are more desirable. In the last few years a substantial number of reports have appeared in the literature demonstrating that improved preparation techniques make studies of human material obtained by excision and needle biopsy more useful. These results have considerably enhanced the potential of the application of electron microscopy in surgical pathology. This optimism is reenforced by the finding that a short time gap between death and fixation does not preclude ultrastructural analysis with the electron microscope. As a result, in recent years an increasing number of publications dealing with the application of electron microscopy to both surgical and autopsy material clearly indicates the potential utility of this technique (see p. 345).

F. DISCUSSION

In the preceding section, considerable emphasis was placed on the rapidity of exposing the tissues to the fixative so as to minimize post-mortem alterations in fine structure. A systematic study of the alteration in cell ultrastructure after death has revealed that many membranous organelles possess a surprising degree of stability.[22] This study suggests that for many tissues extremely rapid specimen fixation is less urgent than formerly was thought. It has been suggested[90] that many of the distortions in fine cytological detail, which were previously attributed to delayed fixation, were not directly due to this cause, but rather to a post-mortem increase in susceptibility to polymerization damage following methacrylate embedding. As a result, when specimens were embedded in Epon, cell structure was found to be nearly normal in appearance even though fixation had been delayed for substantial periods. Thus, while needless delay is clearly not desirable, haste of such a nature as would risk gross mechanical damage during excision may be unwarranted.

Chapter 8

FIXATION

BIOMEDICAL electron microscopy, as it has been applied up to now, involves a study of the ultrastructure of non-living cells and tissues initially treated with a solution that, hopefully, neither shrinks nor swells their constituents. This solution must serve, not only as a killing agent, but must also act as a fixative; i.e., the various tissue constituents must be affected in such a manner that they both maintain their initial form and become capable of resisting the effects of subsequent steps in the preparative procedure.

It was appreciated quite early (even at the light microscope level of resolution[21]) that most classical fixatives and fixing mixtures used in light microscopy were less than ideal. They were found to produce distortions of the cytoplasmic structures relative to their appearance in the living cell. Notable exceptions were neutral solutions containing OsO_4.

A. OSMIUM TETROXIDE FIXATION

The usefulness of osmium tetroxide as a histological fixative was recognized by Schultze about a century ago. While the detailed nature of all the interactions of OsO_4 with cellular constituents has not been fully clarified, its reactions with unsaturated lipids are in part evident by rapid blackening of the tissue. It is also believed that osmium tetroxide probably reacts at the site of the polypeptide side-groups of tryptophane and histidine, thus linking protein chains together. Osmium tetroxide acts to stabilize cellular proteins which form the matrix of the protoplasm. Specimens fixed with this medium yield a cytological picture under the electron microscope, which is very "life-like" at the limits of resolution of the light microscope.

It was found that a number of factors (e.g., *p*H, temperature, duration) are critical for good fixation. Knowledge of these factors gave a substantial impetus to the useful application of electron microscopy in biological research. Subsequently, several other fixation media were reintroduced (see Chapter 8 B), or introduced, which resulted in reducing the complete dependence on OsO_4, although this medium is still used in the majority of studies. For this reason, the bulk of this section is devoted to a discussion of the application of OsO_4 as a fixative for electron microscopy.

1. *p*H of the Fixative and Buffering Media

One of the significant findings that greatly facilitated the successful use of osmium tetroxide was the discovery that the *p*H of the fixing solution is critical. It was found by Palade[40] that unbuffered solutions of OsO_4 result in acidification of the excised tissue, which precedes fixation and causes artifacts. Thus buffering of the fixative, in general, was found to be helpful and a *p*H range of 7.2–7.6 was recommended. With protozoa, invertebrate and embryonic tissues, a more alkaline *p*H going up to about 8.0 has sometimes been found to be more effective.

In addition to pointing out the need for *p*H control, Palade favored the use of acetate-veronal buffer of Michaelis. This buffer was used routinely for a long time, but has gradually been replaced by others which are thought to improve the quality of preservation. The buffering media most commonly employed at present are S-collidine[91] and phosphate.[92] (It is desirable that the buffer have a *pKa* near *p*H 7.0, be non-toxic, and not react with OsO_4. This has restricted the range of choice considerably.)

2. Tonicity of the Fixative

After introduction of the Palade medium, attention began to focus somewhat more on the tonicity of the fixing solution. As a result, various salts were added to the standard medium. A little later it was suggested that sucrose be used since, in addition to controlling molarity, this compound was thought capable of passing freely in and out of the cell. It is now felt that the concern over tonicity of the fixative was exaggerated. Thus, while it appears logical to use a balanced salt solution or sucrose, their presence does not seem essential. Tonicity is perhaps a factor that needs special attention when working with particularly sensitive material rather than in routine fixation.

3. Temperature of the Fixative

Immersion of the specimen in cold osmium tetroxide has been advocated for a long time. This idea stems from the desirability of reducing the rate of cellular changes following excision and prior to the action of the fixative. In addition, it is felt that at low temperatures (0–4°C), an increased proportion of the block will be well-fixed. Thus, routinely, specimens are cut in a few drops of precooled fixative and are then placed in glass-stoppered vials of fixative which have been kept in crushed ice. Equilibration to room temperature takes place during the latter part of fixation or during dehydration (see p. 137).

Chilled fixatives can also be used with animal tissues *in situ* fixation. For plant material, the temperature of fixation is thought to be critical. Some plant specimens appear to be preserved best at 0°, while others require fixation at room temperature or higher.

4. Duration of Fixation

In the infancy of biomedical electron microscopy (1952–1955), it was the practice to fix tissues for periods of 2–6 hours time. Subsequently, it became evident that prolonged fixation resulted in solubilization and subsequent leaching out of proteinaceous substance from the specimen by the buffered fixative.

Currently, the general practice (with OsO_4) is to fix material for from 30 to 90 minutes, the optimal time depending on the size and nature (density) of the specimen. Very short periods of fixation are inadvisable, since then the specimen often is not in a state that will stand up under further processing. Thus the duration of fixation should be adjusted so that a reasonable compromise is reached between the two simultaneous effects of the fixative, namely, the stabilization of the tissue proteins and subsequent solubilization of some of the tissue components.

5. Modification of the Standard Osmium Tetroxide Medium

Several modifications of the standard Palade medium have been introduced for a number of reasons, some of which were noted in the preceding Chapter (8 A 2). In this connection, the following media have been used:

(a) Zetterquist's buffered isotonic osmium tetroxide (for formula, see ref. 119).

(b) Caulfield's[93] buffered osmium tetroxide fixative with sucrose.

(c) Dalton's[94] chrome-osmium fixative. (This medium was suggested in order to reduce extraction. It often, therefore, produces specimens with less contrast among cell organelles than the other OsO_4 fixatives).

The standard osmium fixation medium is the one used in most studies.

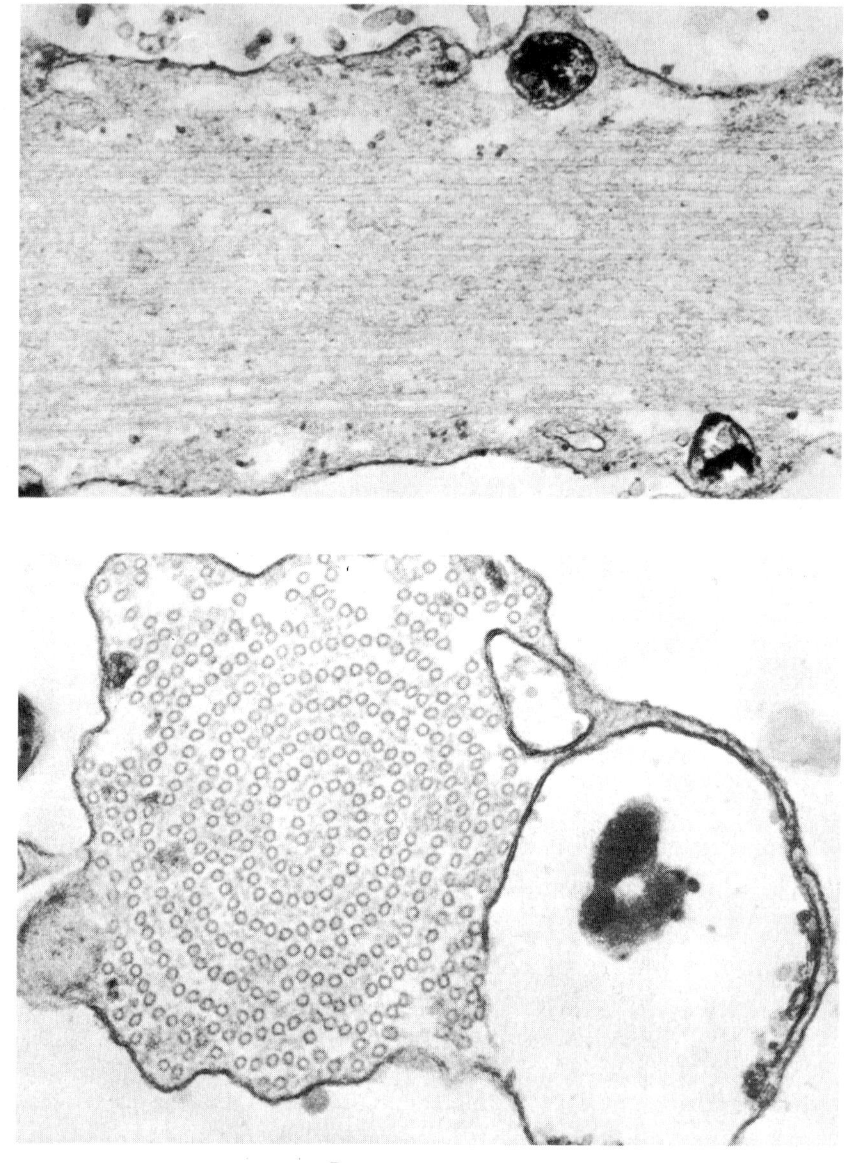

B

FIG. 75. Demonstration of glutaraldehyde fixation of the axopodia of the protozoan *Actinosphaerium nucleofilum*.
(Courtesy of Drs. L. G. Tilney and K. R. Porter.)
(A) Longitudinal section through a recovering axopodium. Dense granules are present just peripheral to the axoneme (birefringent core). The latter is made up of microtubules arranged parallel to each other and the long axis of the axopodium. 42,000X.
(B) Transverse section through a recovering axopodium. Within the plasma membrane is a partially ruptured dense granule and axoneme consisting of a double coiled array of microtubules. 71,000X.

B. OTHER FIXATION MEDIA

The dependence on osmium tetroxide has long been a source of concern to investigators. A number of other media have been introduced but none, as yet, have met with the same general acceptance.

1. Aldehyde Fixatives

Formaldehyde and glutaraldehyde are the most popular aldehydes in current usage as fixatives.

a. FORMALIN

Although formalin by itself, followed by methacrylate embedding, has not resulted in as good preservation as with osmium tetroxide, this medium appears to have some usefulness for the fixation of certain plant tissues.[95] When formalin fixation, however, is followed by Epon embedding, greater structural stability of the specimen under the electron beam appears to occur.[96]

A chromic acid-formalin mixture was suggested as having some limited value.[97] (Improvement in formalin fixation, especially for studies at high magnification, was reported when followed by treatment with buffered osmium tetroxide.[68,69]) Also, formalin treatment following permanganate fixation was reported to be superior to formalin fixation by itself.[98]

Brief fixation in formalin (10–15 minutes) followed by embedding in a water-soluble embedding medium has been used to study the action of hydrolytic enzymes on thin sectioned specimens. In this case, the cellular constituents are being deliberately "understabilized" so as to retain their susceptibility to specific reagents.

b. GLUTARALDEHYDE

In 1959, Luft[99] suggested that acrolein (acrylic aldehyde), because of its penetrating and preserving properties, be used for electron microscopy. Subsequently, a number of dialdehydes were introduced as fixatives for electron microscopy.[83] These substances provide a firmer tissue consistency which permits cutting the specimen into small pieces for postfixation with buffered osmium tetroxide. The latter treatment is desirable in order to insure adequate specimen stability during dehydration and embedding and in the electron beam.

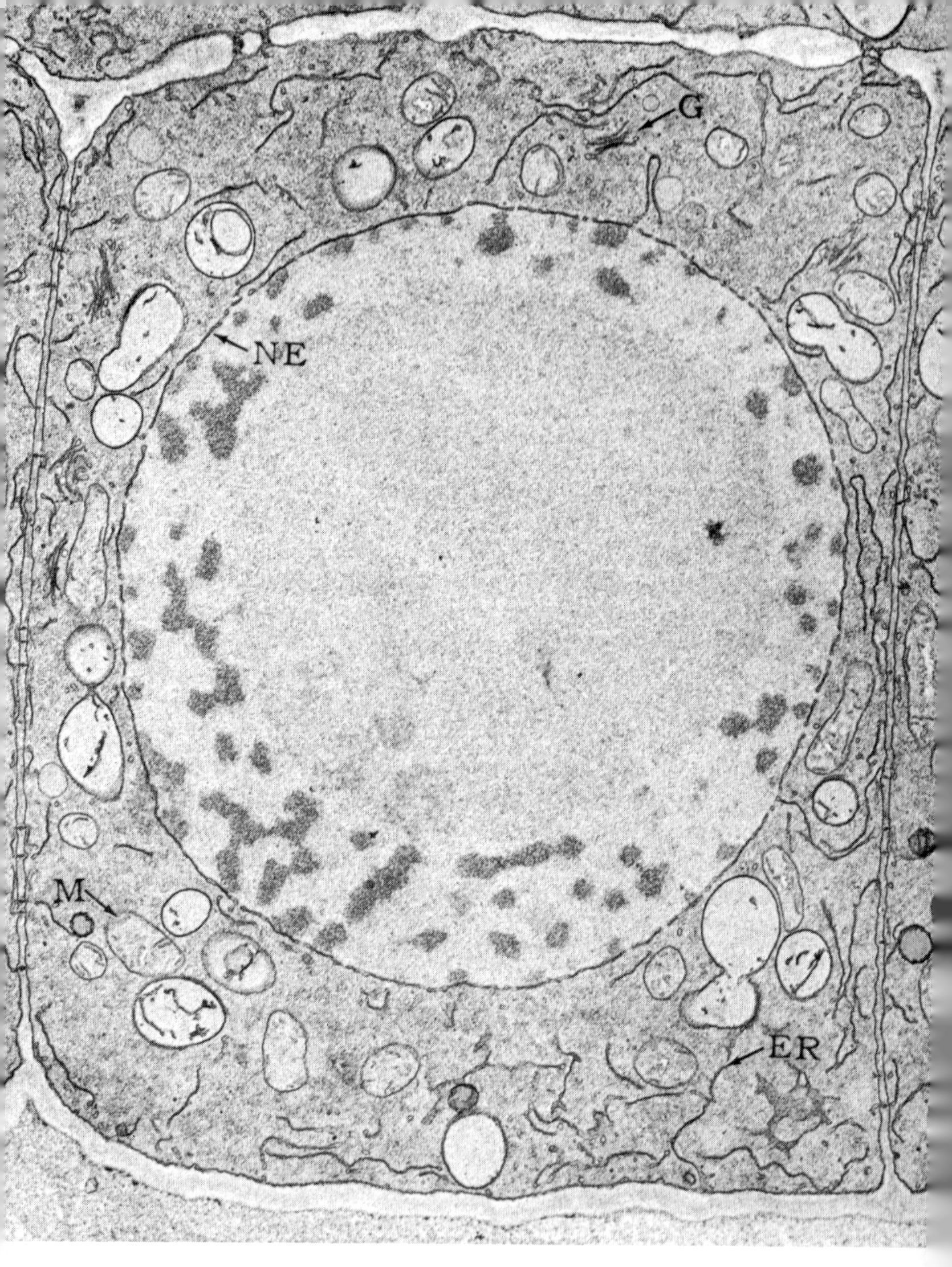

FIG. 76. Demonstration of permanganate fixation of plant tissue. Cells of the cortex of the maize root tip. The cytoplasm contains numerous mitochondria (*M*), clusters of Golgi material (G) and individual strands of endoplasmic reticulum (*ER*). The discontinuous nuclear envelope (*NE*) bounds the nucleus. 15,000X. (Courtesy of Dr. H. H. Mollenhauer.)

Aldehyde fixatives have two advantages over buffered osmium tetroxide:

(1) They penetrate tissues rapidly. By the time the tissue is trimmed, it is firm. As a result, distortion is reduced and structural relationships are maintained.

(2) Cytoplasmic matrix components seem less prone to extraction. As a result, the fidelity of the ultrastructure is improved.

Of the various dialdehydes tested, glutaraldehyde appears to be best. On the basis of electron microscope studies that are currently being published, there is widespread use of glutaraldehyde, especially in combination with postfixation with osmium tetroxide. The use of glutaraldehyde has been recommended since initial observations suggest that its preservation of cellular ultrastructure may be more reliable than even that of osmium tetroxide.[100,101] This seems especially to be the case of the class of organelles called microtubules (Fig. 75). Its value as a fixation medium extends to plant as well as animal material.[102]

2. Permanganate

Luft[54] introduced buffered potassium permanganate as a fixative. This strong oxidizing agent was found to preserve some cell membranes and to "produce" a finely granular cytoplasmic matrix. Both DNA- and RNA-containing elements, however, were not satisfactorily preserved. Improved results were obtained when permanganate was used selectively rather than as a general purpose fixative. Thus, permanganate fixation provided very satisfactory results (Figs. 76 and 77) when, for example, used to study membrane systems in plant material[103] and to determine the ultrastructure of the myelin sheath of the axon.[104]

A further evaluation of the effectiveness of potassium permanganate fixation for electron microscopy was made by Bradbury and Meek.[105] In addition to confirming Luft's findings, they reported that permanganate causes considerable initial swelling as contrasted with the shrinkage resulting from treatment with osmium tetroxide. The rate of penetration of permanganate is, like OsO_4, slow in comparison with that of aldehydes. Permanganate probably forms insoluble, electron-dense, MnO_2 by reaction with tissue constituents, thus adding to image contrast. The deposited non–MnO_2 products are particles about 50 Å in diameter which are insolubilized during the dehydration with alcohol rather than by the initial permanganate treatment. Thus, since permanganate does not preserve most protein or phospholipid constituents of the protoplasm, these authors do not regard it as a true cytological fixative.

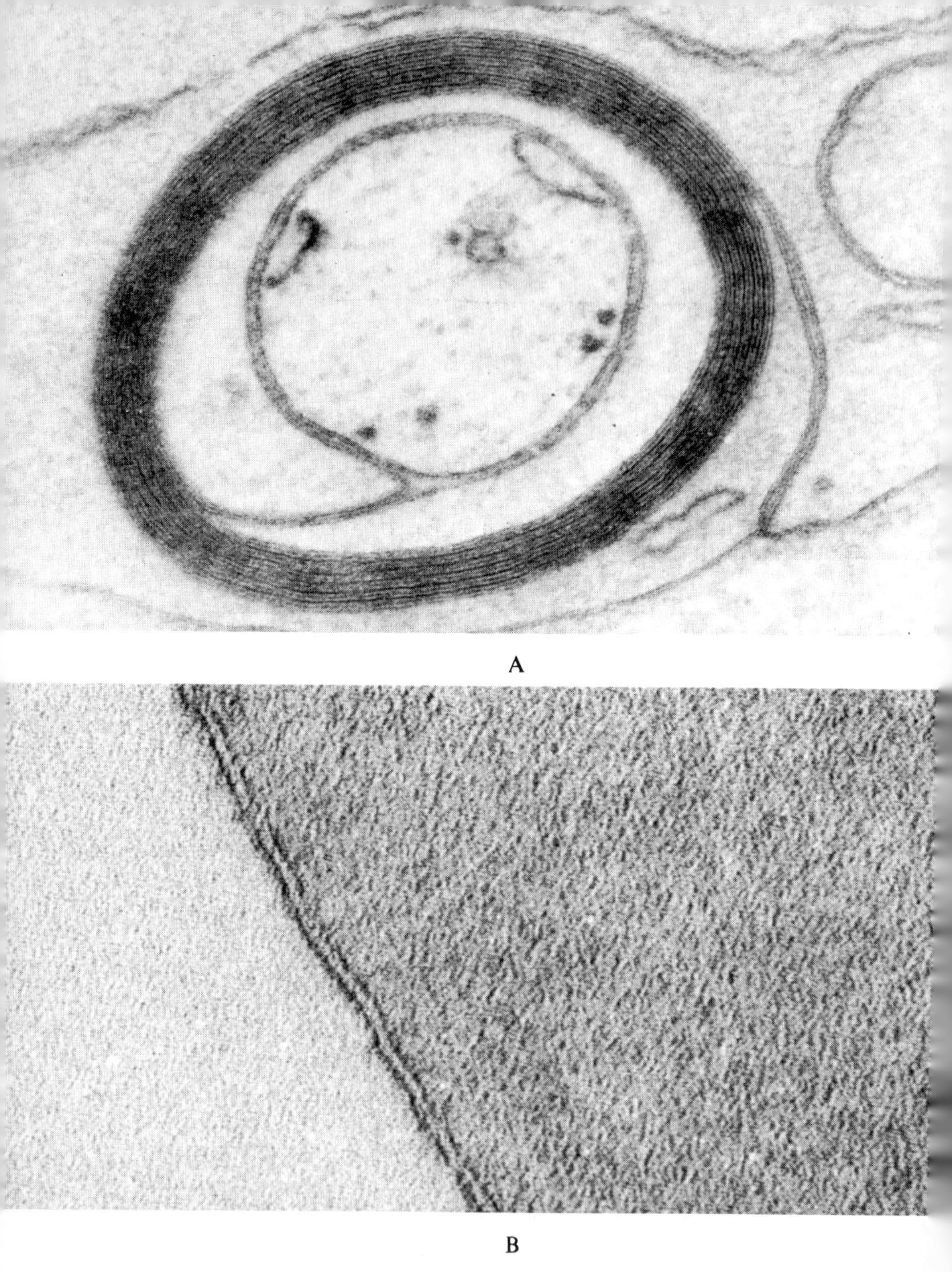

FIG. 77. Demonstration of permanganate fixation of animal tissue. (A) Cross-section of an axon. The individual membranes making up the myelin sheath are delineated. 120,000X. (Courtesy of Dr. J. D. Robertson.) (B) The unit membrane at the surface of a human red blood cell. 380,000X. (Courtesy of Dr. J. D. Robertson.)

C. DISCUSSION

Examination of thick sections under the phase contrast microscope can provide initial information as to the state of specimen preservation. Under the electron microscope, subjective judgment of the state of preservation is based primarily upon overall appearance, as determined by the degree of orderliness and the pattern of distribution of the components and, to some extent, by the aesthetic appeal of the image.

The specific criteria related to the ultrastructural appearance of the cellular elements after good osmium tetroxide fixation are summarized in Table VI and illustrated in Fig. 78.

The basic validity of cell ultrastructure found in osmium tetroxide fixed tissues has been confirmed by finding a very similar, but not always identical, fine structural organization after gluteraldehyde fixation and, in part, after preservation with potassium permanganate.

TABLE VI

CRITERIA FOR ULTRASTRUCTURAL APPEARANCE OF CELLULAR ELEMENTS

Cellular Element	Appearance After Good Fixation
Mitochondria	Few swollen or empty looking; external and internal membranes smooth; no gross distortion in shape.
Endoplasmic reticulum	Membranes intact, essentially parallel. When stacked, cisternae arrangement uniform.
Golgi membranes	Intact; vesicles of various dimensions appearing in a circumscribed area.
Plasma membrane	Intact and smooth around entire cell.
Cytoplasmic matrix	Fine precipitate; no empty spaces generally present.
Nuclear envelope	Both membranes are intact (except at pores) and essentially parallel to each other.
Nuclear contents	Finely granular with denser masses both in the interior and periphery of the nucleus.

Recent studies [106,107] have made it clear that various buffers produce different appearances in ultrathin sections of tissue fixed with osmium tetroxide of identical concentration and *p*H. The effect of the buffer upon the specimen can be recognized by differences in the preservation of nuclear chromatin, endoplasmic reticulum, mitochondria, and background cytoplasm. Thus the choice of buffer can result in differences in the electron

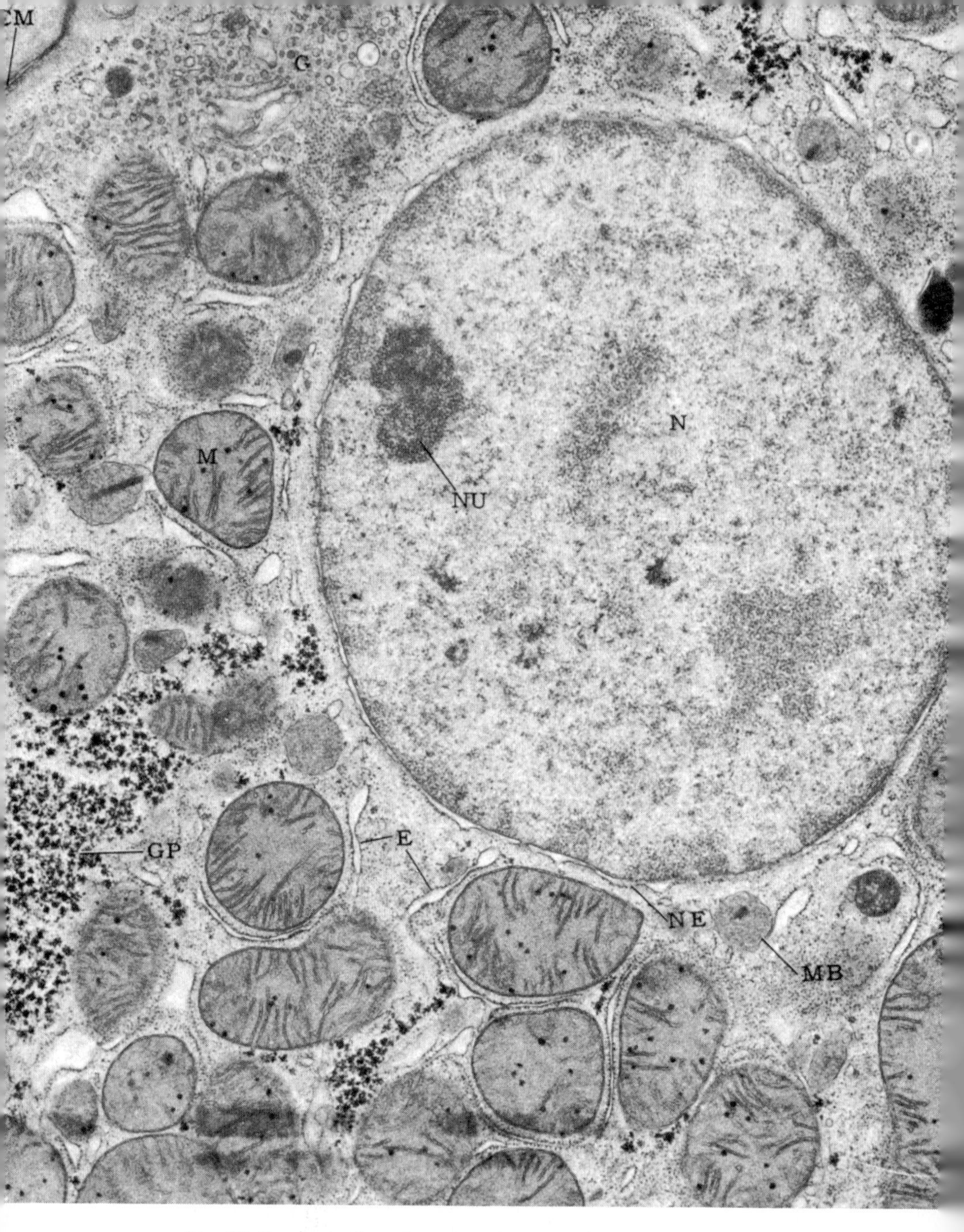

FIG. 78. Demonstration of good osmium tetroxide immersion fixation. This is an electron micrograph of a part of a liver cell of a bat. The nucleus (*N*) contains a nucleolus (*NU*) and is bounded by a nuclear envelope (*NE*). The cytoplasm contains the usual organelles such as mitochondria (*m*), cisternae of the endoplasmic reticulum (*ER*), Golgi complex (*G*) as well as microbodies (*MB*) and glycogen particles (*GP*). The cell membrane (*CM*) is visible in one corner. 9,600X. (Courtesy of Dr. K. R. Porter.)

microscope data obtained. This points up the fact that much remains to be clarified about the "optimal conditions" of fixation.

The reaction of the fixative with cellular constituents takes place at a "macromolecular" level. While many of the details of the chemistry of fixation have, as yet, not been worked out, proteins are best preserved with OsO_4 or formalin, while potassium permanganate does not insolubilize many of the proteins. This selective action of permanganate serves to make the membranes stand out, a feature which is probably further emphasized by its strong reaction with the phospholipid component of the membranes. Osmium tetroxide is largely active by virtue of its reaction with proteins and lipids. Both types of nucleic acids do not react extensively with these fixatives, but their preservation is probably due to their being bound to proteins and by subsequent staining.

Recently, Pease[108,109] demonstrated that for some purposes chemical fixation of tissues for electron microscopy is unnecessary if the tissues are dehydrated with increasing concentrations of inert, water-soluble, permeating, organic substances (Fig. 79). The simplest successful procedure found was to use ethylene glycol as the dehydrating agent and to transfer the tissue directly from this into partially prepolymerized HPMA (see p. 143).

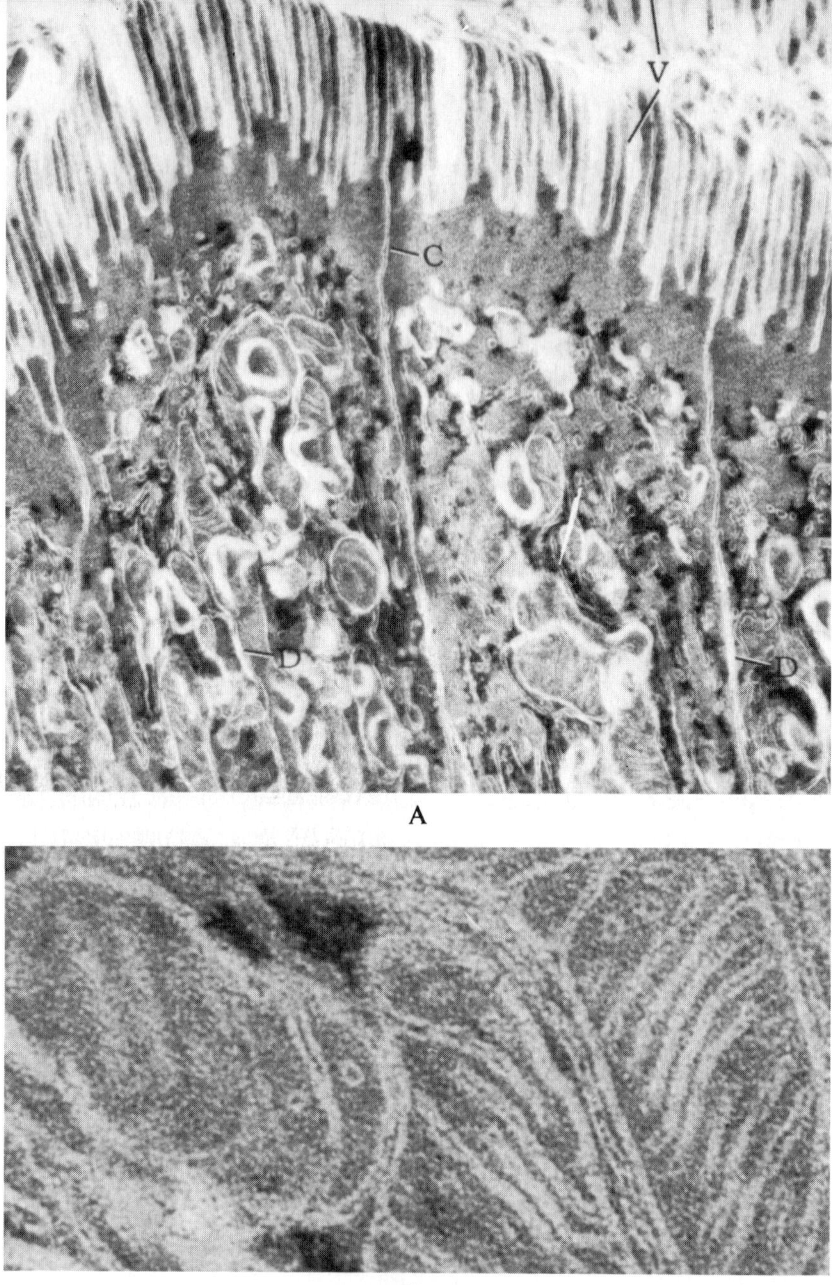

FIG. 79. Demonstration of preservation by dehydration with "inert agents." Both specimens, from the rat, were embedded in HPMA and stained with uranium and lead. (Courtesy of Dr. D. C. Pease.)

(A) Proximal portion of several epithelial cells of the jejunum. Note the cell membranes (*C*) which exhibit desmosomes (*D*) and also the numerous microvilli (*V*). 21,000X.

(B) Three mitochondria from the basal part of a proximal tubular cell of the kidney. Note the well-preserved mitochondrial membranes. 94,000X.

Chapter 9

DEHYDRATION

IT WAS FORMERLY customary to wash the fixed specimen with distilled water, saline or the buffer medium. Later it was found that this step was usually unnecessary and, moreover, that washing after veronal-osmium fixation could have harmful secondary effects. Currently, therefore, washing is often omitted from the processing procedure and the specimen is dehydrated immediately after fixation.

A. DEHYDRATION MEDIA

Because ethyl alcohol was the dehydrating agent traditionally used in histologic and cytologic processing procedures, it was natural for this solvent to be initially selected for use in electron microscopy. It has proven to be a reliable and effective agent. When water-soluble resins are used, however, ethanol (or other agents) are unnecessary since these embedding media can also be used to dehydrate the tissue directly.

Acetone has also been used as a dehydrating agent, mainly for procedures involving later embedding with Vestopal, a polyester resin, not miscible with ethanol. Ethanol can, nevertheless, be used even in this case, if styrene or acetone is interposed as an intermediate step just prior to infiltration with Vestopal.[110]

Propylene oxide, a solvent ordinarily used between alcohol and embedding media, such as Epon, can also simultaneously serve as a dehydrating agent. This is possible because propylene oxide is both partially miscible with water and dissolves in epoxy resins. The use of propylene oxide as both a dehydration and infiltration agent, however, has not been widely accepted.

As noted earlier (see p. 135), Pease demonstrated that inert compounds such as ethylene or propylene glycol can be used effectively as dehydrating agents. They serve both to displace tissue water and stabilize the cell's macromolecular systems. As a result, cytoplasmic proteins are very largely retained and most fine structural relationships remain intact. The cytomembranes in such preparations are seen in negative contrast, outlined by neighboring material (Fig. 79).

B. DURATION OF DEHYDRATION

In the past, it was commonly believed that slow dehydration was necessary to prevent distortions due to large transient changes in osmotic pressure. It was subsequently found that this classical histologic approach often contributed to extraction of tissue components and subsequent shrinkage. At present, the usual routine employed calls for rapid dehydration. Dehydration usually starts with a short treatment with 70 percent ethanol and then a transfer to 95 percent. From the latter the specimen is transferred to 100 percent alcohol for a longer period of time (*circa* one hour). It is common practice to use cold alcohols through at least the 95 percent concentration, at which point the vial containing the specimen is allowed to come to room temperature.[85]

Chapter 10

EMBEDDING

IN THE EARLY days of thin sectioning, paraffin waxes were tested as embedding media, but were found to be too soft to enable sections thinner than about 1 μ to be cut. Subsequently, harder waxes and double embedding techniques were tried, but without substantial overall improvement.

The first embedding medium found to be especially suitable for electron microscopy was poly-butyl methacrylate. This embedding matrix and slight modifications of it were used exclusively despite severe shortcomings for about ten years following its introduction in 1949. Epoxy resins began to be widely accepted by 1960, and have essentially replaced methacrylate mixtures as the medium of choice. Thus our discussion of butyl methacrylate will be primarily of historical interest.

A. METHACRYLATE

The introduction of methacrylate by Newman *et al.*,[36] was one of the three major steps (the others are the introduction of glass knives and the controlled fixation with OsO_4) which resulted in the beginning of the establishment of routine specimen processing for electron microscopy. These innovations stimulated the now classic studies of the ultrastructure of the normal cell.

The suitability of methacrylate as an embedding medium for biomedical electron microscope studies resides in its ease of handling, its good rate of penetration of tissues and its capacity to be cut to thicknesses in the range of 500–1000 Å with relative ease. Varying the ratio of a mixture of methyl methacrylate and n-butyl methacrylate monomers to suit the density and composition of the tissue specimen to be examined permits blocks of proper hardness to be produced. Hardening of the blocks is the result of polymerization of the monomer molecules to form simple linear polymer chains. At any but low beam intensities, unsupported methacrylate sections almost always develop tears. Such sections, therefore, require supporting membranes made of Formvar, Parlodion, or, preferably, carbon.

Methacrylates are esters of methacrylic acid (2-methyl propenoic acid). The conversion of the acid to an ester is by reactions typical of carboxylic acids.

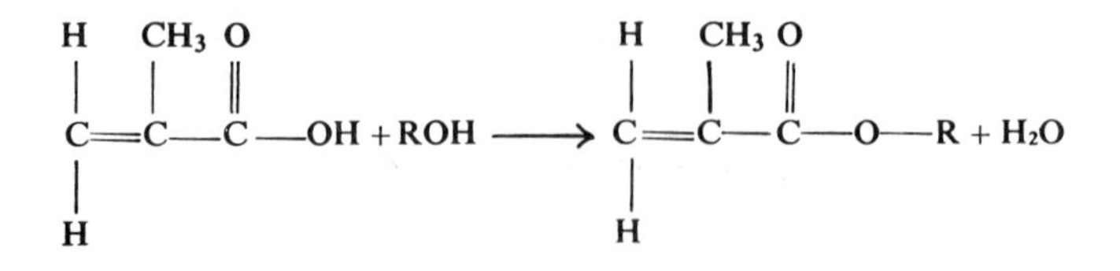

where:

$R = CH_3$ for methylmethacrylate.

$R = CH_3—(CH_2)_3$ for n-butylmethacrylate.

The initiation of polymerization involves the dissociation of the catalyst into free radicals. This dissociation is accelerated by heat or, alternatively, results from the absorption of a quantum of light (e.g., U.V.). The free radical can then form polymer radicals by reaction with monomers. These polymer radicals then, in turn, initiate the chain reaction of polymerization.

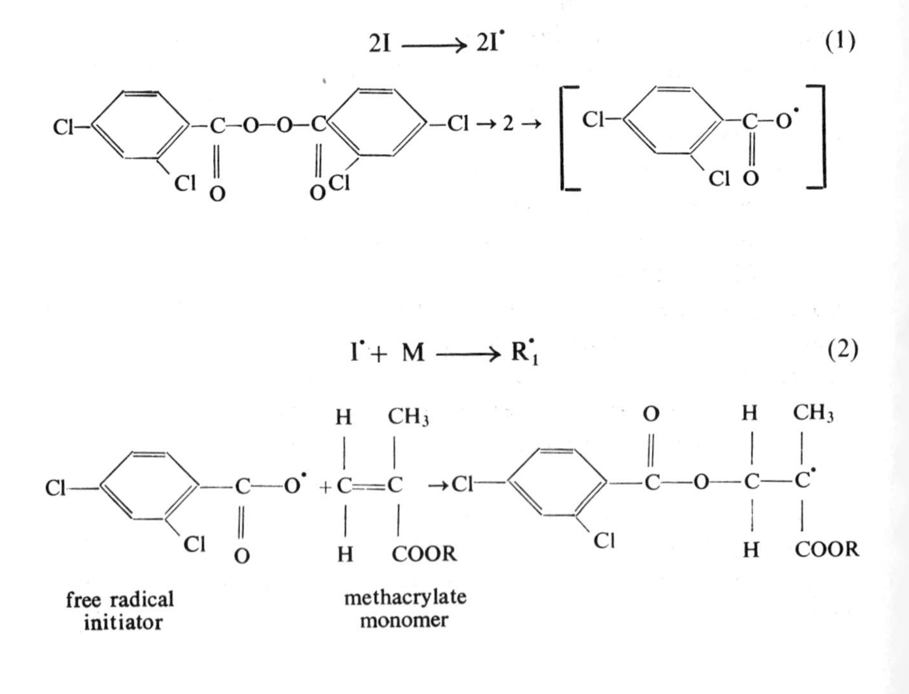

where:

I = initiator.
I$^{\bullet}$ = free radical initiator.
M = monomer.
$R^{\bullet}_1$ = radical formed by combination of monomer with initiator radical.

(where $\cdot$ represents the unpaired electron of a free radical).

Following initiation, the normal propagation step may be represented by the following equations:

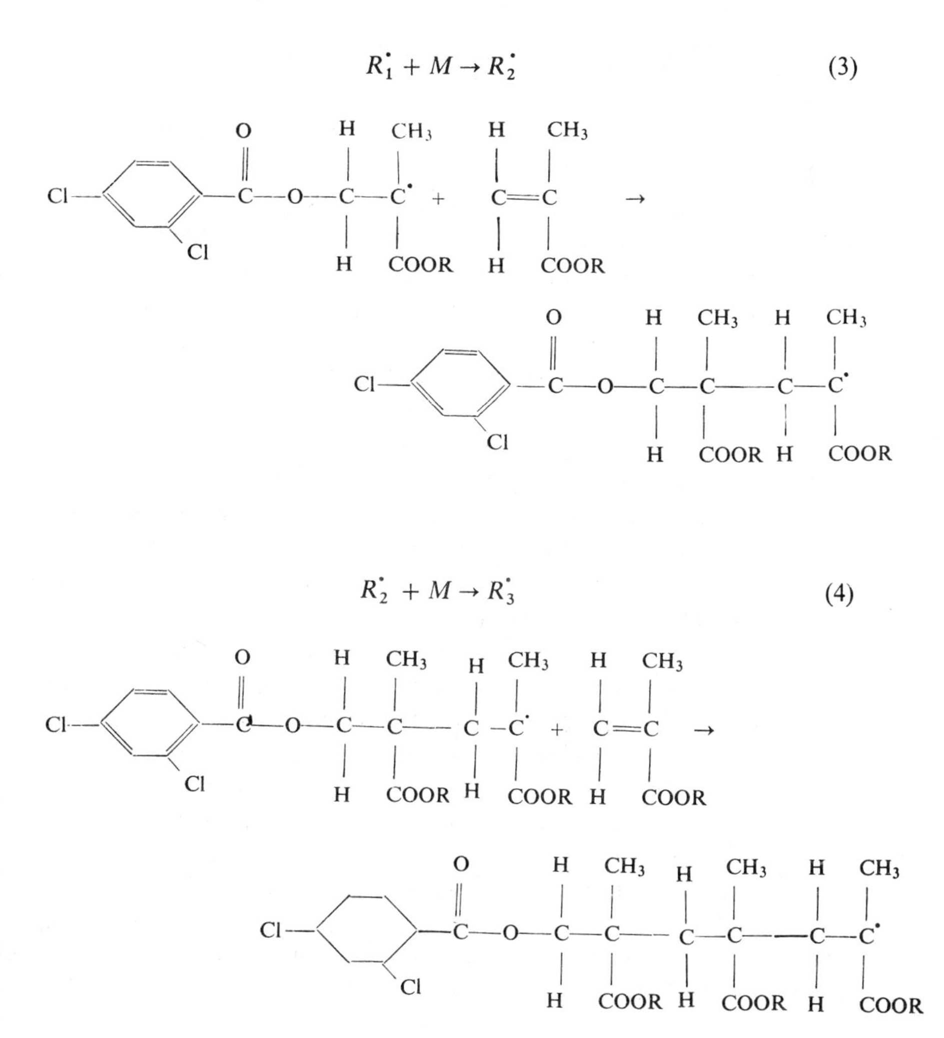

or, in general,

$$R_n^{\bullet} + M \rightarrow R_{n+1}^{\bullet} \tag{5}$$

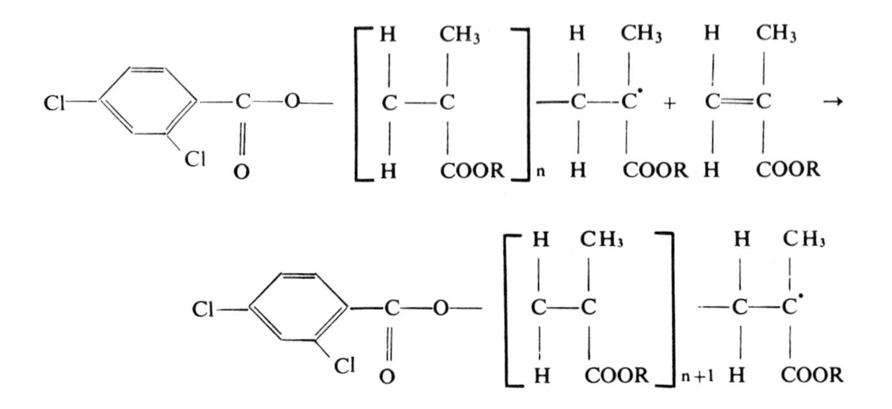

Three major problems were encountered in processing tissues with methacrylate. These were:

1. Shrinkage. A substantial decrease in volume (up to 20%) occurs when methacrylate polymerizes. This may reduce the size of the specimen embedded within it. The use of partially polymerized methacrylate reduces this shrinkage.[111]
2. Polymerization damage. Frequently specimen artifacts were introduced during polymerization.[21,112] Several ways to reduce the effect of this drawback were suggested.[112,113]
3. Beam evaporation. Methacrylate was found to be unstable under electron bombardment. Sublimation of up to 50 percent of the methacrylate in the thin tissue section usually takes place under normal operating conditions.[50] Such losses due to sublimation can be accompanied by a flow of the remaining methacrylate and, as a consequence, the structure of the specimen may become distorted (although the contrast is often improved by this removal of part of the embedding medium). Such loss of methacrylate can be minimized by prolonged irradiation of the sections with a low intensity electron beam prior to viewing at higher intensity. Alternatively, sublimation can be reduced by sandwiching the specimen between protective films of carbon.[114]

While the three problems associated with methacrylate embedding were partially controllable, preparations continued to show unpredictable variability, even in the hands of experienced investigators.

B. WATER-SOLUBLE EMBEDDING MEDIA

Such media have been employed for histochemical studies. They were introduced in the hope that some lipid-containing structures would be better preserved and also that water-soluble histochemical reagents (e.g., enzymes) might penetrate more rapidly into sections. Aqueous solutions of these media are used immediately after fixation, both as dehydrating agents and then in pure form as infiltration and embedding agents.

Leduc and Bernhard,[76,115] have for several years been examining water-soluble embedding media with respect to the preservation of cell ultrastructure and to the digestibility of sections of formalin-fixed tissues by appropriate hydrolytic enzyme solutions. They found that the susceptibility to the enzymes varies with the embedding medium employed. This differential response was suggested as being due to the variable manner in which the media combine with reactive groups of proteins and nucleic acids. These studies point to the possibility that, when technical problems are overcome, a new approach to cytochemical analysis on the ultrastructural level will be available. This subject is discussed more fully in Chapter 9 A.

Three water-soluble agents have been tested.

1. Glycol methacrylate. Thin sections of specimens polymerized in glycol methacrylate tend to swell in water. They also stretch under the electron beam thus resulting in a distorted image. This compound is 2-hydroxyethyl methacrylate (HEMA). The hydroxypropyl derivative (HPMA) has been claimed to be a superior water-soluble embedding medium[116] (Fig. 80). Hydroxypropyl methacrylate, however, is less "water-like" than HEMA.
2. Aquon. This is a water-soluble component of the widely used epoxy resin, Epon 812. It is a resin of relatively low viscosity and sections fairly well. Because tissues embedded in it resist enzymatic digestion, its use for such histochemical studies is limited.
3. Durcupan. This medium is a water-soluble epoxy resin that is commercially available. When it is used without other added epoxies the blocks are very soft, thus producing serious sectioning problems. When sufficient additives are provided to improve its sectioning properties, most of the advantages associated with its water solubility appear to be lost. This has severely restricted its usage.

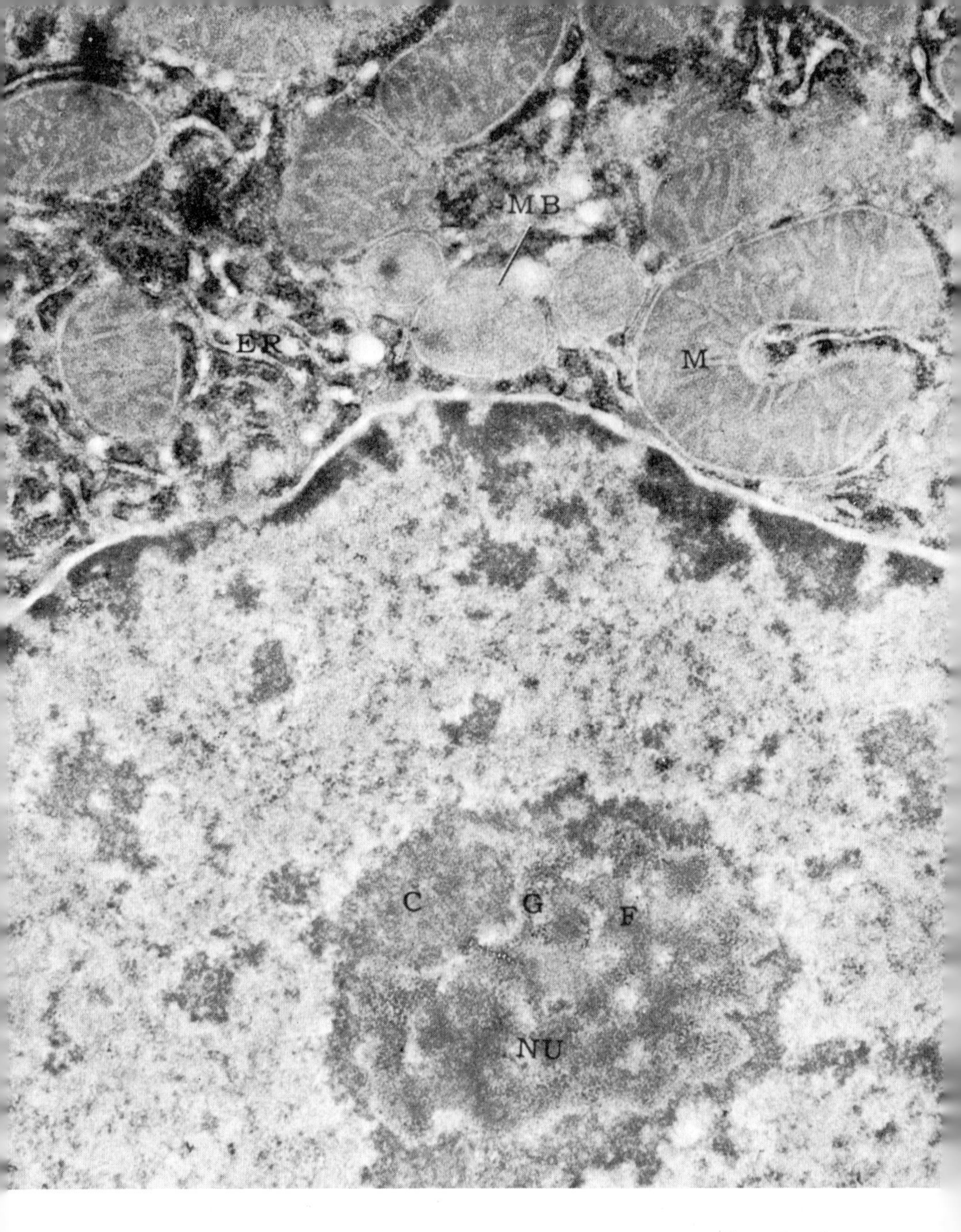

FIG. 80. Demonstration of embedding with a water-soluble medium. Rat liver fixed in gluteraldehyde and dehydrated and embedded in HPMA. (Compare with Figure 78.) The membranes are unstained and appear in negative image. The most heavily stained components of the cells are the chromatin (*C*) and the nucleolus (*NU*) including both its granular (*G*) and fibrillar components (*F*). The cytoplasm contains mitochondria (*M*), microbodies (*MB*), strands of endoplasmic reticulum (*ER*). 30,000X. (Courtesy of Dr. E. Leduc.)

C. POLYESTER RESINS

The polyester, Vestopal W, another acrylic resin, was introduced[58] as an alternative to methacrylate. Its use also requires an initiator and activator. Since the embedding medium is not miscible with alcohol, after dehydration in alcohol, the specimen should be placed in styrene.[109] This liquid is miscible with both alcohol and Vestopal (and is a component of Vestopal) and reportedly assures consistently well-infiltrated specimens. Alternatively, dehydration may be carried out in acetone. Vestopal sections do not require supporting films.

Polyesters may be defined as polycondensation products of dicarboxylic acids with dihydroxy alcohols. Vestopal belongs to the unsaturated group of polyesters. Unsaturation is usually introduced by the use of unsaturated dicarboxylic acids such as maleic or fumaric acid. Thus, for example, the reaction of ethylene glycol (a dihydroxy alcohol) with maleic anhydride would result in a linear unsaturated polyester, ethylene glycol maleate, having the following formula:

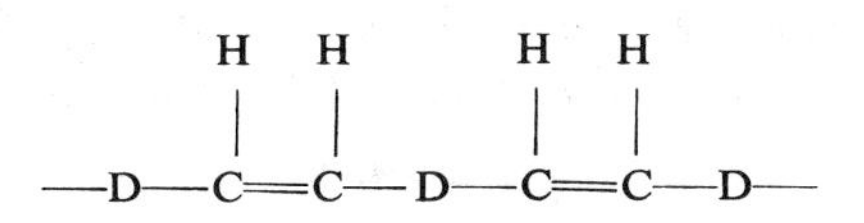

where:

```
        ┌          H    H           ┐
        │          |    |           │
  D =   │  —C—O—C—C—O—C—            │
        │   ||     |    |      ||   │
        └   O      H    H      O    ┘
```

Heating of this linear polyester (especially in the presence of peroxide initiators) will cause some reaction between double bonds of the same sort as in the polymerization reaction for the methacrylates, resulting in the formation of a cross-linked resin.

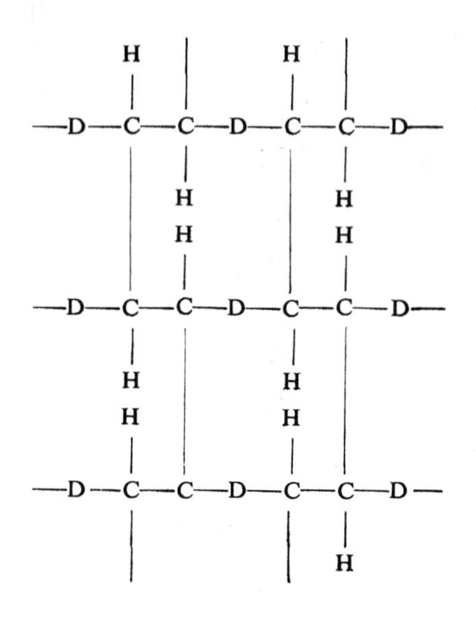

The unsaturated polyester, Vestopal W, used for electron microscopy is usually cross-linked with the unsaturated monomer styrene($C_6H_5CH=CH_2$). One type of cross-linking which can occur is illustrated by the following formula which represents the compound formed when styrene cross-links the double bonds in the linear ethylene glycol maleate of the preceding example.

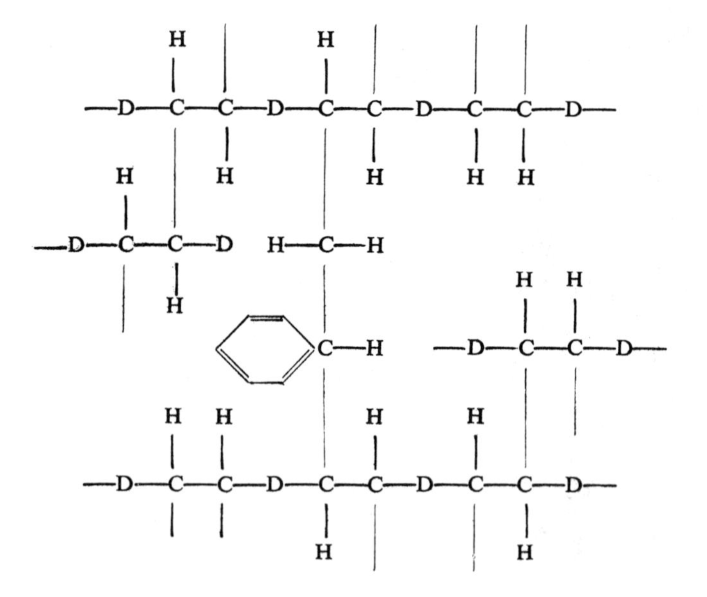

In a similar manner, longer *chains* of polystyrene may also act as cross-links. This cross-linking, by means of saturated chains, is again perfectly analogous to the polymerization of the polymethacrylates (and it is also initiated by peroxides).

D. EPOXY RESINS

In the mid-1950's, attempts were being continuously made to find alternate embedding media that would avoid the aforementioned problems associated with the methacrylates. During that period, epoxy resins were introduced for industrial purposes. In use, a mixture is prepared which consists of the epoxy resin, a hardener, an accelerator (to control the rate of hardening) and a plasticizer (to control the hardness of the block). Maaløe and Birch-Anderson[52] carried out initial investigations along these lines using a Shell epoxy resin (EPO). No polymerization damage (as usually seen with methacrylates) was observed. This resin, however, because of its extremely high viscosity, proved impractical due to its poor penetrability into the specimen. This work was sufficiently promising to stimulate others to search for more suitable epoxy resins. This early work was followed by the introduction of Araldite.

The chemical model of the configuration of typical epoxy resins can be related to the linkage of two glycerol molecules by an ether (-O-) linkage.

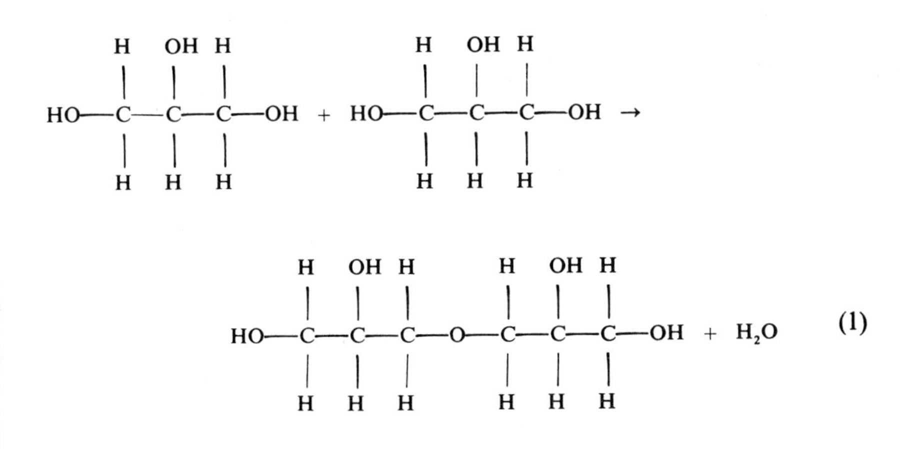

The diglycerol ether is dehydrated, producing the illustrated diepoxide (epoxide group –C—C–):

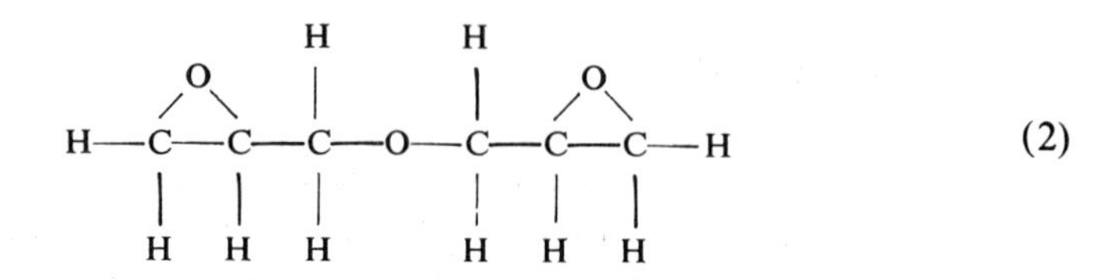

Glycerol can also be dehydrated to form polyethers of glycerol with terminal epoxide groups. Quite commonly, more complex polyethers are synthesized with aryl groups* (R), located at the sites of the ether linkages.

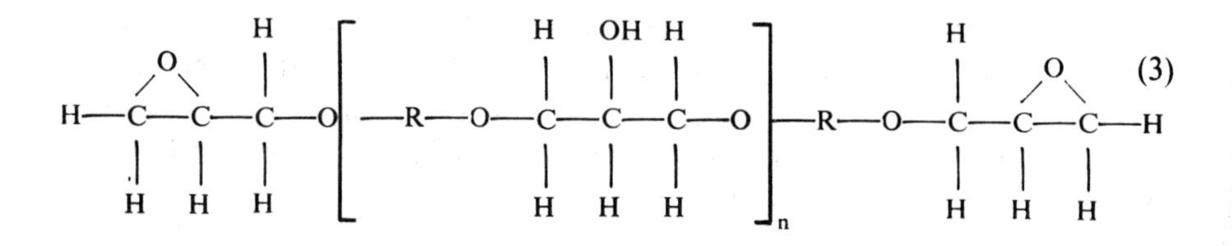

For example, when the aryl group is diphenylpropane, we get the following general formula for a particular epoxy resin, which is a polyether epoxide linear polymer.

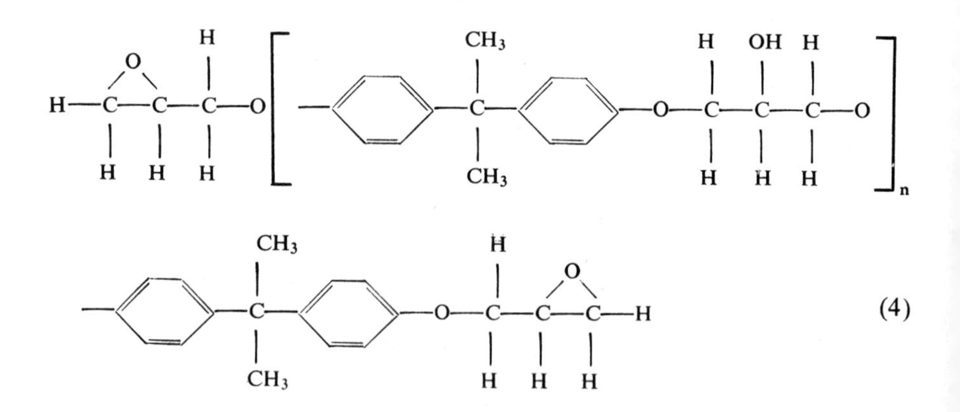

* An aryl group refers to any organic radical obtained from an aromatic hydrocarbon.

By themselves, these liquid resins have little utility. They are, therefore, further polymerized ("cured") by the action of a curing agent and heat. By the addition of such materials, the epoxy resins polymerize, forming cross-links to give clear, tough thermosetting resins with high strength and with little shrinkage.

As curing agents, aromatic anhydrides are used for hardening at elevated temperatures. The further addition of aliphatic polyamines permits room temperature curing.

The addition of an aromatic anhydride [e.g., dodecenylsuccinic anhydride (DDSA) or methyl nadic anhydride (MNA)] to the linear epoxy resin results in the formation of cross-linked polyester-polyethers.

The general formula for a particular linear epoxy resin (see above) can be rewritten in a simpler form as:

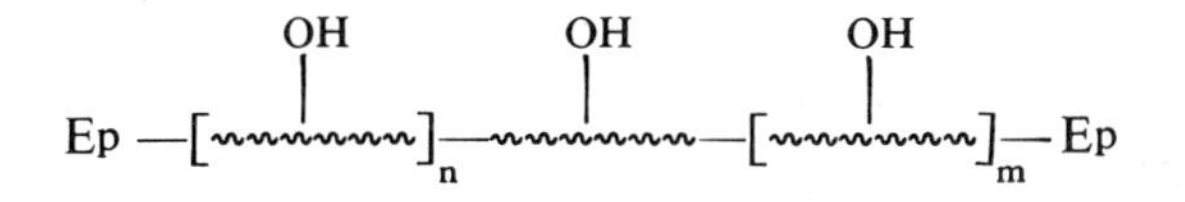

where:

Ep = epoxide group
OH = hydroxyl group
[OH monomeric unit] = monomeric group

When the epoxy resin reacts with an anhydride, the following reaction takes place:

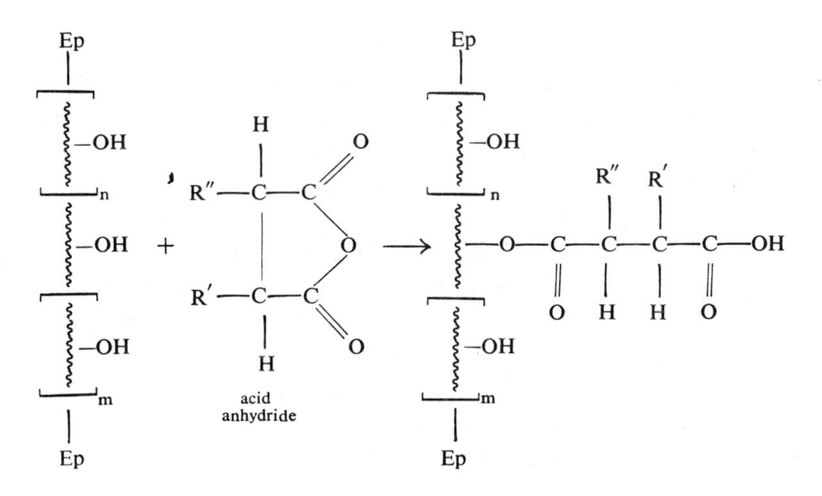

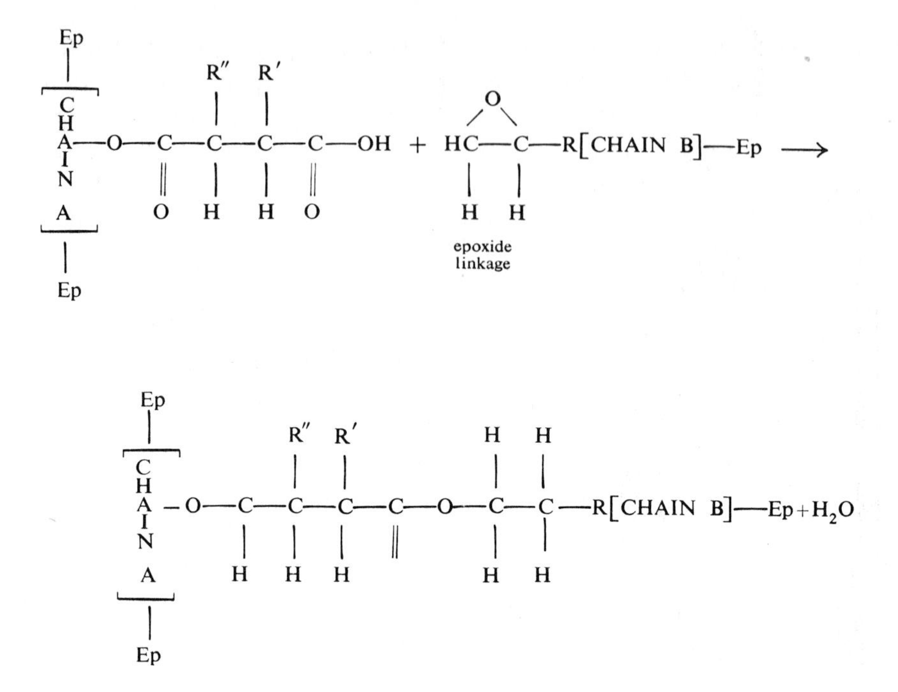

where:

chains A and B = contents between terminal Ep groups of linear polyether epoxide polymer.
R = hydrocarbon portion of epoxy resin molecule.
R′ and R″ = H, alkyl or aryl groups.

Other internally located OH groups and the terminal epoxide groups of the same and other chains react with, and are bridged by, the anhydride in a similar manner, producing a highly cross-linked resin from the original, essentially linear, polyether epoxide polymers.

When curing is completed the epoxy polymer can be viewed in two ways; one tracing a path down the original polyether chain and the other at "right angles" tracing a path from cross-link to cross-link. When viewed down the polyether chain, the epoxy resin is unique among the resins here discussed. When viewed at "right angles," the appearance is like the main chain of the polyesters. (Similarly, in the case of the polyesters, while the main chains are like the cross-links in the epoxy resins, the cross-links are "like" the backbone of the polymethacrylates.)

Thus we see that there are substantial physical and chemical similarities among the set of resins used as embedments for thin sectioning. The differences in numbers of cross-links per unit volume, together with the differences in the physical characteristics of the different kinds of polymer chain backbones, are largely responsible for the different cutting properties of the various resins.

1. Araldite

Glauert *et al.*[53] found that Araldite, the trade name for an epoxy resin from Ciba Ltd., of Basel, Switzerland, could be used successfully as an embedding medium. Because of its solubility in ethanol, and its increased penetrability over EPO, it was adopted in many European laboratories.

Araldite was initially found to have three significant disadvantages which have been substantially overcome. The first was its high viscosity. This made prolonged infiltration (for several days) necessary. If propylene oxide is used as a transitional solvent, the infiltration interval for Araldite can be reduced to about six hours.

A second problem that arose was the difficulty of cutting thin sections from Araldite blocks. It was found that this problem could be reduced by insuring uniform penetration of the embedding medium and by correctly adjusting the hardness of the final block. Such improvements were achieved during a period when Ciba was also modifying and improving the Araldite prepared specially for electron microscopy. Araldite sections do not require special supporting films.

A third problem became apparent when examining specimens under the electron microscope. Such specimens demonstrated considerably less contrast than users had become accustomed to with methacrylate, a feature probably due to the greater stability of Araldite than methacrylate when exposed to the beam. This "difficulty" was compensated for by the development of stains (see Chapter 12B). A search for a more reliable epoxy resin continued. It led to the introduction of Epon.

2. Epon

Luft[77] was the first to use an epoxy-containing mixture known as Epon 812. Because of its low viscosity, Epon has good penetrating capability. It is also superior to Araldite in providing better contrast, even though the level of contrast still usually requires enhancement by staining. The excellent quality of epoxy resins as a group, together with the two aforementioned features, have contributed to making Epon the most widely used embedding medium at the present time.

Impregnation with Epon requires the use of propylene oxide after dehydration. Epon, like Araldite and Vestopal, sections are very stable under electron irradiation and do not require supporting films.

3. DER-334

A recently introduced Dow epoxy resin known as DER–334 has been reported [117] to produce blocks that exhibit superior cutting properties. The same advantage has been described[118] when DER–332 is used in combination with DER–732, which serves as a flexibilizer. Sections prepared from blocks of this mixture exhibit considerable contrast with unstained specimens. Moreover, it was found that different batches of these resins have a closely uniform composition and the ingredients apparently are not modified by prolonged storage at room temperature. These characteristics permit more uniform embedding in comparison with Epon 812, which is subject to change during storage and some variability in composition.[119] In contrast to Epon 812, greater granularity is noted in high magnification micrographs.

4. Maraglas

This epoxy resin was introduced by Freeman and Spurlock.[120] It is processed in the same manner as Epon. Initial difficulties with Maraglas embedding led to subsequent adjustment in the composition of its mixtures.[121]

Two advantages have been claimed for Maraglas: that it requires shorter curing time than Epon and that larger blocks of tissues can be easily sectioned. This medium has not yet been widely adopted. An improved Maraglas mixture, recently suggested,[122] may stimulate wider usage of this embedment. This mixture employs a di-epoxide flexibilizer (DER–732) rather than a mono-epoxide (Cardolite).

E. DISCUSSION

An ideal embedding medium might have the following properties:

1. It would be soluble in ethanol (or acetone) before polymerization.
2. It would not, itself, chemically modify the specimen.
3. It would not cause physical distruption or distortion of the specimen.
4. It would harden uniformly.

5. It would produce a final block hard enough but plastic enough to enable ultrathin sections to be cut.
6. It would be relatively stable under bombardment by the electron beam.
7. It would be easily soluble in an appropriate inert solvent.

None of the embedding media in current use possesses all of these properties, since some of the aforementioned criteria appear to be incompatible with others. Thus, while the newer embedding media are substantially superior to polymethacrylate, the search will continue for even better ones. A more comprehensive understanding of the physical basis of ultrathin sectioning than is presently available would possibly facilitate the development of even more suitable embedding media.[21] Of the media presently in use, the epoxy resins are generally the most satisfactory. Quantitative studies of these resins as embedments have confirmed that they set uniformly, produce very little shrinkage and appear to cause no cellular deformation.[123] The newer epoxy embedding media, however, demand skill in microtomy. They are more difficult to section than methacrylate.

Chapter 11

MICROTOMY

IMPROVEMENTS in the method of thin sectioning were needed to take full advantage of the increased resolving power of the electron microscope. The full range of application of microtomy presently extends almost over four orders of magnitude of specimen thickness (see Table VII).

TABLE VII

SECTION THICKNESS REQUIRED AT VARIOUS MICROSCOPIC LEVELS

Subject	Section Thickness Required in μ
Histology, embryology, histochemistry	10–40
Cytology, cytochemistry	1–10
Phase contrast microscopy	0.2–4
Electron microscopy	0.01–0.2

As is evident from this table, it has been necessary to extend the lower limit of section thickness suitable for light (phase) microscopy considerably in order to meet the special requirements of electron microscopy. Sections in the 0.01–0.2 μ range are thin enough for the electron beam to easily penetrate. This is also critical since, in general, resolution diminishes as specimen thickness increases. Over the years, a number of ultramicrotomes have been built which are capable of reproducible advance in the 0.05–0.5 μ range.

A. PRINCIPLES OF MICROTOMY

In order to efficiently cut successive thin sections, and to do so reproducibly, an ultramicrotome must structurally meet the following requirements:

1. All of its movements must be free of vibrations of magnitude of the order of 0.01 μ.
2. The advance mechanism should be sufficiently free of static friction so as to permit evenness and continuity of the knife's cutting movement.

3. The incremental advance of the specimen to the knife should be adjustable down to about 0.01 μ, to secure the desired thin sections.
4. The specimen should pass the knife edge only once, i.e., during the return phase of the sectioning cycle the clearance between the knife and the specimen should be such as to insure that the face of the block is not compressed by rubbing against the back of the knife on the return stroke.

Basically, an ultramicrotome consists of a horizontal bar, to the front end of which is attached the specimen holder (Fig. 81). This bar is moved forward by means of an advance mechanism. A knife mount is positioned in front of the specimen. Such an instrument cuts sections by repeatedly moving the specimen past the knife edge, with a very small advance of the specimen towards the edge made between each successive cut. The thickness of the specimen is determined by the magnitude of its forward advance.

The advance mechanism of an ultramicrotome usually moves the specimen forward at a preset increment during each cutting cycle by means of a micrometer screw and a suitable high-reduction lever [or its mechanical analogue (see details in Chapter 10 B)]. Some mechanical advance microtomes are hand-operated. They also have been adapted for motor drive operation. In thermal advance microtomes the cantilever bar is heated and consequently expands. The current to the heater is controlled so as to provide a steady linear expansion of the rod over a substantial period of time. Since the forward movement of the bar in thermal advance systems takes place at a steady rate, a regular rate of cutting is required to produce sections of uniform thickness. This is achieved with a motor drive.

The condition of the thin section that is cut depends largely on the response of the specimen block to the strains to which it is subjected while actually being cut. The ideal embedding medium is one which absorbs all extraneous strains elastically and recovers completely after the section is cut.

The position of the knife, relative to the tissue block, is critical if uniformly thin sections are to be cut. The positional relationship of the knife edge to the specimen involves several angles (Fig. 82) which can be defined as follows:

rake angle (γ) is the angle between the line perpendicular to the front face of the block and the upper facet of the knife edge.

knife angle (β) is the angle subtended by the rear and upper facets of the knife edge. It is also known as the bevel angle.

clearance angle (α) is the angle between the rear facet of the knife edge and the vertical plane of cutting.

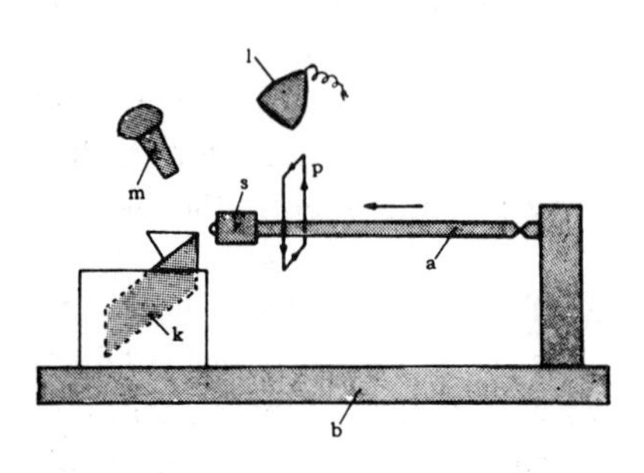

FIG. 81. The basic elements of an ultramicrotome.
(A) Cantilever arm (a) that moves in the direction indicated by the arrow; specimen holder (s); knife (k) mounted in a holder; parallelogram trajectory (p) of the cantilever arm; lamp (l); microscope (m); and microtome base (b).

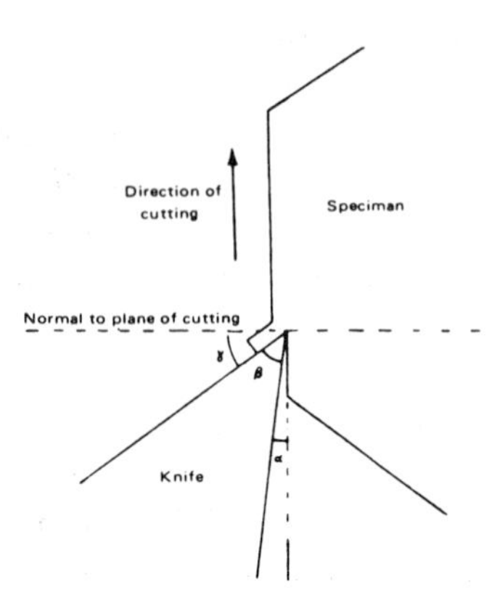

FIG. 82. A diagram of the knife edge beginning to cut through a specimen. It presents the various angles involved in sectioning: α — clearance angle; β — knife angle; γ — rake angle.

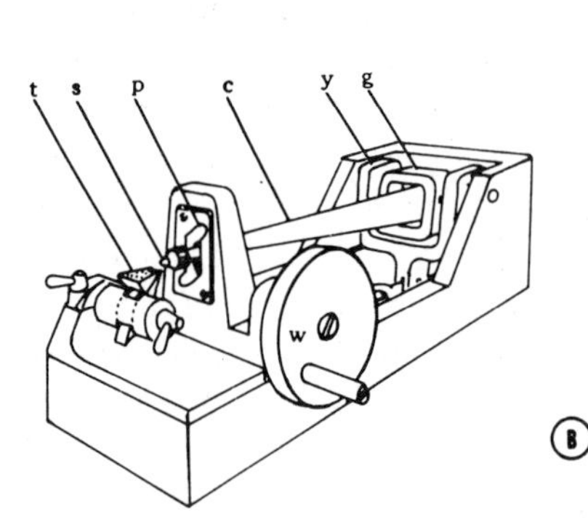

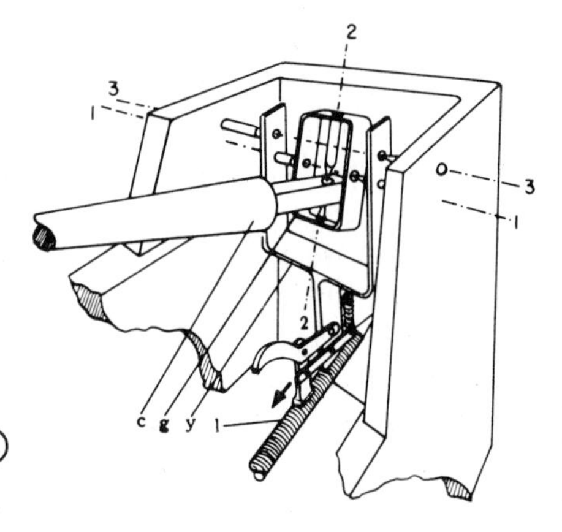

FIG. 83. Sorvall (MT–1) ultramicrotome.
(A) The microtome. The cantilever arm (c) is fixed in a gimbal (g) which is suspended in a yoke (y). The arm extends throughout the parallelogram-shaped opening (p) and a specimen is mounted at its front end. The specimen (s) will move past the knife edge which is enclosed on three sides by a metal trough (t). Movement of the arm will take place by rotation of the spindle wheel (w).
(B) The advance mechanism (enlarged). The cantilever arm (c) is suspended in the rectangular frame of the gimbal (g). The latter consists of two sets of pivots, one horizontal (1–1) and one vertical (2–2). The gimbal is suspended by a forked yoke (y) which is mounted on the support frame by a pair of horizontal pivots (3–3) that are positioned a little above the horizontal pair of the gimbal. The yoke joins the gimbal to a nut on a precision lead screw (l).

The usual knife angle is 45°–80° (see section C), and the clearance angle is commonly adjusted to 2°–5°. For successful cutting, the rake angle is made as large as possible without reducing the clearance angle below about 2°.

B. ULTRAMICROTOMES

One of the first practical approaches to thin sectioning involved modifying the conventional microtome so that the specimen block was mechanically advanced by substantially smaller increments than the usual limit of 1 micron. Subsequently, the principle of thermal advance was introduced.[36] This was followed by quick appreciation of the importance of inclusion in the microtome design of a bypass mechanism to avoid specimen-knife contact on the up or return stroke.[34] Development of satisfactory mechanical advances incorporating bypass features led to the production of commercially manufactured ultramicrotomes. Wachtel *et al.*,[21] have reviewed in detail the stages in the development of ultramicrotomes.

At present there are a considerable number of ultramicrotomes which are commercially manufactured. Of these, four have performed so satisfactorily that they have received widespread adoption.

1. Sorvall MT–1.[41] This instrument, introduced in 1953 by Porter and Blum, and then improved by them, has been the most widely used of the commercially manufactured microtomes (Fig. 83A). In the improved model, the specimen holder, which is a collet-type chuck, is mounted on one of the ends of a horizontal aluminum bar, the cantilever arm. The latter is pivoted at the opposite end in such a way that the specimen end of the arm is free to move both vertically and horizontally.

The specimen end of the cantilever arm protrudes through a parallelogram-shaped opening in a guide plate, and the edge of this opening is responsible for directing the arm through the sectioning cycle. During each cycle the arm is first drawn vertically downward past the knife edge and then is directed laterally to the left, away from the knife holder. The arm is then returned to its original position by moving upward and then sideways to the right. By inscribing such a path, possible damage to the face of the specimen block during the upstroke phase of the cutting cycle is prevented by moving the block away from the position of the knife.

The mechanical advance mechanism of this ultramicrotome (Fig. 83B) consists basically of three elements, a gimbal, a forked yoke, and a precision screw. The gimbal is made up of a rectangular frame that is held in place by a pair of horizontal pivots extending from the yoke. Thus the gimbal can rotate around a horizontal axis. The back end of the cantilever arm passes through the gimbal frame and is attached to it by two vertical pivots. Thus the arm can rotate around a vertical axis. The yoke is attached to the supporting frame by a pair of horizontal pivots, positioned slightly above the pair that attaches the gimbal to the yoke. These two pairs of horizontal pivots enable the yoke to act as a lever with large mechanical advantage. The lower leg of the yoke is coupled to a midline precision screw by an attaching nut whose grooves engage the threads of the screw.

When the operating handle is rotated during the return phase of the cutting cycle, the precision screw turns through an arc preset by the operator (see below) and the yoke is advanced. Forward movement of the yoke, in turn, advances the cantilever arm. This advance system permits the specimen block to be moved forward in increments of 0.025 μ, permitting any section thickness between 0.025 and 0.5 μ to be obtained. The section thickness is dialed in by turning a thickness control knob which sets the arc through which the precision screw turns.

This ultramicrotome has enjoyed wide popularity. Its mechanical simplicity has resulted in it being a remarkably reliable cutting instrument.

2. Sorvall MT–2. The suspension and feed mechanisms of this ultramicrotome are similar to those in the MT–1. It, however, has a motor drive that provides a rhythmic motion of variable speed. The range of section thickness extends from 0.01 μ to 4 μ. Rapid change-over from thin to thick sectioning is provided. The operational advantages incorporated into this model have contributed to improved efficiency of sectioning and the quality of the sections.

3. LKB Ultratome. Introduced by Hellstrom,[124] this electrically operated microtome makes use of a hollow specimen arm as the essentially moving element (Fig. 84). This arm, which supports the specimen block holder at its front end, is attached to the foundation of the instrument by means of flat-blade springs which limit movement of the specimen arm to a short vertical arc. The specimen arm is suspended near its point of equilibrium by a cord and pulley to the shaft of an electrically controlled moving coil, to regulate cutting speed.

During ultrathin sectioning, the specimen is moved forward by a thermal advance mechanism which involves the placement of heating coils around the back end of the specimen arm. By coupling this efficient linear thermal feed system with an extended range of cutting speeds, this ultramicrotome is capable of automatically yielding advances of from 0.005–0.13 μ in reliable sequence. During the return stroke of each sectioning cycle, the knife holder assembly is pulled back about 25 μ by an electro-magnet which enables the specimen to bypass the knife without rubbing against it.

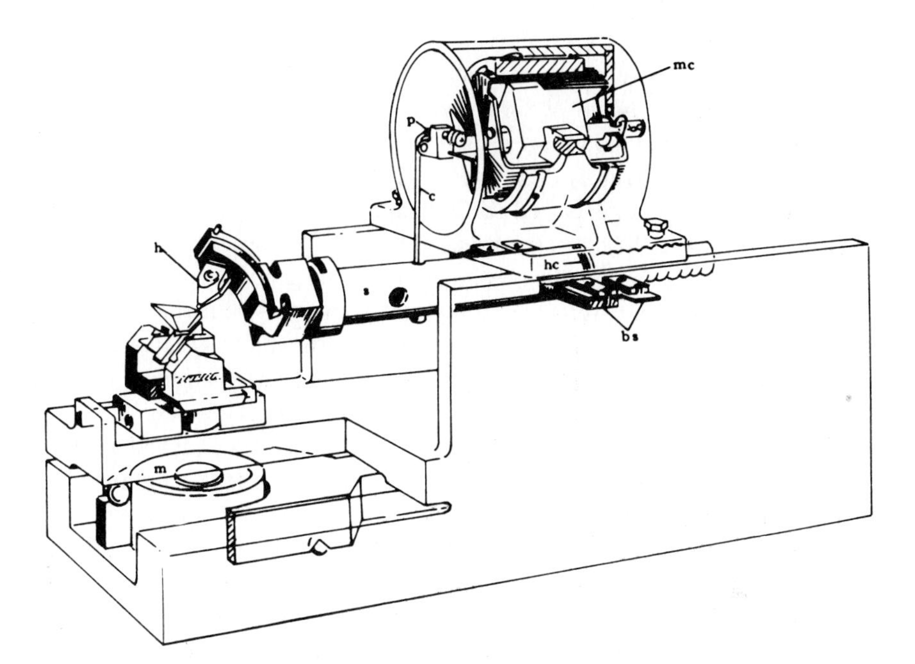

FIG. 84. A diagrammatic presentation of the LKB ultratome microtome. Specimen arm (*s*); specimen block holder (*h*); bladesprings (*bs*); cord (*c*); pulley (*p*); moving coil (*mc*); heating coils (*hc*); retraction magnet (*m*).

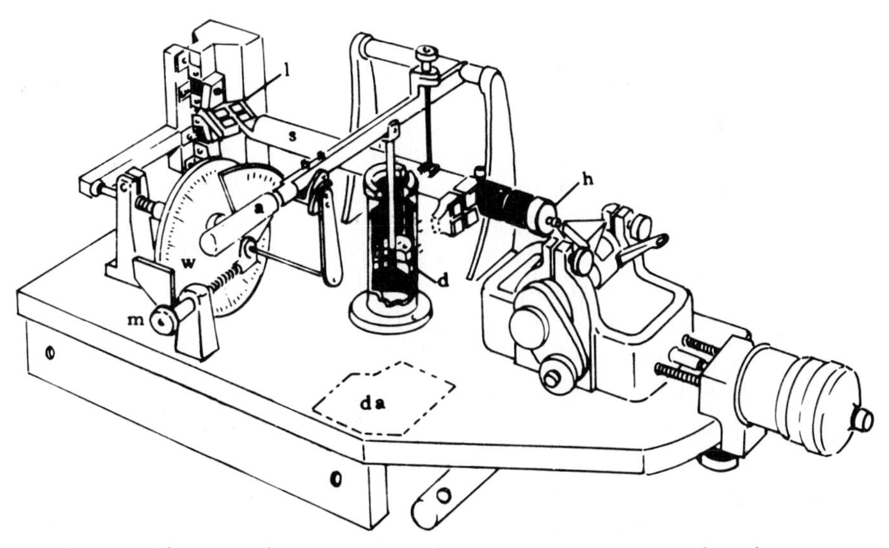

FIG. 85. Diagrammatic presentation of the Cambridge-Huxley ultramicrotome. Specimen arm (*s*); operating arm (*a*); reducing lever system (*l*); micrometer screw (*m*); calibrated wheel (*w*); oil-filled cylinder (dashpot) (*d*); site of displacement and advance mechanism (*da*).

The overall design of the microtome provides operational convenience and avoids fluctuations in section thickness. This microtome is especially popular in European laboratories.

4. Cambridge-Huxley Microtome. This ultramicrotome, introduced by Huxley[125] and manufactured in England, also has as its essential moving element a cantilever-like specimen arm (Fig. 85). The tissue block holder is attached to its front end. The specimen arm is suspended from the end of an operating arm which is located transversely above it. At its back end, the specimen arm is hinged to a reducing lever system which is activated by a micrometer screw. The thickness of the sections is determined by presetting a calibrated wheel which is coupled to the micrometer screw.

During the sectioning cycle, the operating arm is picked up and allowed to fall by gravity. Cutting of the specimen takes place as the arm falls. The rate of fall can be adjusted by the valve of a piston, the latter being suspended from the operating arm in an oil-filled cylinder. This insures a constant and vibrationless downstroke of the specimen arm, even at very slow cutting speeds. During the return stroke, a second system of levers insures the necessary lateral displacement of the specimen arm from the plane of the knife. The special features of this ultramicrotome provide for excellent precision cutting.

C. KNIVES

The ideal knife for use in ultrathin sectioning might have the following characteristics:

1. The edge would have a radius of curvature considerably smaller than the thickness of the thinnest section required.
2. It would be resistant to chemical decomposition.
3. It would have a degree of hardness and toughness that makes it impervious to cleavage or chipping on impact, even with hard blocks.
4. It would be made of a homogeneous material with the edge of the same quality everywhere along its length.*
5. It would be physically stable so as not to be subjected to net molecular migration near room temperature, due to Brownian motion.

---

* Uniformity of sharpness of the edge can be determined by examining it by the light microscope at about 400 X (N.A. $\sim$ 0.85), with the light beam directed horizontally at the knife edge. The amount of light which is scattered from the edge is noted, the edge being sharpest where the scattering is at minimum. Breaks or nicks along the knife edge are detected as either bright scattering points or interruptions of an otherwise evenly scattered edge.

Single-crystal (diamond) knives approach this ideal, but obtaining diamond knives of good quality has proven, in practice, to be a substantial problem. The first knives used for ultramicrotomy were either specially sharpened razor blades or those made from selected hard steel that were lapped and polished. These knives, in practice, had a relatively short life and their edges were subjected to corrosion by the trough liquid and room atmosphere. Thus, especially in view of the considerable effort expended to prepare them, steel knives proved to be of limited usefulness for thin sectioning.

1. Glass Knives

The glass knife was introduced by Latta and Hartmann[38] and has proven to be the most popular of the cutting edges. The reason for this is that the glass knife is relatively easy to make, is inexpensive and is convenient to use. A mechanical glass knife maker, based on a prototype described by Fahrenbach,[126] is now commercially available.

The procedure used in preparing glass knives is dependent upon the shape of the knife required which, in turn, is determined by the microtome's knife holder. Thus, where a simple triangularly-shaped knife is needed, a large piece of glass may be cut up into uniform squares of glass which are then diagonally bisected.[119] The knife edge to be used represents the intersection of a diagonal break with one of the original fracture faces of the glass square, very near a corner. Knives are also frequently broken from strips of glass.

The cutting edge should be straight and even and the front knife surface (i.e., that which will face towards the specimen block) very flat and smooth. An arc, or *fracture ridge*, formed by stress generated during knife making, is present beneath the edge on the back surface of the glass knife (Fig. 86). It provides a guide to the best segment of the cutting edge. This is usually located near the top of the ridge. The knife angle is greatest at the top of the ridge (near 90°) and, therefore, the best portion of the edge is nearer to the center. The end of the edge opposite the ridge frequently presents a spike of varying height. The edge may also exhibit saw-tooth irregularities.

For a given clearance angle, the stress in the sections during cutting decreases as the knife angle decreases, but at the same time the knife edge becomes more fragile. A compromise between these conflicting requirements is usually obtained with a knife angle of about 45°. The exact angle at the knife edge, however, is difficult to determine. Generally, it will be found that the angle at the edge is greater than the original score mark as the

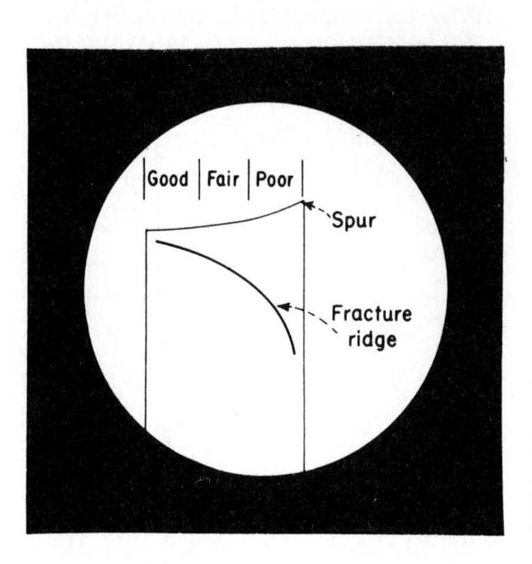

FIG. 86. A diagrammatic presentation of the edge of the average glass knife. The location and cutting properties along its length with respect to the fracture ridge are indicated.

fracture tends to break away from a straight path. At the top of the fracture ridge, the angle can approach 90°. It is usually smallest at the opposite end of the ridge. It appears that,in practice,[119] the most acute knife that one can hope for has an angle of approximately 52°. More acute angles are rarely produced by trying to bisect angles of much less than 90°.

The resistance to impact of the glass edge varies considerably with the type of glass used. For some grades, the edge becomes nicked after just a very few cuts while for others, the edge remains sharp for many sections. The rate of deterioration of the knife edge will also depend on the hardness of the block, the size of the block face and section thickness. To prolong the "life" of a good knife, the middle third of the knife should be used for trimming, after which the outer third nearer the top of the stress ridge can then be used for thin sectioning. Knives usually are broken a short time before sectioning and are kept in a dust-free container.

2. Diamond Knives

The relatively short useful life of glass knives prompted a search for more long lasting and less fragile materials. An obvious choice would be single-crystal knives. An unsuccessful attempt was made to produce a synthetic sapphire knife as early as 1950. Fernàndez-Moràn[42] was the first to produce sharpened diamond knives applicable to thin sectioning.

Diamond knives are currently commercially available. Their durability and sharpness have been variable and there has been no adequate assurance that a really good knife can routinely be secured. A satisfactory diamond knife, however, provides a means of sectioning material of considerable hardness and can be used repeatedly before resharpening is required. (Some knives have been used steadily for more than two years.) The cost of diamond knives is high and varies according to the length of the cutting edge.

D. THE TROUGH

Thin sections, as might be expected, have little physical strength. If thin sections are cut on a dry knife, the static friction between the facet and the cut surface almost always causes the section to crumple up on itself as it is cut. In the rare cases when an uncrumpled section might be obtained,

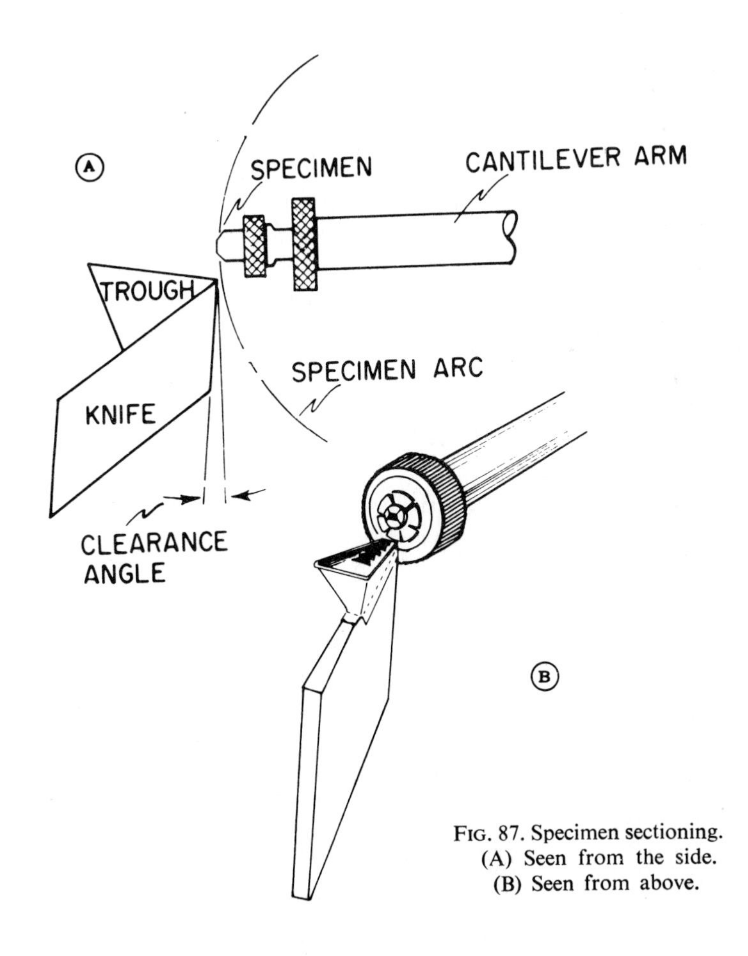

FIG. 87. Specimen sectioning.
(A) Seen from the side.
(B) Seen from above.

the chances of picking it up undamaged and undistorted are very slim. For this reason, little useful thin sectioning was reported until Claude[34] and again, independently, Gettner and Hillier,[37] introduced the use of a trough or "boat" of liquid at the cutting edge to provide a low friction surface (the liquid surface) for support of the freshly cut section and to coincidentally simplify picking up the section on specimen support grids (Fig. 87). The addition of acetone or alcohol (typically about 25%) to the water in the trough can produce slight softening of (some) plastics used in embedding and thus release some compression strains and distortions introduced during sectioning. The surface tension force acting on the cut section is then able to stretch the section out to the original dimension of the block face. The latter processes can also be aided by exposing the freshly cut section, while floating on the surface of the trough fluid, to the heavy vapors of a solvent (e.g., applied by holding a xylol wet camel's hair brush about a millimeter above the surface of the section[127]).

The simplest way to prepare a trough is with the use of some waterproof adhesive or masking tape.[21,119] It is positioned so that it does not overlap the cutting edge or front face of the knife. The base of the trough, especially at its "heel", is often sealed with melted wax to insure against leakage. Troughs made of aluminum or sheet copper or malleable rubber are preferred by some workers. Prepared metal troughs made of bronze or stainless steel are also available. Holders for diamond knives are made so that the trough is an integrally cast or machined part of the knife vise.

E. SECTION THICKNESS

A judgment as to section thickness needed can be made by taking into account the desired resolution. With the light microscope, using aberration-free objectives of suitable numerical aperture, maximal resolution can be obtained when the specimen thickness is many times thicker than the depth of field of the lens, if the object detail of interest displays sufficient contrast relative to its surroundings.

With the electron microscope the situation is more complex since electron lenses are uncorrected monochromatic lenses and, in addition, contrast is dependent on density differences rather than differences in the absorption of light by an object and its surroundings. Wachtel *et al.*,[21] noted that contrast is dependent on the "effective density":

$$c \approx \frac{2d\Delta p}{V\alpha_0{}^{1/2}}$$

where:

c = contrast.
d = object thickness (in Å).
Δp = effective density (i.e., difference in density between the object and its surroundings).
V = accelerating voltage.
α_0 = objective aperture angle.

This equation suggests that a decrease in voltage and objective aperture angle and an increase in object thickness, which all favor increased contrast, are to be desired. However, as voltage decreases and object thickness increases, chromatic aberration rather than contrast sets the limit of resolution. Therefore, depending on the particular properties of both the specimen and the microscope (i.e., specimen thickness, objective aperture angle, objective focal length and accelerating voltage*), different specific compromises may be required for maximal resolution.

For example, the aforementioned authors have calculated the maximal specimen thickness for 20 Å resolution for a variety of microscope lenses. They have demonstrated that even with a 0.3 μ thick specimen, resolution better than 30 Å can be attained photographically with a microscope equipped with a 1.7 mm focal length lens operating at 100 KV.

Even if very thick sections would not degrade resolution (e.g., very much higher voltages and shorter focal length lenses were routinely available), the low numerical aperture produces a great depth of field in electron microscopes (see p. 61) and all detail at all depths in the specimen would be in focus at the same time (in contrast to the typical case with a high aperture light microscope). Therefore, unless stereoscopic views (see p. 104) were recorded, interpretation of such very thick sections would usually be quite difficult.

In practice, section thickness is usually judged by the interference color exhibited by reflecting light from the surface of a cut ribbon that is floating on the liquid in the trough (see Table VIII).

The quality of the section cut is, in practice, primarily dependent on the quality of the knife edge and the dimensions and shape of the specimen block. A dull knife, inadequate clearance angle and too large a block face may result in faulty sectioning. With a good knife edge that is properly positioned and with a well-trimmed block, the quality of the section then

* Since as noted earlier, (see p. 55) the limit of resolution due to chromatic aberration is $d_c = k_c \cdot f \cdot \alpha_0 \cdot \Delta V/V$, therefore, the shorter the focal length, f, and the higher the accelerating voltage, V, the larger can be the section thickness for a given chromatic limit of resolution, d_c (since ΔV increases with thickness[21]).

becomes dependent on the physical properties of the embedded block which affect its behavior when it is under strain during the actual cutting process. The ultramicrotome usually is the least likely source of sectioning problems.

TABLE VIII

INTERFERENCE COLORS OF ULTRATHIN SECTIONS

Section Thickness in μ	Interference Color
0.010–0.060	Gray
0.060–0.090	Silver
0.090–0.150	Gold
0.150–0.190	Purple
0.190–0.240	Blue
0.240–0.280	Green
0.280–0.320	Yellow

(For a discussion of methods of measurement of section thickness see Wachtel *et al.*[21])

F. DISCUSSION

Ideally, reproducibility of performance by the ultramicrotome should be assured at any setting of the specimen advance. In practice, however, reproducibility as in any other instrument is affected by the total random fluctuations which occur during the time interval needed for the incremental advance of the specimen arm. A number of possible variables that may affect the orderly sequential sectioning pattern, several of which can be minimized, will be briefly discussed.

1. External temperature fluctuation. One source of error for all microtomes is thermal contraction and expansion of the instrument's structural components. This may be the result of currents of air varying from the average ambient temperature passing both through and around the instrument. To minimize this factor, the components of ultramicrotomes, where possible, should be short and massive in design. The entire instrument should, as far as is convenient, be thermally insulated from room temperature fluctuations.

2. Forced motion stress. Another variable may be introduced by bending and changes in length of the structural components of the microtome as a result of stresses of forced motion. This source of error in an ultra-

microtome can be corrected partially by reducing the number of its moving parts and contiguous surfaces, and by massive construction.

3. External stress. External forces (e.g., the hand motion applied to drive the instrument), operating during the sectioning procedure, may introduce stress to all parts of the microtome. These influences are distribued differently depending on the structure and functional contacts of the parts of the microtome. They may result in distortions in the sections which can be minimized by smooth sectioning movements.

4. Static friction. Static friction is another source of error. An advancing force greater than that of the static friction inherent in the system must be used. This may produce an erratic movement cycle of the advance mechanism because sliding friction is always less than static friction. Erratic movements may result in uneven specimen advance, thus producing skipping of some sections and cutting of other sections thicker than intended, due to fluctuations in forced motion stress (variable 2 above).

5. Sliding friction. In all microtomes where one part may move against another part, sliding friction will vary both with the speed of motion and from point to point. This variation is a potential source of error and can be reduced by the use of oil films. The most efficacious use of oil films, however, requires that all lubricated moving parts operate at constant speed. Since this is not possible in hand operated instruments, oil films are avoided in those parts of the advance mechanism where the incremental movements are of the same order of magnitude as the specimen advance. When, however, a constant speed motor is used to drive the ultramicrotome, many critical oil films can be kept at more nearly constant equilibrium thickness and thus may be profitably used.

6. Motor vibration. Where motors are used their vibrations may introduce large discontinuous stresses on the sensitive structural components of the microtomes. Special precautions can usually be taken to insulate the microtome against such vibrations.

In the light of such possible sources of error, it is rather surprising that satisfactory sections can be cut with regularity. Reproducibility of section thickness will be dependent upon the set increment of advance and on the root of the sum of the squares of all random changes that might occur when an average section is cut during an average sectioning interval. On some occasions these random changes may cancel each other, while at another time they might complement each other. The fact that thin sectioning with current ultramicrotomes is a relatively routine procedure testifies to the fact that the sources of error introduced into the system are relatively small down to thicknesses of 500 Å.

Aside from using a suitable ultramicrotome (p. 154) to attain satisfactory ultrathin sections, the following summary of additional conditions needs to be met.

1. The specimen must be embedded in a suitable medium which is polymerized so that a tissue block of high degree of homogeneity is attained (p. 152).
2. The specimen block should, when possible, have the form of a truncated pyramid with a block face that is very small (0.1 mm or less).
3. The cutting edge should be free of nicks.
4. The rake angle should be about 30°, the clearance angle about 3° and knife angle less than 57°. (For hard blocks, both rake and knife angles should be decreased.)
5. The sections should be collected on a liquid surface.

Chapter 12

STAINING

A. PRINCIPLES OF STAINING

With the light microscope, the observed differences in the various parts of a specimen that are accentuated by staining are due to differential absorption of visible light. Such differences in light absorption can make the components of a specimen readily resolvable and visible in an optical image. In electron microscopy, the portion of the electron beam absorbed is extremely small. To appreciate the effect of staining on ultrathin specimens, which also increases resolution, it is desirable to clarify the mode by which images are formed with the electron microscope.

As described in the first part of this book (p. 59), image formation is the result of the "subtractive" action of the objective lens. Increase in image contrast can be attained by the use of small apertures which enhance the subtractive action of the objective. To supplement this mechanism, use is made of electron stains. Staining increases the mass density differences at specimen points, and thus increases scattering outside of the objective aperture angle.

The relationship between image contrast and specimen properties has been quantitated.[128] It has been established that with a typical electron microscope containing a standard objective aperture, an object, irrespective of its chemical composition, will be reasonably well-defined *in the imaging screen* if the product of the thickness of the object (in Å) and its weight density (in $g/cm.^3$) is more than about 400 above its surroundings. For the object to be visible at all, the value of this difference must be more than about 100. Biological specimens have a density of about 1, which means that in unstained specimens the minimum thickness for visibility is about 100 Å for such an unsupported specimen element. Positive staining involves treatment with a chemical that will increase weight density.

Substantial improvement in tissue contrast (and thus in potential resolution) can sometimes be attained by the staining of thin sections with solutions of heavy metal salts. The metal ions of the staining solutions form complexes with certain components of cells, thus increasing their density.

In many instances, such staining has little chemical specificity, but the contrast of such components as membranes, ribosomes and glycogen is increased relative to their surroundings. The maximum increase in specimen density that can take place, without creating additional problems of interpretation due to the resolution of the stain deposit itself, is usually a few times the original density. For example, with aberration-free lenses, resolution of structures on the viewing screen* at about the 100 Å level would be possible within a 100 Å thick section of an object of density 2 gm/cm^3 in an embedding medium of 1 gm/cm^3 and of 25 Å in a 400 Å thick section where a stained object (e.g., a membrane) of density 2 gm/cm^3 extends through the thickness of a section of density 1 gm/cm^3.

Increased resolution down to the 10 Å level has been attained with cell fragments and viruses with microscopes with low chromatic and spherical aberration by the use of negative staining (see Chapter 7 C). This involves surrounding the particle with a structureless layer of material having a high density. When an object is embedded in a substance, the difference in density between the biological material (1) and its surrounding stain can be very great. For example, in the case of phosphotungstate, the difference in density will be about 3 gm/cm^3. Thus particles about 30 Å thick are visually resolvable on the screen. Staining has, therefore, served as a very practical means of exploiting the resolution potential of the electron microscope.

B. POSITIVE STAINING

A variety of stains are available at the present time. The most useful employ lead or uranium ions. These have made epoxy and polyester resins especially effective embedding media by compensating for the intrinsic low difference in contrast between the plastic and the contained biological material.

In preparation for staining, thin sections are placed on grids in the usual manner and permitted to dry. Each grid is then floated, with the thin sections downward, on the surface of the staining solution. After staining for the desired interval, the sections are washed with distilled water by floating the grid on the surface of distilled water or by gentle agitation in distilled water. The grid is then allowed to dry, with the sections upward, on filter paper.

Watson[129] suggested various staining media such as saturated uranyl acetate, phosphotungstic acid, phosphomolybdic acid, ammonium molybdate and saturated sodium uranate. A short time later he also suggested[130] the

* Higher resolutions are attainable photographically because of photographic contrast enhancement.

use of lead and barium hydroxides. The lead stain was widely used at first but the original procedure had the drawback of bringing about contamination of the specimen with precipitates of lead carbonate particles due to the absorption of atmospheric carbon dioxide by the staining solution. Several procedures have been devised to reduce contamination.[119]

Currently, the most popular stains are uranyl acetate and various lead salts, namely, lead tartrate,[131] lead cacodylate,[132] lead citrate,[133] and lead ammonium acetate.[134] Of these, lead citrate has the advantage of greater stability, while the others lose their staining properties after a few days or weeks.

Studies on improving uranyl acetate staining[135] resulted in the recommendation that the stain be used in a methanol solvent rather than being dissolved in a water-ethanol solution. As a result of this modification, it was found that the staining time was reduced, image contrast was increased and less contamination from the stain occurred.

Recently, it has been suggested[136] that double staining of OsO_4 fixed and Epon embedded tissues with uranyl acetate and lead hydroxide provides greater contrast than either stain alone. It has also been recommended[136] that uranyl magnesium acetate be used as a means of decreasing contamination. For combination with lead staining, the tartrate salt is most frequently used. A number of other stains have been suggested for use but these, however, have not received widespread popularity (see p. 353).

C. NEGATIVE STAINING

As was noted previously, the contrast of the image is one of the decisive factors in determining the resolution level that will be attained. Although the study of fragments or mechanically disrupted cells dried onto a supporting film is a technique that was widely used prior to the development of thin sectioning methods, the availability of the thin sectioning methods resulted in the fragmentation technique being largely neglected until its recent exploitation with negative stains.

The theoretical basis for negative staining has been discussed previously (Chapter 12 A). One of the techniques of negative staining involves suspending the particles or fragments in an approximately one percent solution of the negative stain. Then, fine droplets of this suspension are sprayed onto a carbon-coated grid and allowed to dry (in air or in a vacuum). An alternate procedure is to place a single large drop of the suspension on the grid and "blot" its edges so that only a thin film remains. The stain penetrates into the interstices of the particle to bring out its details. In both cases, the

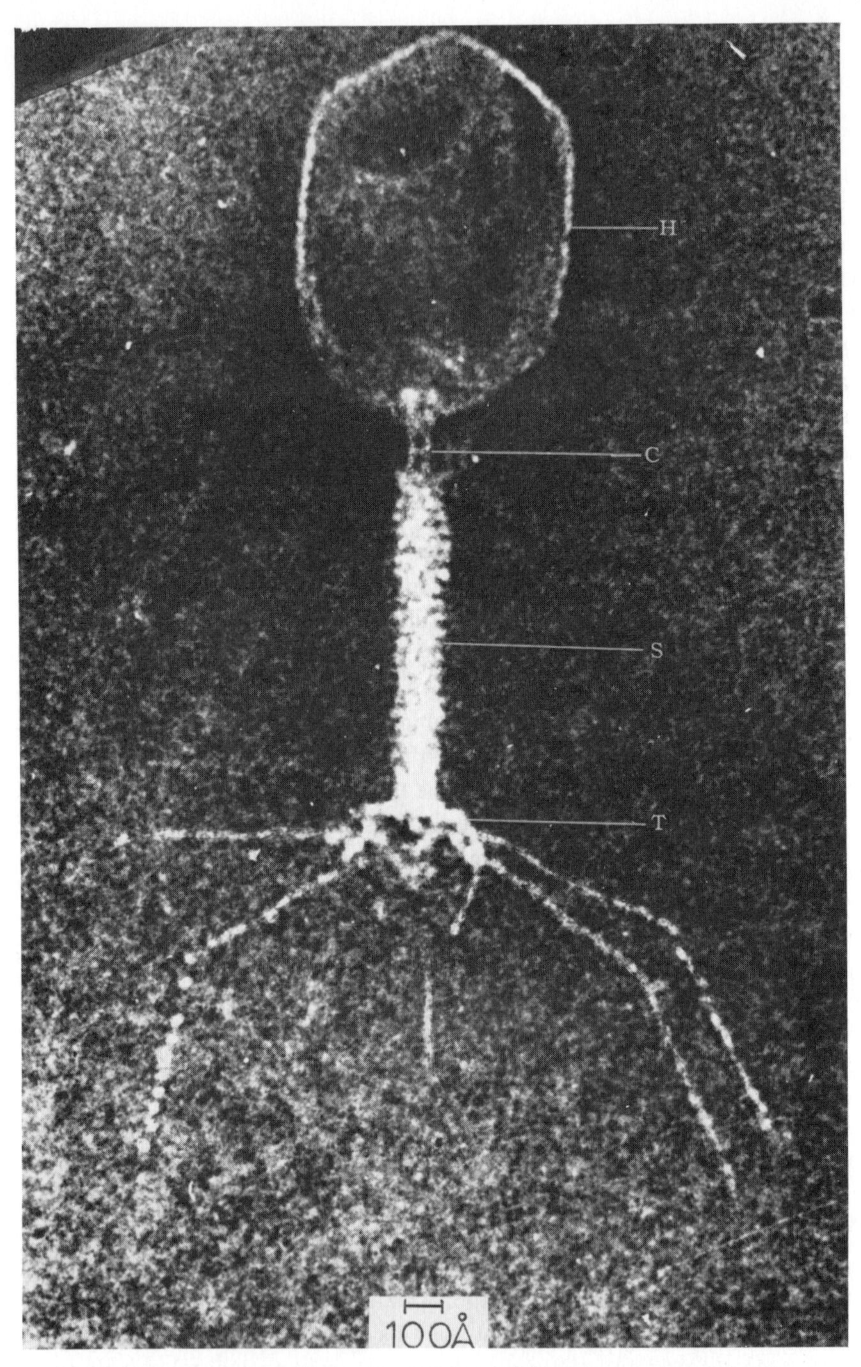

FIG. 88. Demonstration of negative staining. High resolution electron micrograph of T2 bacteriophage showing complex organization of head (*H*); sheath (*S*); and tail (*T*) components. Improved preparation procedures and microbeam illumination with a single-crystal pointed filament electron source reveal new details of the periodic sheath striations (35–40 Å spacing), collar type formation (*C*) at the head-sheath-core junction, and of the tail-plate with attached kinked fibers. Modified PTA negative staining. 400,000X. (Compare with Figure 98.) (Courtesy of Dr. H. Fernández-Morán.)

preparation dries rapidly and the dissolved substance precipitates out of solution in an amorphous condition at the 10 Å level and is deposited over the supporting film and exposed surfaces of the specimen. The method has provided new information relative to the surface configuration and fine structure of various biological entities (Fig. 88).

Valentine and Horne[137] evaluated various substances as potential negative stains. The theoretical requirements of a good negative stain are that (1) it be a substance of high density to provide high contrast, (2) it have a high solubility (at least 80 g/100 ml) so that it does not come out of solution prematurely but only in the final stages of drying, (3) that it have a high melting and boiling point so that the material does not evaporate at temperatures induced by the electron beam, and (4) that the precipitate be essentially amorphous down to the limit of resolution.

Substances having all of these properties are few and only one, cadmium iodide, was found to have potentially some slight advantage over phosphotungstate or sodium tungstate, the media that are most often used. Thus a substantial increase in resolving capacity beyond that attainable with a tungstate medium cannot be anticipated by altering the negative stain that is used.

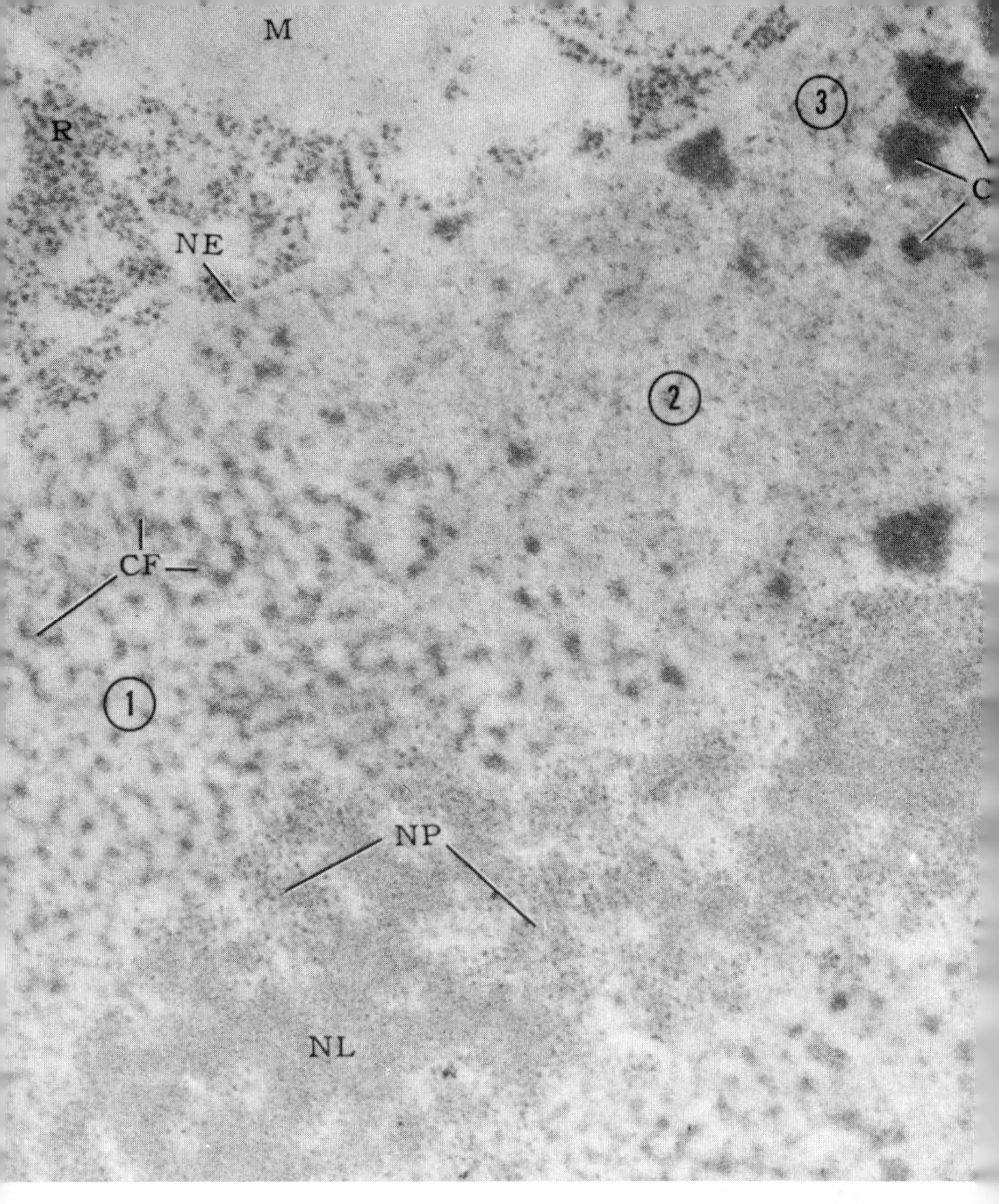

FIG. 89. Demonstration of nucleoprotein staining with indium trichloride. A micrograph of a portion of the replication band in the macro-nucleus of *Euplotes* fixed with glutaraldehyde. Three regions of the band are apparent: from left to right, forward zone (1) of 300 Å chromatin fibrils (*CF*), "homogeneous" trailing zone (2) in which DNA replication presumably occurs, and post-replication area (3) in which chromatin masses (*CH*) are being reconstituted. Note nucleolus (*NL*) with 100 Å nucleic acid containing particles (*NP*), and cytoplasmic ribosomes (*R*). Note non-stained nuclear envelope (*NE*) and mitochondria (*M*), the latter containing small indium positive granules. Magnification 73,000X. (Courtesy of Drs. M. J. Moses and W. Phegan.)

Chapter 13

SPECIALIZED ELECTRON MICROSCOPE TECHNIQUES

IN RECENT years, a number of techniques have been developed which significantly extend the potential of electron microscopy. These approaches have been aimed at attaining a better understanding of the functional significance of the morphological findings. These techniques are being gradually refined.

A. ULTRASTRUCTURAL NUCLEOPROTEIN LOCALIZATION

Progress in morphological and chemical studies of the nucleus on an ultrastructural level has been slower than comparable investigations of the cytoplasm. This probably results from the fact that localization of (nuclear) masses that neither contain lipids nor are enclosed by a membrane presented difficult problems of definition of boundaries.

Three basic approaches have been used in an attempt to localize nucleoproteins. The first involved the use of heavy metals. Initially, iron was suggested[138] as a possible stain for nucleic acids. Later, uranyl salts were tried.[139,140] More recently, bismuth has been recommended.[141] While these heavy metals certainly act as general contrast-enhancing stains, and to an extent appear to have some degree of specificity, it is difficult to evaluate nucleoprotein localization adequately with osmium fixed material.

A second and more successful approach[142] begins with fixation with acrolein rather than OsO_4. Acetylation of the protein amino groups is then carried out to free all potentially reactive phosphate groups of the nucleic acids (Fig. 89). These are then identified by staining the tissue blocks with the highly specific trivalent indium ion, utilizing indium trichloride in acetone. Nucleic acid-containing structures, as seen in thin sections, are found to be quite dense while other areas are unstained. This method is particularly successful[143] because the total mass of indium bound per mass of nucleic acid is quite high (i.e., up to 114 gms. of indium are bound per approximately 340 gms. of nucleic acid).

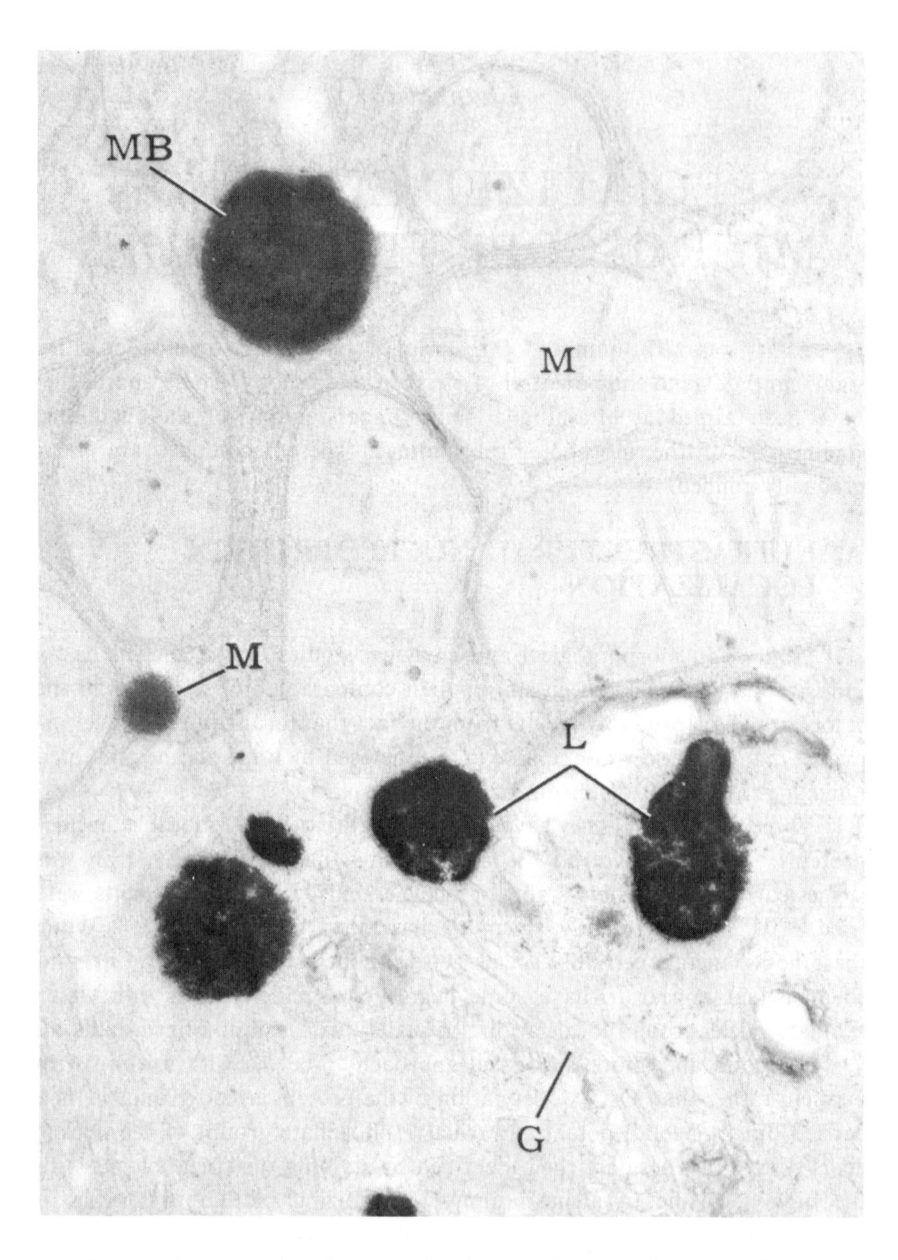

Fig. 90. Demonstration of enzyme chemistry for electron microscopy. Granular reaction product from acid phosphatase activity in lysosomes (L) and homogeneous reaction product from diaminobenzidine oxidation in microbodies (MB) of rat liver. Also labelled are: mitochondria (M) and Golgi complex (G). 41,000X. (Courtesy of Dr. A. Novikoff.)

The third approach, introduced by Swift,[144] involves the use of specific digestive enzymes or nucleic acid solvents in an attempt to remove nucleic acids. Then the preparations are stained with heavy metals. Nucleic acid sites are thereby identified by comparison with undigested stained controls. Small blocks of tissues are fixed in cold buffered formalin or acrolein and their nucleic acids extracted with nucleases, or perchloric acid. (Alternatively, thin sections are extracted.) The tissues are then embedded in Epon and stained with heavy metals. The results from this method, although promising, are variable in nature, especially where high resolution studies are concerned. Better results are claimed when this method is used with water-soluble embedding media.[76] This permits more reproducible enzymatic digestion to be carried out on thin sections and eliminates the variable introduced by uneven enzyme penetration.

B. ULTRASTRUCTURAL ENZYME CYTOCHEMISTRY

One of the approaches which may make it possible to determine the functional significance of cellular ultrastructure is the application of cytochemical staining methods to localize enzymes at the subcellular level. In such studies the technical problems are more complex than in the application of similar methods to light microscopy. This is due to the fact that not only must enzymatic activity be preserved but the cellular structure has to be maintained at a finer level during the first stages of preparation, fixation and incubation of specimens. In addition, the stain must also be able to resist dissolution or displacement by the dehydrating and embedding media, must be relatively stable in a vacuum and under electron bombardment, and must exhibit sufficient contrast in the electron microscope. The initial success in meeting these demanding conditions has been somewhat limited by the fact that compromises have often had to be made between faithful preservation of ultrastructure and meaningful localization of functional activity.

A general class of approaches has been used in electron cytochemical investigations. The tissue is fixed in an aldehyde medium, after which small blocks or frozen sections are cut and then incubated with the cytochemical medium. This is followed by postfixation in osmium tetroxide. This method is the outgrowth of the work of Wachtel *et al.*;[68] and Essner *et al.*[45] Gluteraldehyde, introduced by Sabatini *et al.*,[83] now often replaces formaldehyde, and the incubation of small tissue blocks rather than of relatively thick frozen sections has been reintroduced following the recommendation of Goldfischer *et al.*[145] These two changes in procedure help to provide good

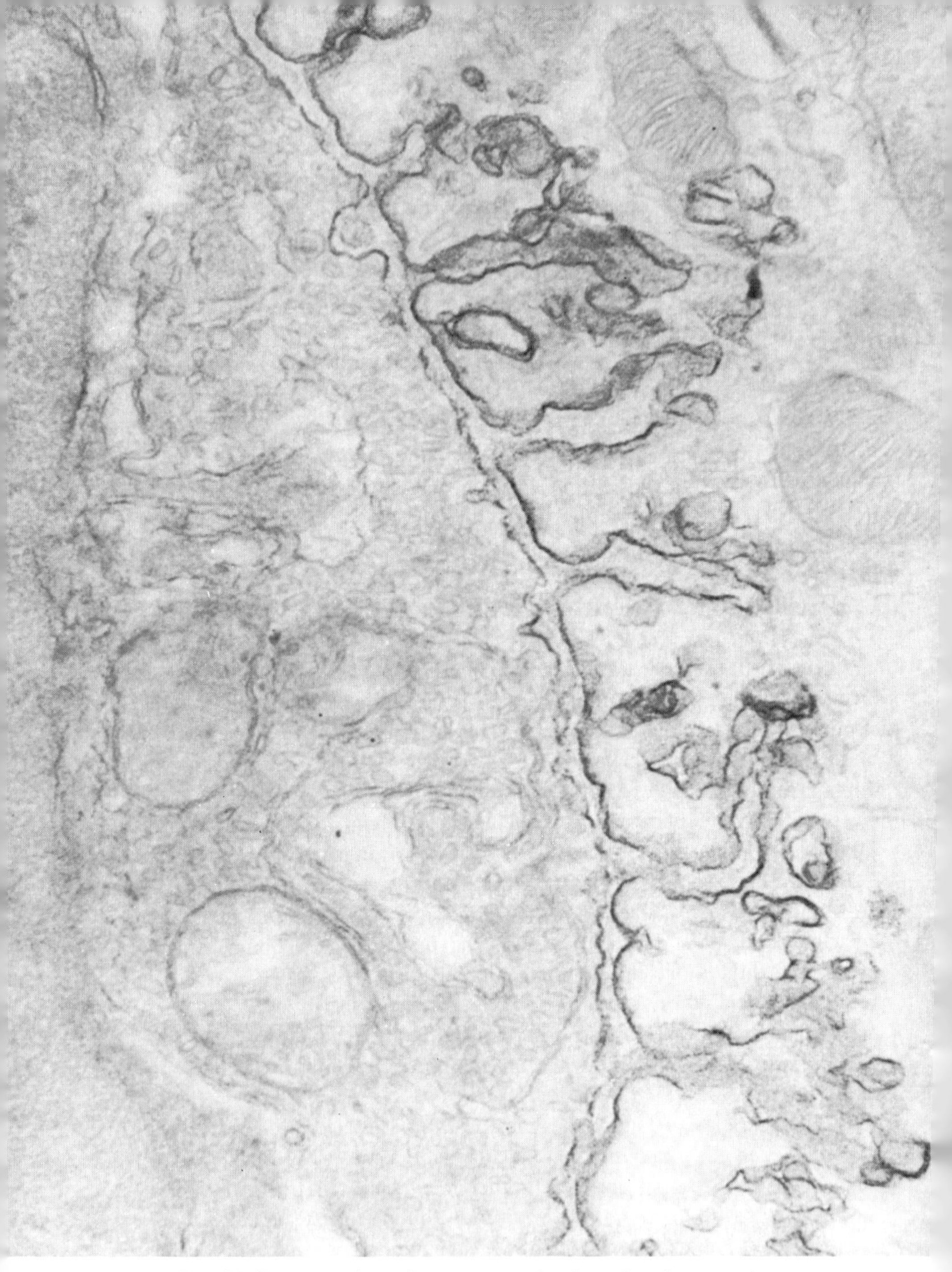

FIG. 91. Demonstration of enzyme cytochemistry for electron microscopy. The motor endplate of mouse intercostal muscle stained for acetyl-cholinesterase by the gold-acetyl disulfide method. Acetylcholinesterase activity is represented by lines of fine precipitate of Au_2S at pre- and postjunctional membranes. 50,000X. (Courtesy of Drs. G. B. Koelle, R. Davis and M. Devlin.)

preservation of cellular morphology and, at the same time, often maintain a high level of enzyme activity. A recent modification[146] suggested the use of gluteraldehyde fixed, non-frozen 10-50 μ thick sections cut with a special tissue "chopper" (Sorvall Model TC-2 tissue sectioner). In tissue sections resulting from all of these methods of treatment, the ultrastructure is often well-preserved and sites of enzyme activity are clearly identifiable.

Enzymes have high turnover numbers, i.e., they can process large numbers of substrate molecules in a given amount of time. If a product of the enzymatic reaction is or can be made insoluble in the incubation media and all succeeding media through which the tissue is processed and if it has a density which exceeds that of the embedding medium, the deposition of the products of activity of relatively few enzyme molecules should be evident in an appropriate electron micrograph. A number of kinds of histochemical reactions for a variety of enzymes have been developed in an attempt to satisfy these conditions. These can be grouped into three types.

1. Insoluble metal salts of enzymatic reaction products (Figs. 90 and 91). Select metallic salts (e.g., of lead) of reaction products which, though almost insoluble and of high contrast, are often semicrystalline and have slight residual solubility. Therefore, the interpretation of fine details of distribution demonstrated with such techniques is sometimes open to question.
2. Non-metallic organic compounds of enzymatic products. The most successful of these involve the use of modified azo dye methods which yield amorphous products. The density of these, however, is not as great as with metallic salts.
3. Organic compounds which can bind substantial concentrations of metals. Introduced recently by Hanker *et al.*,[147] this approach utilizes enzyme substrates which contain a group capable of reacting relatively selectively with OsO_4. This results, after post-treatment with OsO_4, in deposition of "osmium black" at the enzyme reaction sites. The sites of deposition of the insoluble reaction product are not altered by embedding and this product has a fine amorphous electron opaque character.

Another method in this category was introduced by Karnovsky.[85] Using formaldehyde-gluteraldehyde fixation and frozen sections, peroxidase (injected as an ultramicroscopic tracer) was localized by using 3, 3′-diaminobenzidine and H_2O_2 in the substrate medium. The reaction product is non-crystalline, insoluble and is also extremely electron opaque after post-treatment with osmium tetroxide (see Fig. 93).

As has been emphasized by investigators in this field, the localization of enzymatic activity on a subcellular level may be influenced by the fixative, the method of tissue preparation, possible redistribution of very slightly soluble reaction products and of the amount of reaction product deposited. Thus, while the potential relative to functional interpretations of subcellular activity that may be derived from this research area is great, considerable room remains for decreasing errors of interpretation and for developing a more reliable methodology.

C. TRACERS IN ELECTRON MICROSCOPY

Wissig[66] was the first to introduce the use of tracers in electron microscopy and suggested gold, thorotrast and ferritin (Fig. 92). Of these, the latter has also become useful in immuno-electron microscopy, as will be described in the succeeding section (p. 181).

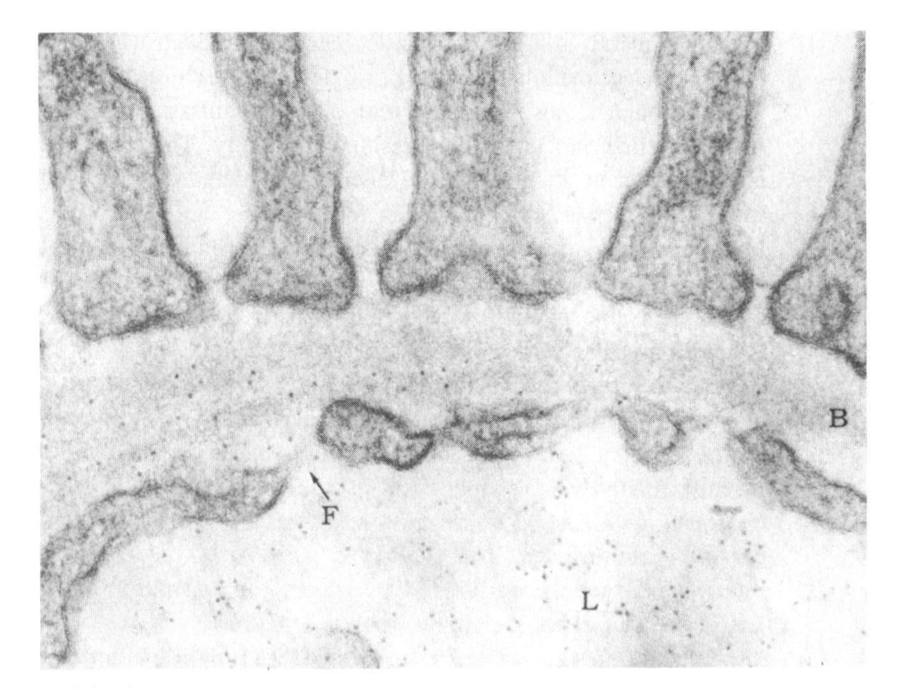

FIG. 92. Demonstration of the use of ferritin as an extracellular tracer. Portion of a glomerulus taken from a normal rat given IV ferritin 30 minutes prior to sacrifice. The ferritin molecules are seen at high concentration in the capillary lumen (*L*), and a few are present within the endothelial fenestrae (*F*). Some accumulation of ferritin is seen in the luminal part of the basement membrane (*B*), whereas its deeper part contains relatively few molecules of the tracer. 90,000X. Compare with Fig. 121. (Courtesy of Dr. M. G. Farquhar.)

The use of horseradish peroxidase as a tracer in electron microscopy was introduced by Karnovsky.[88] It has the advantages of being a relatively small protein (m.w. 40,000) and also that very sensitive localization is obtained by virtue of the amplification of visible markers through enzymatic activity. Thus the presence of the enzyme at a particular site can be detected by allowing the enzyme to act on a suitable substrate to yield an electron-dense reaction product in sufficient amount to be easily visualized, because a few molecules of enzyme yield a large number of molecules of electron-dense reaction product (Fig. 93). The use of horseradish peroxidase in combination with immunochemical methods will also be noted (p. 185).

Recently Karnovsky and Revel[87] introduced a method of impregnation of tissue blocks with lanthanum for tracing out the extracellular space in electron microscope specimens. They found that the tracer can permeate spaces 20 Å wide or less (Fig. 94), filling them with a uniformly dense matrix. This is a kind of negative staining.

D. ULTRASTRUCTURAL IMMUNO-ELECTRON MICROSCOPY

By using fluorescein-labelled antibodies as specific reagents for localization of antigenic macromolecules, it has been possible to substantially clarify the cellular location of a number of proteins. To extend this method so that localization could be carried out at the ultrastructural level required the development of an electron-opaque label for the antibody. Singer[70] developed a method for conjugating of ferritin, single molecules of which are directly resolvable in the electron microscope,[148] to the antibody to serve as a marker for the antibody-antigen reaction sites (Fig. 95). His method was essentially a two-step reaction. In the first, ferritin was mixed with an aromatic diisocyanate to obtain a ferritin-xylene isocyanate complex. This was then mixed with rabbit gamma globulin. Up to one-third of the antibody molecules could be conjugated to ferritin. Subsequent improvements in the preparation method were reported.[149,150,151]

Sternberger introduced the immunouranium technique.[152] In this technique, immuno-specific purified labelled antibodies are prepared. This was done with the aim of securing electron-opaque antibodies with a minimal concomitant increase in the size of the antibody molecule. To achieve this, the antibody was protected by saturating it with specific antigen prior to exposure to uranium. In the reaction of uranium with the antigen-antibody complex, both the antigen and the non-specific areas in the antibody ligated with uranium, but the specific combining sites in the antibody did not react

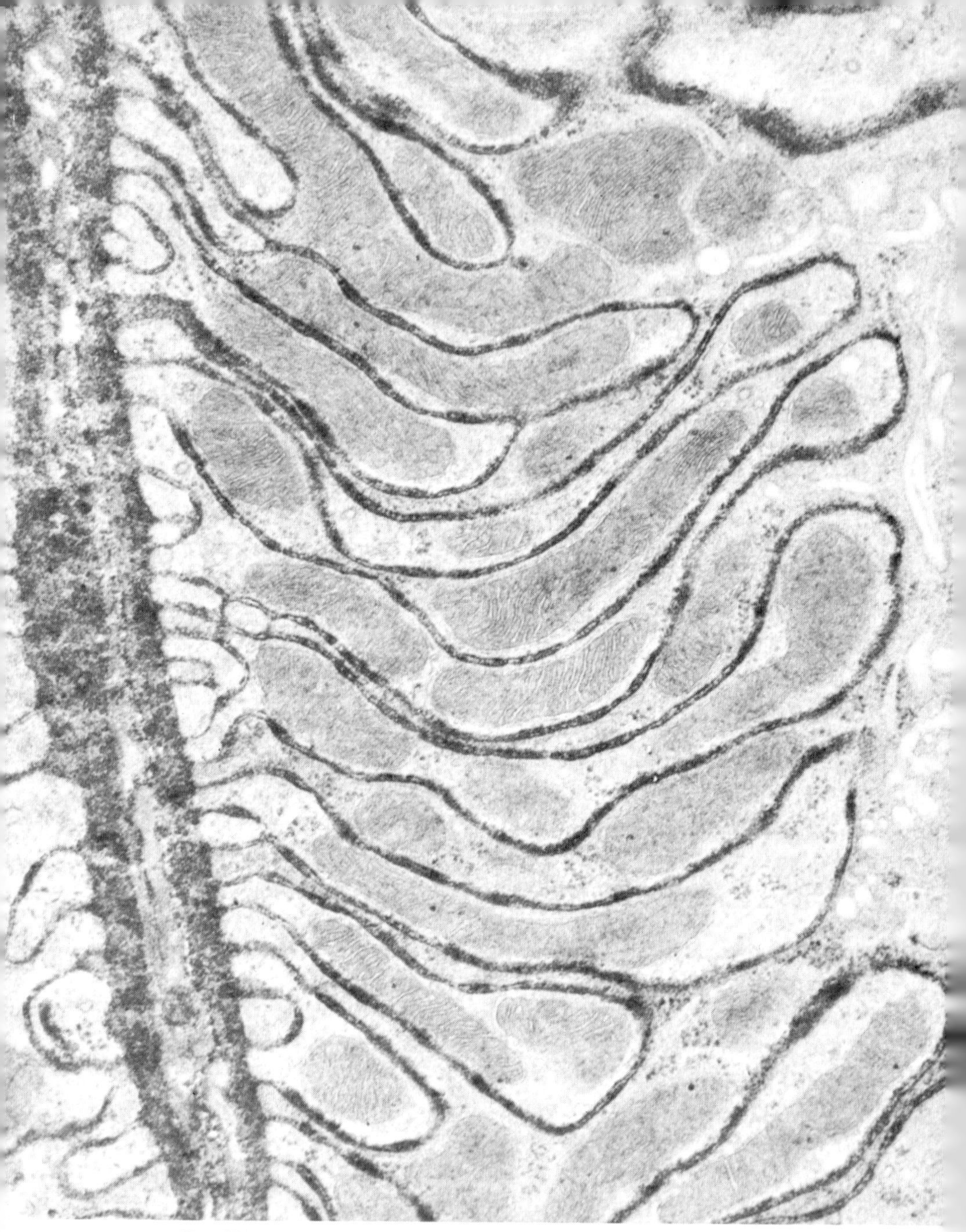

FIG. 93. Demonstration of exogenous peroxidase as an ultramicroscopic tracer. Renal tubular cell from mouse injected intravenously with horseradish peroxidase. Sites of peroxidase activity are demonstrated with 3, 3′-diamino-benzidene as electron donor. The peroxidase has escaped from the peritubular capillaries, and has permeated the basement membranes and the extracellular space between the basal infoldings of the plasma membranes. Formaldehyde-glutaraldehyde-osmium fixation. 33,500X. (Courtesy of Dr. M. J. Karnovsky!)

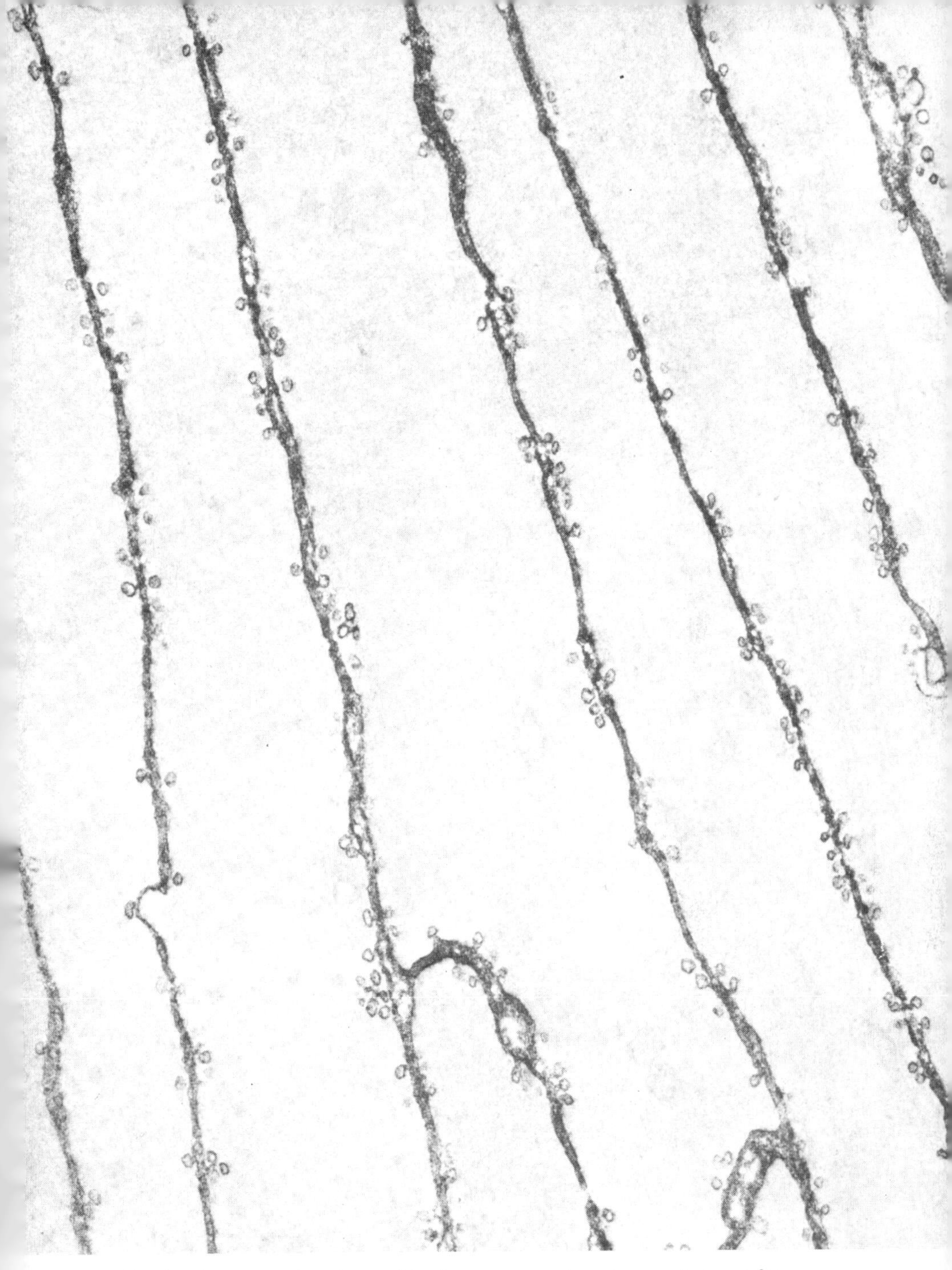

FIG. 94. Demonstration of use of lanthanum as an ultramicroscopic tracer for the extracellular space. The lanthanum fills the extracellular space between the smooth muscle cells. Superficial vesicles which open to the surface are also filled. Formaldehyde-glutaraldehyde-osmium fixation. 24,000X. (Courtesy of Dr. M. J. Karnovsky.)

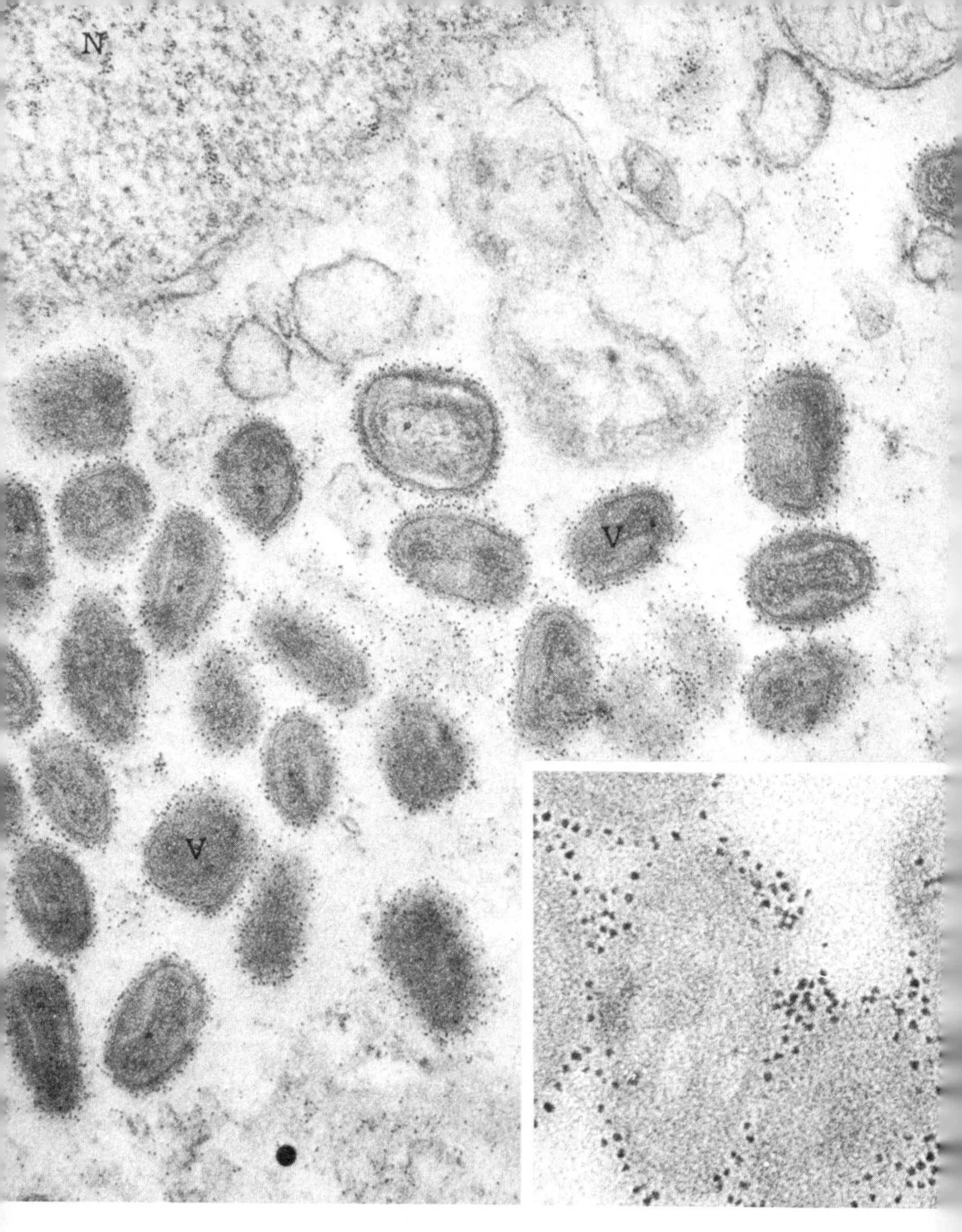

FIG. 95. Demonstration of the use of ferritin as an intracellular tracer. A thin section through the cytoplasm of a cell injected with vaccinea virus. Part of the nucleus (*N*) can be seen in one corner and a mitochondrion (*M*) at the other. Ferritin-conjugated antivaccinea antibody can be seen attached to the surface of the viral particles (*V*). 95,000X.
Inset: The tagged vaccinea virus are even more evident. 225,000X.
(Courtesy of Dr. C. Morgan.)

because they were occupied and protected by antigen. In the next step, the antigen was separated from the antibody, and the antibody, which contained uranium in the non-specific areas of the molecule, was recovered, but possessed specific combining sites devoid of uranium and hence still capable of reacting with antigen.

It has been found that purified uranium-labelled antibody does not localize non-specifically on methacrylate or Epon sections. However, it unfortunately also fails to adhere consistently to specific antigen sites in otherwise untreated sections. Etching of ultrathin methacrylate sections with suitable solvents makes the antigen more accessible to the antibody. No solvent suitable for use with Epon sections is as yet available. The immunouranium technique was improved by bridging uranium in antibody with thiocarbohydrazide, leaving one of the groups of the latter free to react with osmium tetroxide.[153] Subsequent exposure to osmium yields "osmium black" deposited on sites of uranium antibody. It was found that stability of the sections under the electron beam, which was poor for those treated with the original immunouranium technique, was improved with the use of the thiocarbo-hydrazide-osmium technique.

Recently, the ultrastructural immunochemical and cytochemical methods have been combined in an attempt to localize enzyme-labelled antibodies.[88,154] In this procedure, horseradish peroxidase was conjugated with immunoglobulin using p, p′-difluoro-m, m′-dinitrodiphenyl sulfone under mild conditions. Localization of the conjugate was detected with H_2O_2 as substrate and 3, 3′-diaminobenzidine as capturing agent (see p. 179). Following enzyme action, exposure to osmium tetroxide resulted in the reaction of osmium with oxidized diaminobenzidine, giving a high contrast insoluble product which is stable in the electron beam. The aforementioned method has recently been improved upon by the use of methacrylate as the embedding medium and then processing for peroxidase labelled antibody.

Very recently[154a, 154b] another improved immunochemical technique for localizing cellular and extracellular antigens has been developed. The new technique does not require the chemical conjugation of an antibody to a fluorescent dye or enzyme label as in the Nakane method; rather, it involves the binding of a histochemically demonstrable enzyme, horseradish peroxidase (HPO), to an antigenic site through the antigen-antibody reaction, producing an immunoglobulin bridge.

The procedure involves four basic steps with washings in between:

1. Thin sections are placed in rabbit antisera specific to the antigen of interest.

2. Then exposure takes place to an excess of bifunctional sheep antiserum to rabbit immunoglobin (one antibody site combining while the second site remains free.).
3. Then exposure takes place to immunospecific purified rabbit anti-HPO (which reacts with the second site on the immunoglobin.)
4. Then exposure takes place to HPO which reacts with the immunospecific site on the anti-HPO.
5. Incubation then follows for peroxidase using Karnovsky's 3, 3′-diaminobenzidine and H_2O_2 as substrates, followed by post-treatment with OsO_4.

The full potential of ultrastructural immunochemistry has not yet been realized. It could serve as a means of localizing and identifying normal and abnormal cell products as well as providing an approach to a molecular interpretation of the phenomenon of antigenicity.

E. ELECTRON AUTORADIOGRAPHY

Autoradiography involves the incorporation of a radioactive material into a specimen and its detection by photographic techniques. The latter depends upon contact between the specimen and a photographic emulsion during an exposure interval when the preparation is kept in the dark. During this time the radioactive material decays and the resulting emission of ionizing radiations produces latent images in the emulsion. After photographic processing, the developed image appears as an accumulation of silver grains. With the light microscope these appear as round black spots which may be as large as 1 μ.

If isotopes, such as tritium and I^{125}, which "emit" very low energy electrons with *maximum* paths in tissue of the order of 1 μ were used, and if both sections and emulsions were made thinner, and if the developed silver grains were kept small, and if the specimens were examined with the electron microscope, it seemed reasonable to expect that useful resolution of silver grains of much smaller dimensions, much closer to the site of the radioactive decay than could be visualized with the light microscope, could be attained. The initial efforts to combine electron microscopy with autoradiography resulted in resolution not significantly better than that which could be obtained with the light microscope.

Caro,[79] on the other hand, succeeded in improving the techniques so that silver grains appeared under the electron microscope as small filamentous particles which were resolved to within about 0.2 μ of the source of radioactivity (Fig. 96). His technical procedure, however, left the gelatin matrix

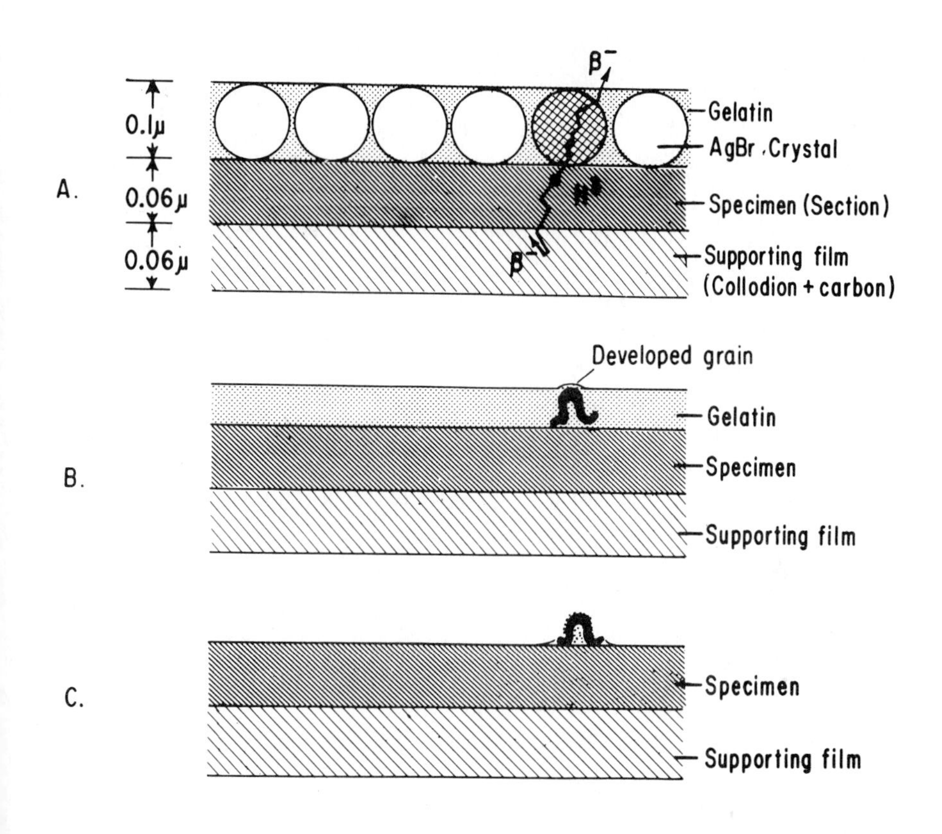

FIG. 96. Diagrammatic representation of the mode of preparation of an electron microscopic autoradiogram.

(A) During exposure the beta particle from a tritium point source in the specimen has hit a silver bromide crystal (cross-hatched). This causes the appearance of a latent image on the surface (black speck on side of this crystal).

(B) After development. The exposed crystal has been developed into a filament of silver and the non-exposed crystals have been dissolved.

(C) After fixation, the emulsion or gelatin layer has been removed.

of the emulsion on the section and this resulted in impairment of resolution and contrast in such areas. Revel and Hay[155] found that by simply exposing the autoradiograph to high pH solutions it was possible to remove most of the emulsion from the exposed and developed autoradiograph. This was accomplished incidental to the use of the lead staining technique recommended by Karnovsky[132] which makes use of a highly alkaline lead solution. Further modifications of and improvements in the autoradiographic technique have been reported (see p. 355).

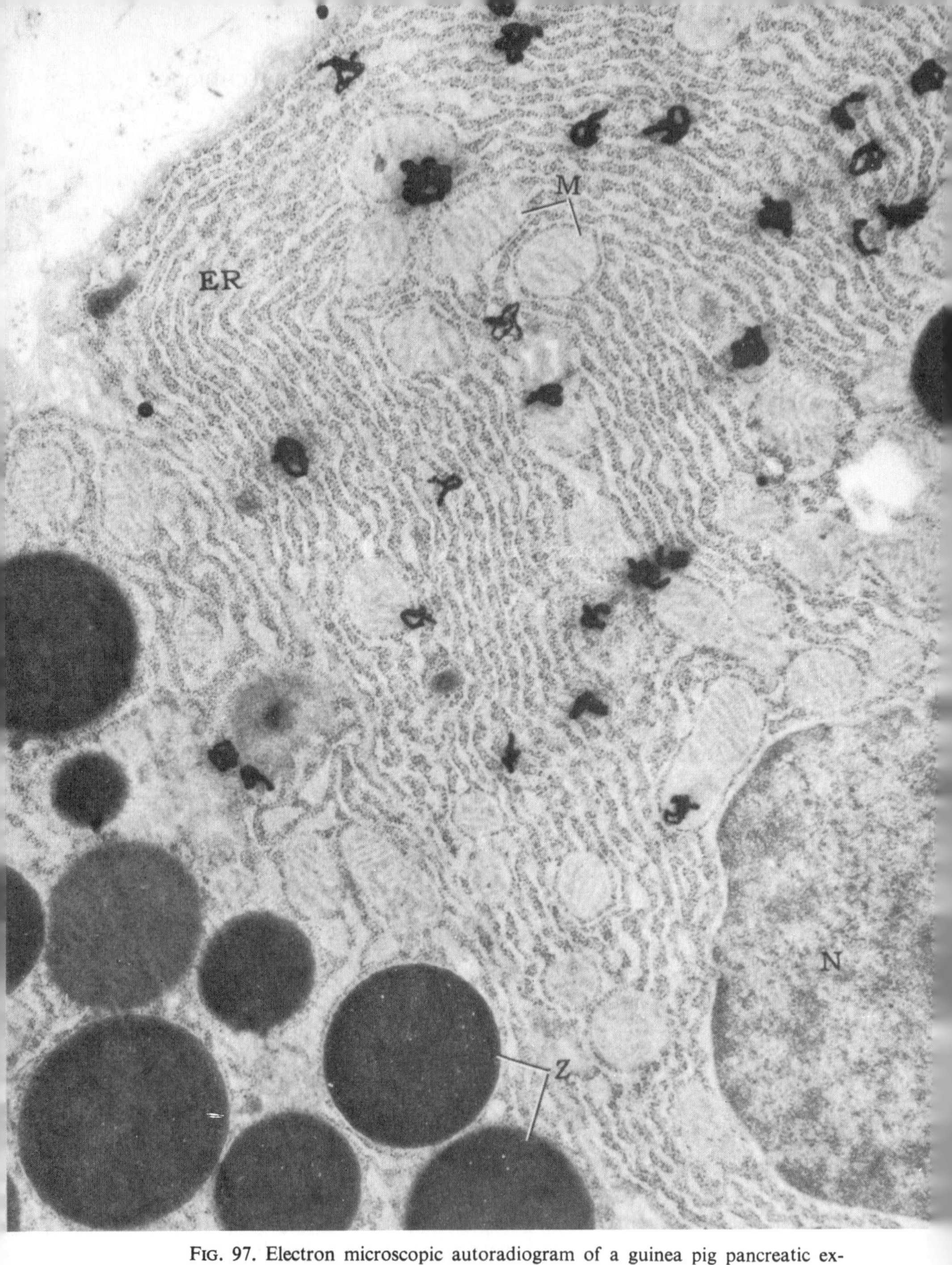

FIG. 97. Electron microscopic autoradiogram of a guinea pig pancreatic exocrine cell at the end of pulse-labelling *in vitro* for 3 min. with ^{3}H-leucine. The label marks predominantly radioactive secretory proteins over elements of the rough-surfaced endoplasmic reticulum (*ER*), rather than other cellular elements such as the nucleus (*N*), mitochondria (*M*) or zymogen granites (*Z*). 18,000X. (Courtesy of Drs. J. D. Jamieson and G. E. Palade.)

Electron autoradiography has proven to be superior to autoradiography with the light microscope in that it provides greater resolution, and permits (because of the great depth of focus in the electron microscope) the simultaneous photographing, in the same focal plane, of both the specimen and silver particle, and enables labelled cell components to be readily identified.

By means of electron autoradiography, it has proved possible to localize metabolic events and still maintain much of the high level of resolution attainable with the electron microscope (Fig. 97). Such studies provided confirmatory evidence as to the function of the various cellular organelles previously derived solely from morphological studies as well as some entirely new correlations of structure with function.

Another potential application of autoradiography that remains to be developed is in the field of cytochemistry. Thus, by using stains, substrates or precursors labelled for specific compounds or enzymes, discrete sites of newly synthesized material may be detected and movement of such material and its ultimate fate determined. An investigation along these lines was recently reported by Salpeter.[156] In this study, tritiated diisopropylfluorophosphate was used to phosphorylate acetylcholinesterase in the motor end plates of mouse sternomastoid muscle, and its distribution was evaluated quantitatively by electron microscope autoradiography.

F. HIGH RESOLUTION ELECTRON MICROSCOPY

High resolution microscopy can be considered to refer to work dealing with structural elements in the 20 to 5 Å range (especially closer to the latter). This ultrastructural level contains many components of biological systems including small viruses as well as both natural and synthetic polymers. These objects often have structures with at least one dimension which may be less than 10 Å. (Fig. 98).

The problem of examining such small structures present in this range is usually not one of resolution, but rather one of contrast. Three basic means can be used to increase specimen contrast: staining, shadow casting and instrument adjustments.

1. Staining. Positive staining, while increasing resolution, usually is limited to work in the 20 Å range or higher (see Chapter 12). Negative staining has greatly facilitated studies in the 20 Å range (see Chapter 12 C).

2. Shadow casting. The shadow casting technique (see Appendix L) has permitted some studies of structures in the 10 Å range—but presents a major difficulty in finding a suitable smooth substrate for supporting the particles. To overcome this difficulty, the *mica substrate technique* was

FIG. 98. High resolution electron micrograph of the head and sheath of a T2 bacteriophage. The fully extended sheath shows approximately 25 striations. Each striation shows a periodic fine structure on the order of 10–15 Å. Uranyl acetate-dialysis staining. (Compare with Fig. 88 of negative stained comparable structure). 680,000X. (Courtesy of Dr. H. Fernández-Morán.)

developed by Hall[157] (Fig. 99). The principle behind this approach was to obtain a smooth hydrophilic surface support from cleaved crystals. Mica is readily available and is easily cleaved over large surface areas and was an obvious choice. In practice, Hall found that spraying a suspension of protein molecules on freshly cleaved mica results in the droplets spreading readily over the hydrophilic surface and drying very rapidly. Polystyrene latex spheres are added to provide a reference standard for the measurement of shadow-to-height ratio as well as an aid for focusing. Shadow casting is usually carried out using platinum (since it is readily stripped from the mica surface). The platinum film is then backed with carbon and may be further strengthened with collodion. The final mica-film sandwich is cut into squares and the film is floated free from the mica onto the surface of water to be picked up and mounted on grids. By these means, globular proteins, collagen, nucleic acids and microsomal particles have been studied.

3. Instrument adjustment. Given a fixed instrumental resolution, the preceding two approaches seek to come close to that limit by modifying the specimen. An alternate approach might be to modify the electron microscope.

In principle, resolution is governed by Abbe's equation which, in simplified form for the electron microscope, can be written as:

diffraction limit $$d = \frac{0.61\,\lambda}{\alpha_0} = \frac{k_\lambda}{\alpha_0\sqrt{V}}$$ (see pp. 11,15) (1)

where:

d = resolution.
λ = wavelength.
α_0 = half angle of the objective lens aperture.
k_λ = 7.5 Å /steradion-volt$^{1/2}$.
V = voltage.

Thus, when $\lambda = 0.37$ Å (at 100 KV) and $\alpha_0 = 1 \times 10^{-2}$ radians, a theoretical resolving capability of about 2 Å would be possible. This level may not be attained because lens aberration, ac fluctuation in lens current and accelerating potential, stray fields, vibration, etc., all act to distort the phases at the image plane. To achieve maximal resolution, therefore, elimination of all disturbing external influences as far as possible is necessary. Reducing the effects of lens aberration is more complex. From an examination of the various equations governing resolution, namely those determining lens aberrations, image contrast and the diffraction limit (eqs. 1–4):

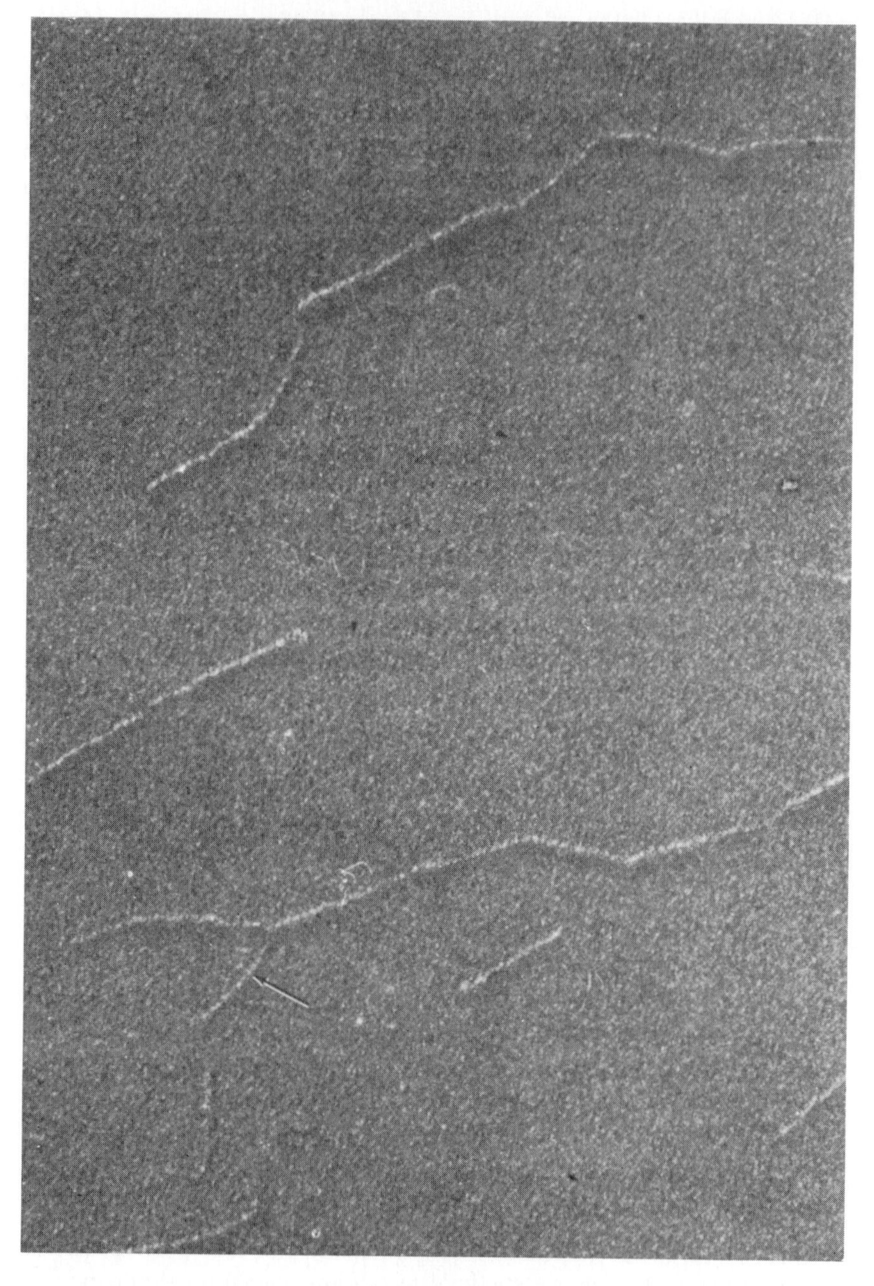

FIG. 99. Demonstration of the mica substrate technique. An electron micrograph of DNA from salmon sperm fragmented by sonic vibrations. Note the DNA strands lying on a smooth background. At one site (arrow) note that the thick strand seems to have split near its end into two separate and thinner strands which is consistent with the double-helix hypothesis. 172,000X. (Courtesy of Dr. C. E. Hall.)

spherical aberration $d_s = k_s \cdot f \cdot \alpha_0{}^3$ (see p. 54) (2)

chromatic aberration $d_{c_v} = k_c \cdot f \cdot \alpha_0 \cdot \frac{\Delta V}{V}$ (see p. 55) (3)

contrast $c \approx \frac{2d\Delta p}{V\alpha_0{}^{1/2}}$ (see p. 160) (4)

it can be seen that there is an interplay of many of the same factors whose alteration may produce contradictory results.

Thus among the variables that affect resolution are lens focal length (f), lens numerical aperture (α_0), and the accelerating potential (V). These variables go into determining the extent of lens aberration and image contrast. In addition, contrast is affected by section thickness. The interplay of some of these factors can be, in practice, seen by examining the effects of changes of the variables on resolution.

1. Focal length of objective lens. As the focal length is decreased, and all other factors remain the same, resolution increases because spherical and chromatic aberration are decreased. But to decrease the focal length (for other than superconductor lenses[20]), the bore of the pole pieces must be decreased and astigmatism, due to inhomogeneities in the pole piece material and/or changes on the walls of the bore, may become limiting factors. The spacing of the pole pieces must also be decreased to reduce the focal length, leaving less and less room for specimen holder, objective diaphragm, cold baffle, etc.

2. Aperture angle of objective lens. By reducing the aperture angle, all other factors staying the same, resolution is increased because spherical and chromatic aberration are decreased and contrast is increased. Since, however, Abbe's law determining resolution indicates that decreasing aperture angles decreases resolution because of diffraction, there is a direct conflict involved which, in practice, is settled by a compromise angle, α_0 (see p. 58).

3. Accelerating potential. Increasing accelerating potential does improve resolution as would be expected from de Broglie's equation and the reduction in chromatic aberration. However, at higher KVs, specimen contrast drops and microscope design rapidly becomes more complex and massive.*

4. Specimen thickness. Since specimen thickness determines the magnitude of change in voltage (ΔV) of the electrons emerging from the

* When one is prepared to sacrifice contrast to improve penetrating power or resolution, it is then useful to consider employing a high voltage electron microscope (see p. 107).

specimen, it has a direct effect in determining chromatic aberration and thus resolution. Thus excessively thick sections are to be avoided. On the other hand, excessively thin sections will have low contrast.

Given the *utilization of scattering* and the "subtractive" action of the objective aperture as the basis for developing contrast in the electron microscope, we find that little room remains for microscope modification to increase resolution. It has been estimated [17,158,159] that to optimize resolution what is needed is an electron microscope which typically has:

(a) voltage stability on the order of 10^{-6} (i.e., 1 part per million),
(b) lens current stability on the order of 10^{-6},
(c) objective lens astigmatism corrected to be less than $0.01/\mu$.,
(d) the ability to hold the specimen position stable during the exposure period to about 2 Å,
(e) no environmental disturbances larger than 2 Å at the specimen level,
(f) no contamination resulting in introducing astigmatism beyond the limiting value,
(g) the ability to operate at about 200 KV[17],
(h) a short focal length objective (*circa* 2 mm or less) having a medium size aperture (1×10^{-2} rad.).

An instrument having these specifications used in examining sections that are not excessively thin or thick (*circa* 500 Å), would be expected to be able to resolve structures with adequate contrast and stability down to the 2 Å level.[17,18]

Such instrument improvements which have been achieved in themselves are not fully meaningful, for while they may result in improving resolution, the important consideration is to secure images of *interpretable* contrast at high resolution levels. There are two mechanisms for producing image contrast in transmission electron microscopy. These are diffraction contrast and phase contrast. *Diffraction contrast* is accomplished by either the removal of scattered electrons by the objective aperture resulting in bright field density variation between individual points (p. 59) or the imaging of only the scattered electrons (dark field imaging). To secure diffraction contrast beyond that provided by the aperture, it has been suggested that the accelerating voltages be drastically reduced (V appears in the denominator of the contrast equation). Unfortunately, while this will increase contrast, low voltages also result in decreased resolution as would be expected both from Abbe's law and the chromatic aberration equation. This type of instrument adjustment, therefore, leaves very little room for improvement of resolution.

The other type of contrast is *phase contrast*. In this case, elastically scattered or diffracted amplitudes must pass through the objective aperture and be recombined with the undiffracted (zero order) transmitted wave at the image plane. Phase contrast is dependent on the phase difference introduced into the scattered wave by the scattering object (relative to that of the zero wave) and by lens aberrations. The phase difference might be controlled, for example, using a $\frac{1}{4}\lambda$ "electron phase plate"[160] during operation of the microscope. The use of phase plates may now become feasible since the development of anti-contamination devices (see p. 100).

In a recent evaluation of phase contrast, Heidenreich[17] has concluded that while present day electron microscopes are capable of producing phase contrast images with detail in the molecular size range, down even to 2 Å, "the change in appearance and contrast with absolute defocus raises problems in interpretation since the detail can vary rather drastically." The ability to interpret phase contrast in this range from films 40–75 Å thick is not yet developed and "probably the situation will be the same for improved microscopes of higher resolving power." He noted that the major unsolved problems in developing "molecular microscopy" are specimen preparation (e.g., to prepare regions of unobstructed separated chains of interesting biopolymers) and electron damage. The damage will be reduced if work is done at low temperatures with an anti-contamination device in the specimen stage and at accelerating voltages near 200 KV (since the ratio of contrast to damage is maximum at about 200 KV),[17] but an irreducible minimum of damage seems unavoidable.

A detailed consideration of the problem of specimen change due to exposure to the electron beam has been discussed previously (p. 97). This subject is significant for (as also noted there), *under optimal operational microscope conditions, specimen change will be the limiting factor determining resolution.*[161]

G. ELECTRON MICROSCOPIC ANALYSIS OF CELL FRACTIONS

The usefulness of the technique of differential centrifugation in cell biology has long been established. The purity of separated cell fractions is determined by their degree of homogeneity. The nuclear fraction can be usefully examined by phase contrast microscopy, while purity of the mitochondrial fraction has been evaluated by virtue of the supravital staining specificity of these organelles for Janus Green. These light microscope methods are much less useful when applied to smaller particles.

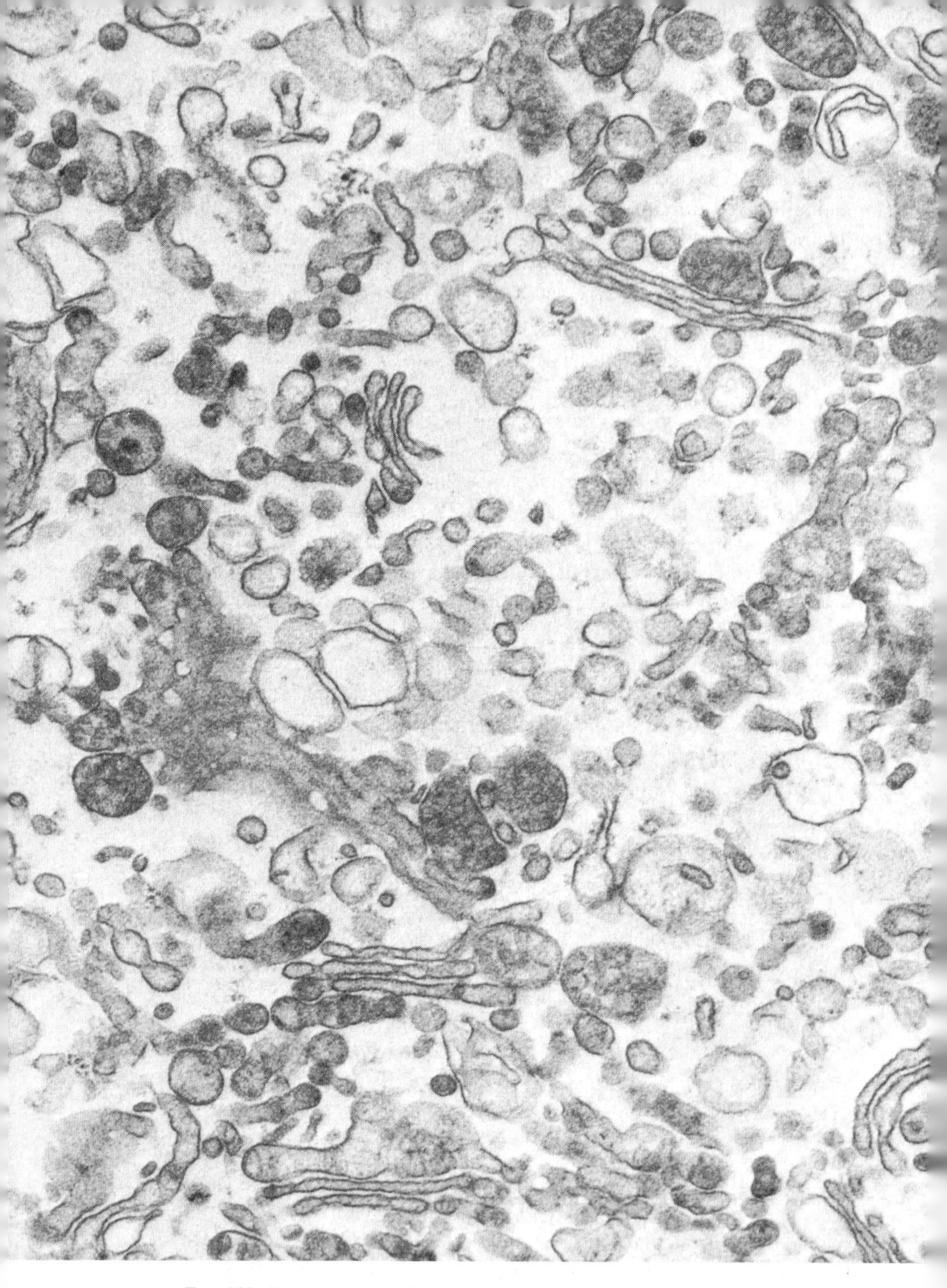

FIG. 100. Demonstration of electron microscopy of cell fractions. The Golgi fraction from the mouse liver consisting of cisternae and vesicles. A few mitochondria that were not separated out are also present. 69,000X. (Courtesy of Drs. D. J. Morre and H. H. Mollenhauer.)

The utility of testing fractions for homogeneity by electron microscopy was elegantly demonstrated by Palade and Siekevitz,[56] who worked with the microsomal layer of liver cell fractions and Novikoff,[57] who made pellets of centrifuged mitochondria which were fixed, embedded and then sectioned. This approach demonstrated the usefulness of electron microscopic analysis as a control when determining the extent of contamination of the fraction being examined (Fig. 100).

This approach is also useful in following the morphology of cellular organelles and their response to chemical treatment. Such observations can serve as a means of evaluating the validity of data obtained by physical and chemical methods which are purported to have relevance to intact cellular organelles (even for the evaluation of tissue sections).

H. CRYOFIXATION FOR ELECTRON MICROSCOPY

While the standard method for immobilizing cytoplasm involves the use of chemical fixatives, this approach, as can be inferred from what has been said earlier, has several disadvantages. First, the penetration of the fixative may result in diffusion of solutes and water as well as in osmotically induced flows, all of which may produce disruption or serious dislocation of delicate cytoplasmic structures. A second disadvantage is that chemical cross-linkages which are formed during fixation may result in macromolecular rearrangements which may find morphological expression. The sensitivity of the cytoplasm and organelles to minor modifications of the fixation medium is well-documented with recognizable structural differences.[162,163,164] In attempts to avoid these problems, freezing has been experimented with as a fixation technique for quite a number of years in the hope that it would reduce the alterations occurring during fixation, and might also serve as a basis for establishing new methods for cytochemical localization. Moreover, it was hoped that it might serve as an additional method for assessing the fidelity of chemical fixation.

The earliest freezing technique used was that of freeze-drying. It was found that artifacts were introduced by such preparations, which may have been brought about partially by surface tension forces acting at the time of drying and/or impregnation. To eliminate these forces, two approaches have been experimented with, namely, freeze-substitution and freeze-etching.

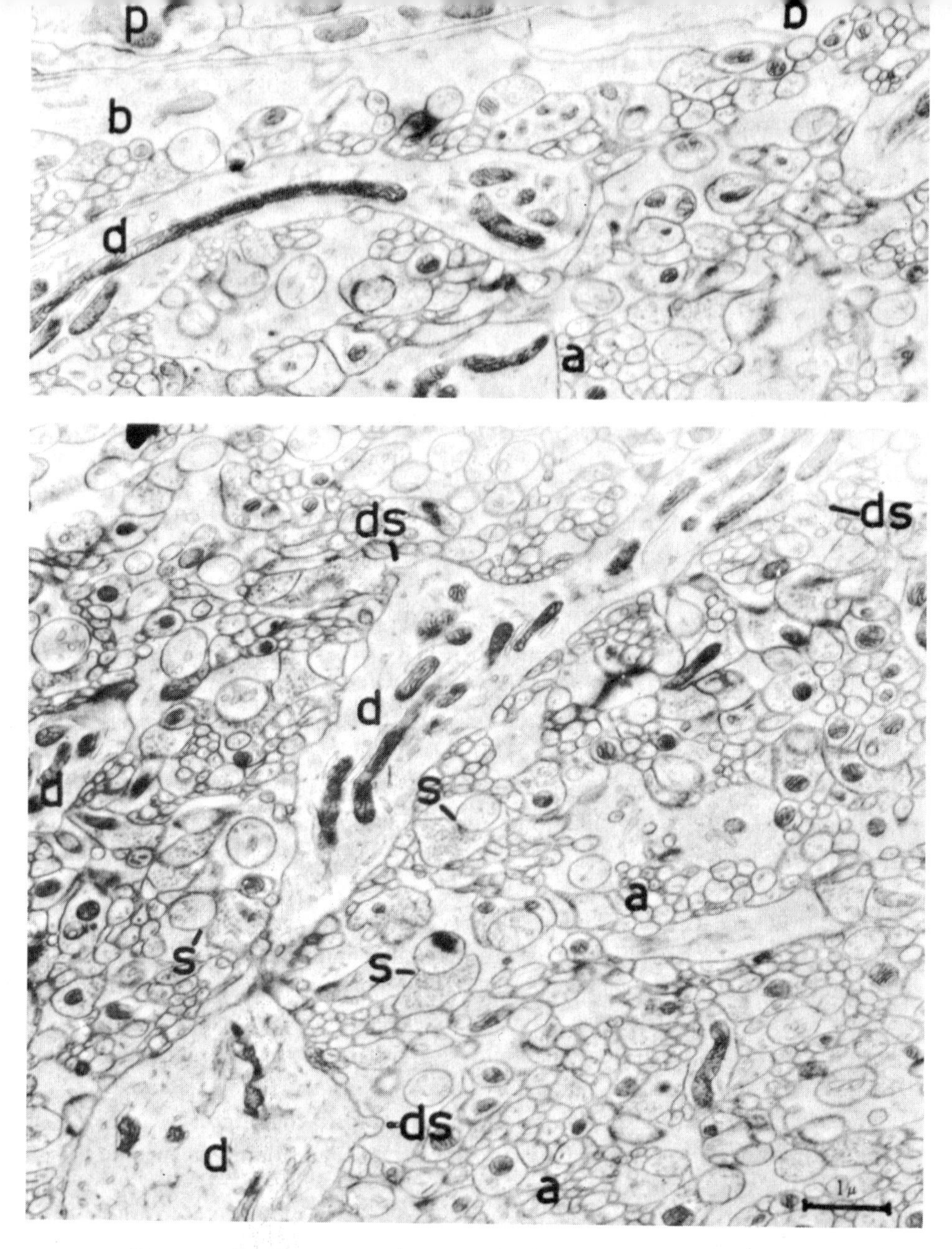

Fig. 101. Demonstration of freeze-substitution. A large field of the molecular layer of asphyxiated cerebellar cortex of a white mouse frozen 8 minutes after decapitation. The dendrite (*d*) in the upper part of the figure is a continuation of the one seen in the lower part. The dendrite is characterized by the presence of spines (*ds*). Immediately underneath the pia (*p*) are end-feet of Bergmann fibers (*b*). Profiles of axons are present in groups (*a*), and numerous synapses (*s*) are recognizable. Ghosts of ice crystals are recognizable as light areas surrounded by relatively electron-opaque material in the lower part of the dendrite. Note the absence of intracellular space. 23,000X.
(Courtesy of Drs. A. Van Harreveld, J. Crowell and S. K. Malhotra.)

1. Freeze-Substitution

Freeze-substitution for the electron microscope[71,72] was a natural outgrowth of the corresponding techniques for light microscopy. It is based on rapid freezing of tissues followed by slow dissolving ("substitution") of the ice within the tissue with an organic solvent at a temperature well below 0° C. The latter step is based on the fact that ice is soluble in many solvents at temperatures far below its melting point. The specimen, free of ice and permeated by the cold anhydrous solvent or solvent solution of fixative (e.g., 2% OsO_4 in acetone[165,166]), is then brought to room temperature, put through additional changes of an embedding medium compatible with the solvents and embedded, sectioned and stained in conventional fashion.

Theoretically, instant freezing and subsequent removal of the ice should insure maintenance of spatial relationships. This has rarely proven to be the case. The major problem arising from freeze-substitution has been wide variability in the appearance of similar structures present in different samples frozen under apparently similar uniform conditions. Analysis suggested that much of this variability was probably brought about by events occurring between the time the specimen was removed from its natural environment and the time it was frozen.[167] The accidental removal of water from the sample by uncontrolled evaporation in the short period of time during which the minute samples pass from their site of removal through air to the freezing bath seemed a potential source of variability.

The extensive experimental work by Rebhun[167] (who plunged the tissues into liquid N_2 cooled propane or Freon which provide high rates of heat transport) has led him to conclude that such freezing technique will almost invariably introduce artifacts because the rate of heat transport is insufficient to prevent appreciable ice-crystal growth, and hence structural variability, unless the cells have been modified prior to freezing. This can be effected, for example, by evaporation, or by "hydrogen-bonding" of water through the use of solvents such as glycerol and dimethylsulfoxide. The findings suggest that, with some further improvement to insure properly standardized freezing techniques, this method may be developed as a supplement to chemical fixation.

A recent variant of the freeze-substitution technique by Van Harreveld *et al.*,[73] has produced very favorable results (Fig. 101). The method involves rapidly freezing the tissue by bringing it into contact with a silver plug of substantial heat capacity, polished mirror smooth and cooled to the temperature of liquid nitrogen (*circa* 207° C) under reduced pressure.

FIG. 102. Demonstration of freeze-etching. A portion of an intestinal absorptive cell from the mucosa of the small intestine of the mouse. The numerous microvilli (*MV*) can be seen projecting into the lumen (*L*). Mitochondria (*M*) and small vesicles are very evident in the cytoplasm. Shadowing is from the lower left. 30,000X. (Courtesy of Drs. S. Bullivant and A. Amer, 3rd.)

The silver surface is protected against the condensation of water and air by a stream of cold dry helium gas flowing over it. The tissue to be frozen is plunged onto the polished silver surface by a guiding system which makes it possible to let it descend through the cold helium at a controlled velocity (30 to 40 cm/sec). The tissue is then transferred to 2% OsO_4 in absolute acetone at 86° C, and substitution takes place.

Pioneering work involving freeze-substitution and specialized instrumentational approaches were developed by Fernández-Morán.[158] By progressive improvements he has succeeded in carrying out high resolution studies of the macromolecular organization of T2 phage particles, ribosomes and isolated cell constituents.[168]

2. Freeze-Etching

Freeze-etching,[59,60] is a special kind of freeze drying which allows the investigation of objects in the frozen state (Fig. 102). There is no chemical treatment of the object during the entire procedure until the replica (see p. 200) is formed on an "etched" fractured surface of the frozen specimen. By means of snap-freezing (as in the freeze-substitution technique) and/or aqueous glycerol impregnation followed by snap-freezing, yeast cells have been freeze-etched so as to preserve the organism in a life-like state.

Specifically, the freeze-etching method involves three steps, all carried out with a specially designed apparatus consisting of a microtome and a freeze-drying and shadow casting installation, all in the same vacuum evaporator.[169]

a. Freezing. Two different approaches may be used.

(1) A small droplet of a very concentrated suspension of cells or very small mass of tissue is placed on a special precooled object table which provides quick freezing, or

(2) A very small droplet is placed on a copper disc which is immersed in propane cooled by liquid N_2. The disc is then mounted on a special precooled object table.

b. Sectioning. The frozen object, mounted on the object table, is cut, etched, (i.e., the cut surface is freeze-dried at -80 to -100° C to a depth of a few hundred Angstroms) and platinum and carbon films are sequentially deposited under high vacuum to produce a replica of the etched surface.

c. Detachment. The vacuum is broken after the deposition of the films on the frozen-etched face of the object. The object table is detached from the stage and warmed to room temperature. The table is dipped into water and the replica floated off. Adhering cells are washed off and the replica is taken up on a Formvar-covered grid.

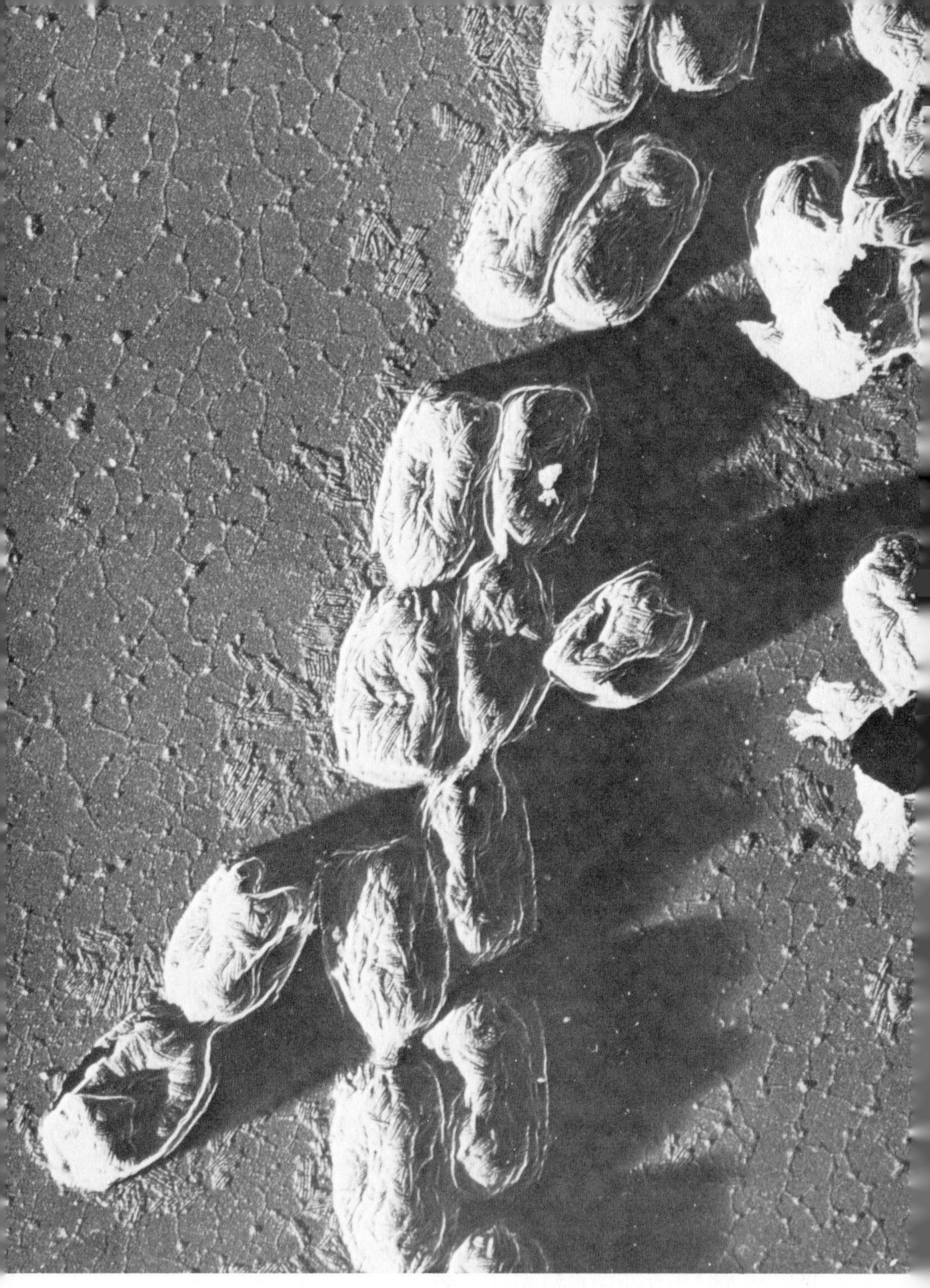

Fig. 103. Demonstration of the replica technique. Surface replica of *Streptomyces venezuleae*. 28,000X.
(Courtesy of Dr. E. Ellis and Mr. R. A. Schlegel; Eli Lilly and Co.)

A suggested simplification in the freeze-etching technique has recently been published.[170]

I. SURFACE EXAMINATION OF SMALL SPECIMENS

The electron microscope can provide significant information concerning the external form and structure of small particles such as bacteria, viruses and protein macromolecules.

There are three basic methods of specimen preparation for the handling of small particles. One, the dispersal technique, involves obtaining good separation of particles on a solid surface from a liquid suspension. Another involves preparation of casts of specimens. The third involves surface spreading on a liquid surface and trapping of the material in a monomolecular layer of surface denatured protein. All three techniques are usually followed by shadow-casting (see Appendix L), or by positive or negative staining.

1. Dispersion Methods

There are two approaches used in the dispersal technique. One, known as the *droplet method*, involves placing a drop of liquid suspension from a platinum loop or pipette onto a film (substrate). After a very brief interval the liquid is drained almost completely from the film. Once the remaining thin aqueous film has dried, the preparation can be shadow cast or positively or negatively stained with a heavy metal salt. The droplet method has two potential disadvantages. First, surface tension forces acting during drying may result in distortion or dissociation of the particles. Secondly, poor dispersion of the particles may result.

The alternate procedure is known as the *spray method*. This involves the deposition on support films of droplets, each containing one or very few particles, by means of a spray gun. Different methods and several gun designs have been described [171] to meet specific needs. When material is sprayed onto a hydrophobic surface (e.g., collodion), the droplet supposedly rides on a vapor pad which gradually evaporates. As a result, shrinkage of the droplet takes place until it finally collapses in a small area. This method, therefore, provides well-dispersed particles.

The spray technique also has two disadvantages. First, since the final specimen area representing each droplet is small, impurities and suspended molecules are also concentrated there. This necessitates working with suspensions that are very highly diluted and using very pure reagents. Secondly, many droplets riding on vapor pads tend to migrate to the grid wires (that serve as the film support) during drying rather than remain on the open filmed areas. For this reason cleaved mica has been used, for it is a hydrophilic surface over which the droplets spread readily and dry very rapidly (see p. 189).

When evaporated metal is applied to a dispersed specimen everything is coated except the areas immediately beyond the particle under investigation. The height of the particle is indicated by "shadows" (metal-free areas) resulting from its blocking the vapor beam for a distance directly proportional to its height. This gives a three-dimensional effect to the micrograph of the specimen (Fig. 103).

2. Replica Technique

The preparation of casts or replicas of specimens which are opaque to electrons also provides a very suitable means for studying the surface structure of small particles to be examined with the electron microscope. Replicas consist of thin films of electron-transparent and electron-dense material, usually carbon or formvar, and a metal corresponding exactly to the surface topography of the specimen.

There are two basic replica techniques. The first, known as the *single-stage* replica method (Fig. 104A) consists of depositing the replicating material directly on the specimen, separating the two, and examining the replica with the electron microscope. The second, or *two-stage* replica method (Fig. 104B), consists of making a preliminary impression of the specimen surface in one material, coating the structure surface of this impression with the final replicating material, separating the two and examining the replica with the electron microscope.

When applied to a replica or casts of specimens, metal shadowing will result in deposition of metal where there are high points in the replica and this region will scatter electrons, taking them out of the beam. Where there are depressions, no metals will reach the surface and this region will easily transmit electrons. These variations in scattering result in the contrast difference of the electron image. The image is analogous to that provided by metal-shadowed specimens and provides similar information.

3. Surface Spreading Technique

Kleinschmidt and Zahn[74] have developed a highly effective method of preparing very long DNA or RNA molecules for electron microscope observation. The method usually involves the use of a shallow rectangular

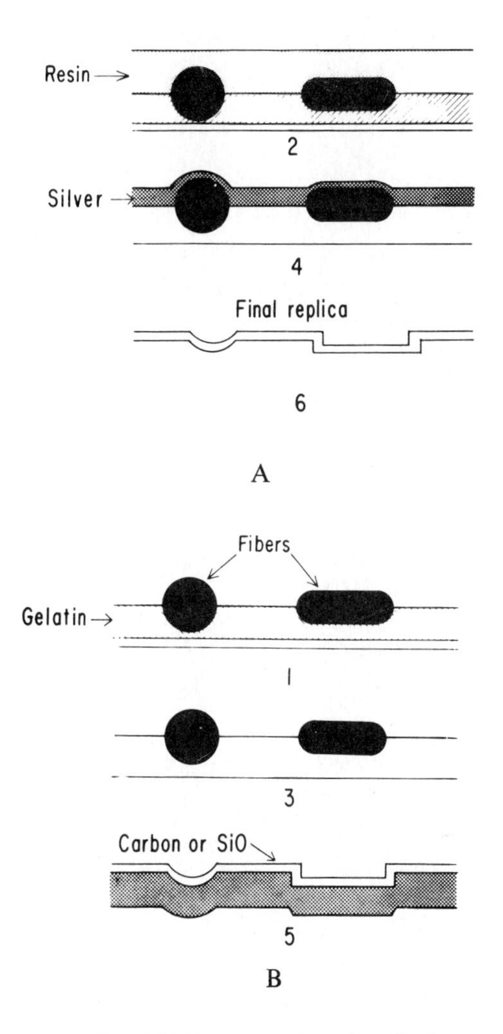

FIG. 104. Demonstration of replication.

(A) The one-stage replica method using a plastic substrate.
(1) Particles mounted on Formvar film on grid.
(2) Particles coated with carbon.
(3) Formvar substrate removed.
(4) Final replica after removal of the particles.

(B) The two-stage replica method for fibers or small specimens.
(1) Fibers lying in swollen gelatin.
(2) Resin backing applied.
(3) Resin parted from gelatin, exposing part of fibers.
(4) Silver layer on exposed fiber surfaces.
(5) Carbon layer on surface of film.
(6) Final replica after dissolution of the silver.

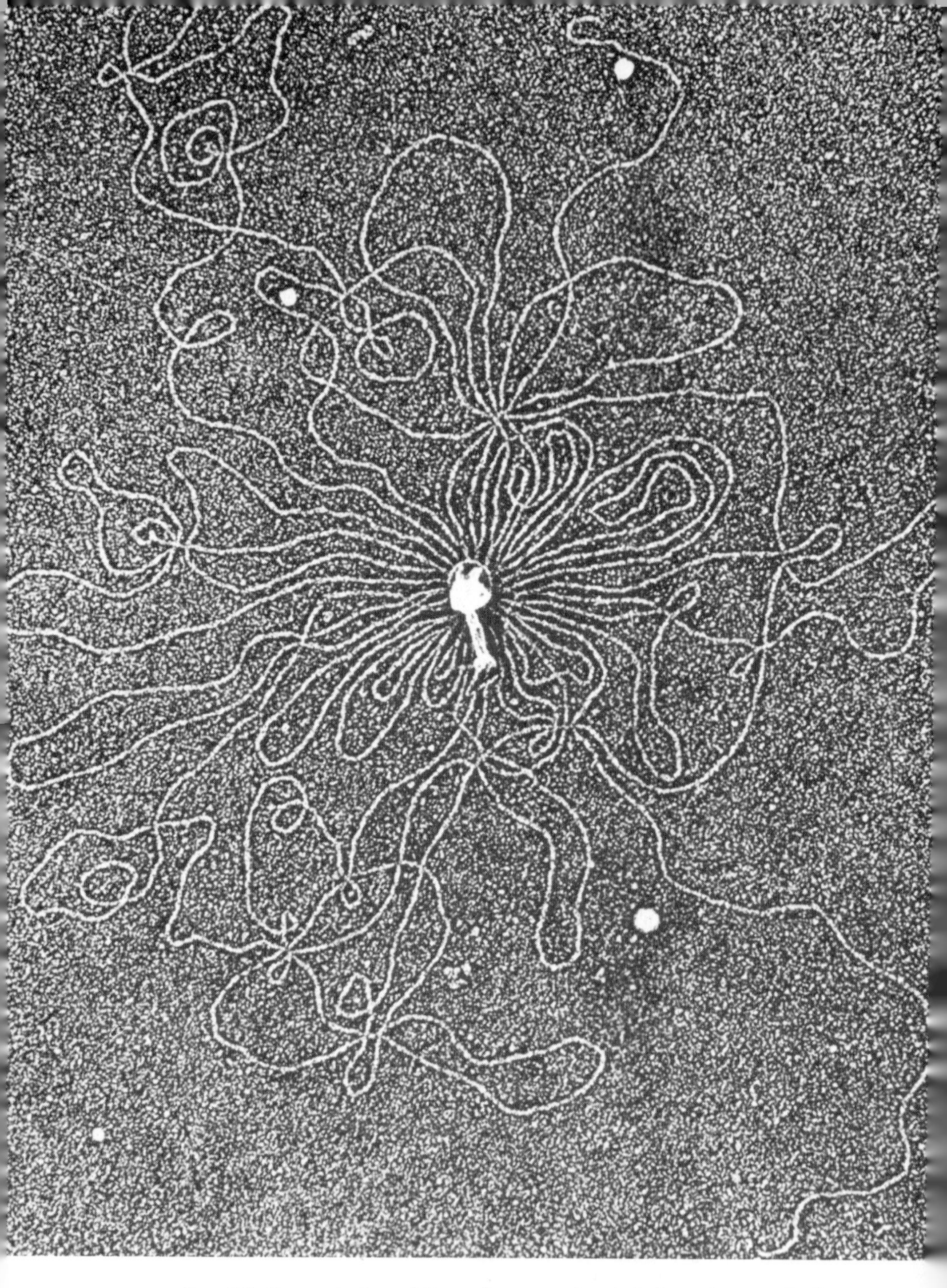

FIG. 105. Demonstration of the surface spreading technique. An osmotically shocked T2 bacteriophage whose DNA content has been spread out as fine strands from the core of the virus particle. 90,000X.
(Courtesy of Dr. A. K. Kleinschmidt.)

glass tray called a *Langmuir trough*. This is filled with a supporting liquid such as double distilled H_2O or, more usually, 0.15 M ammonium acetate. The nucleic acid containing specimen (e.g., virus or nucleic acids released from virus or other sources) is suspended in a protein-salt solution (e.g., 1% cytochrome C in 1 M ammonium acetate). The surface of the liquid is cleaned by placing two water-repellent (e.g., Teflon) bars side by side across the width of the trough and drawing them slowly apart. Talc is spread on the cleaned area to serve as a marker for the spreading film The protein-nucleic acid solution is applied slowly to the surface of the water. The protein immediately spreads out in a monolayer leaving a clear area in the talc. If suspensions of whole virus are placed on the trough surface, osmotic shock occurring during spreading of the protein-monofilm often ruptures the nucleic acid-containing particle and release of molecules in an untangled form may take place. The monolayer is mounted on coated support grids and the excess fluid is replaced with ethanol which is then drained away. The grid is then positively stained with uranyl salts or shadowed with platinum at a low angle while being rotated in the plane of the specimen in the vacuum evaporator. Shadowing in this manner causes metal to impinge from all directions (Fig. 105) and avoids the difficulty that segments of molecules tend to become invisible when they are in line with the shadow direction. The surface spreading method has made it possible to follow extremely long DNA molecules as they twist and turn through the monofilm.

J. QUANTITATIVE ELECTRON MICROSCOPY

As indicated on pages 60 and 164, the optical density of a uniform elemental, i, area of the specimen measured in the image plane [i.e., the logarithm of the ratio of the incident intensity, I_0, (the intensity measured in an image plane with the specimen removed) to the incident intensity minus that which is scattered outside the angle of the objective aperture by the specimen (the "apparent" transmitted intensity in the image plane), I_i] is very nearly proportional to the mass per unit area, M_i, (and, therefore, to mass density for a fixed section thickness) independent of the atomic composition of the specimen (eq. 20). This provides a basis for quantitatively "weighing" various parts of an electron microscopic specimen.

As a result of the very fortunate "accident" that the amount of light transmitted by a portion of an electron micrograph T_{n_i} is directly proportional to the optical density of the portion of the specimen recorded on that part of the *negative* (and this is not the case for light micrographs) (eq. 21), Bahr and Zietler [172,173] have developed a simple quantitative technique which provides directly the total dry mass M_{spec} of any area of an electron microscopic specimen. In principle, two photographs of equal exposure (one with the specimen in place, and one with a standard of known mass per unit area) are recorded on a plate. After development, the area of the negative to be measured is circumscribed by an opaque mask and is illuminated with a uniform light source, and the amount of light transmitted, T_{spec} (eq. 22), is measured with a photometer. The amount of light transmitted through a mask of equal area fitted to the image of the reference object, T_{ref} (eq. 23), is then also measured. The ratio of the first measurement to the second is then multiplied by the known mass of the circumscribed reference object to give the mass of the unknown (eq. 24).

$$M_i = K \log \frac{I_0}{I_i} \tag{20}$$

$$T_{n_i} = J \log \frac{I_0}{(I_0 - I_i)} \tag{21}$$

therefore:
$$T_{spec.} = \sum_{spec.} T_{n_i} = \frac{J}{K} \sum_{spec.} M_i = \frac{J\,M_{spec.}}{K} \tag{22}$$

and
$$T_{ref.} = \sum_{ref.} T_{n_i} = \frac{J}{K} \sum_{ref.} M_i = \frac{J\,M_{ref.}}{K} \tag{23}$$

therefore:
$$M_{spec.} = \frac{T_{spec.}}{T_{ref.}} \times M_{ref.} \tag{24}$$

where:

i = uniform elemental area of the object, of the image of the object or of the negative.
J = constant of proportionality.
K = constant of proportionality.
n = this subscript refers to the transmittance of the negative.

Section Two

SCANNING ELECTRON MICROSCOPY

Part One

INSTRUMENTATION

Chapter 14

INTRODUCTION

THE TRANSMISSION light microscope has been employed by investigators as a research tool since long before the turn of the twentieth century. In most cases this microscope is used with visible light for its source of illumination. Because light has a relatively long wave length it significantly limits microscope resolution (see p. 13). However, since there is an interaction of visible light with the specimen, especially those that have been specifically stained, a wealth of information on its morphology and chemical organization can be obtained. The transmission electron microscope (TEM), an instrument which began to be exploited in biomedical research in the 1950s, has enormously extended our resolution capabilities (down to ~ 5Å). But the informational content of its image is limited because what is usually observed is the two-dimensional, internal organization of a specimen. Specialized preparatory techniques must be employed to secure information about the three-dimensional topography and cytochemical nature of the specimen.

The scanning electron microscope (SEM), which is especially useful in studying surfaces, has become a popular laboratory investigational tool in the 1970s. It falls into an intermediate position between the light and the TEM insofar as the two basic parameters of resolution and image information are concerned and it is similar to the reflecting light microscope. While the resolving capacity is usually less than that of the TEM, the SEM image may be richer by virtue of the variety of information it potentially can provide. For example, such an instrument offers a better than 300-fold increase in the depth of field when compared with the highest quality light microscope which is reflected in the superb "three-dimensional" images it provides. In addition studies of beam-specimen interaction using the SEM can provide useful information about the chemical composition at the specimen surface as well as the crystalographic, magnetic, and electrical characteristics of the specimen.

The SEM, thus, is a very versatile instrument for the examination and analysis of the microstructural characteristics of biological objects (Fig 106). It therefore both complements and supplements the light and transmission

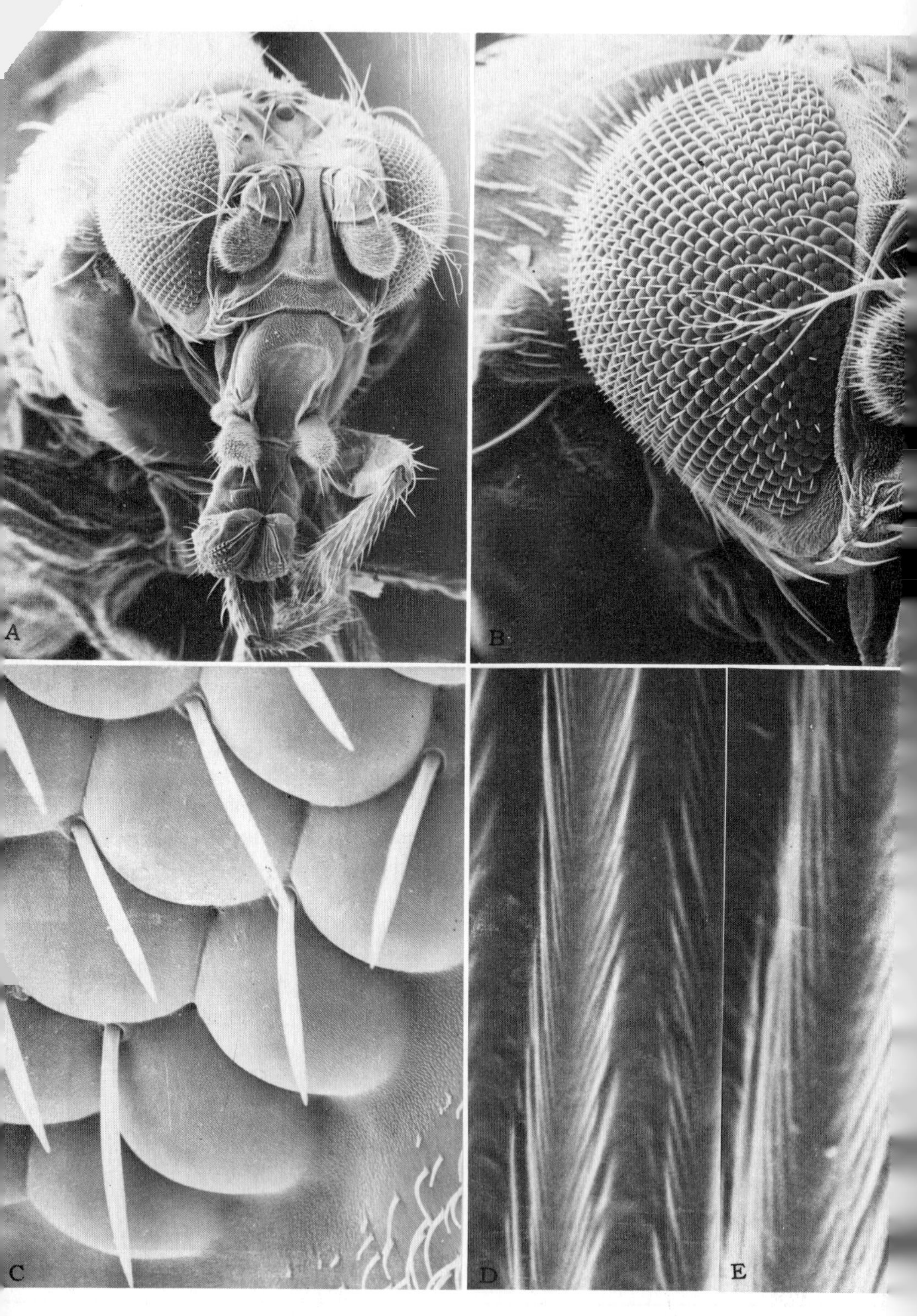
A
B
C
D
E

electron microscopes, and has become an increasingly important instrument in biomedical investigations. The SEM, having achieved the potential of a high resolution stereomicroscope, clearly merits the allocation of a distinct section within this book for a discussion of its basic construction and mode of operation.

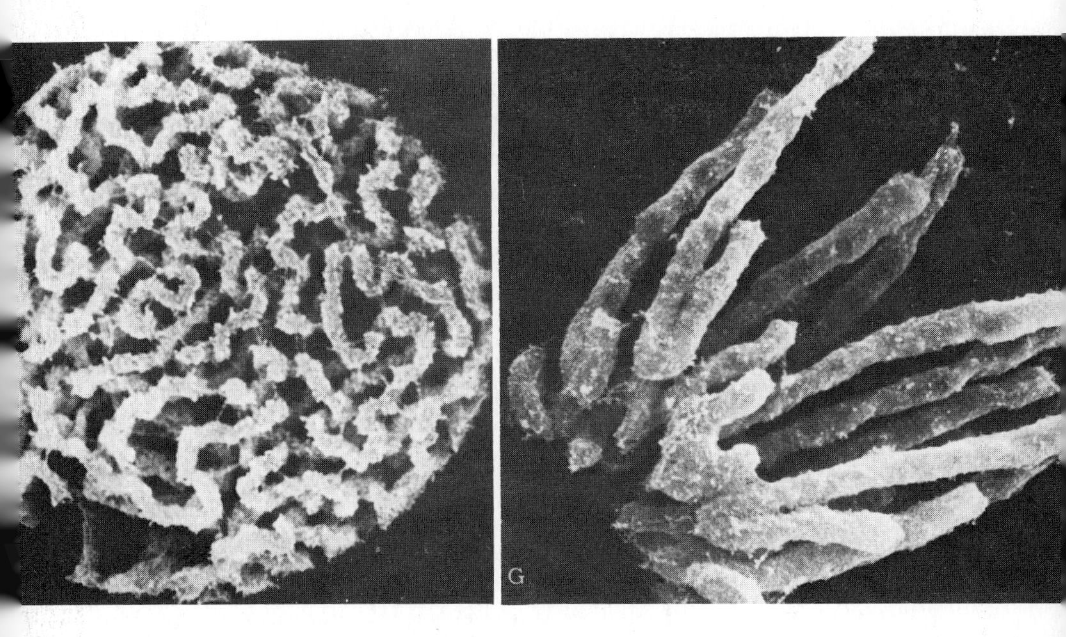

FIG. 106. Scanning electron micrographs of animal and plant material

Animal material (Courtesy of Dr. A. Wachtel)

(A) Drosophila: head on view. 73X
(B) Right eye from the above. 172X
(C) Ventral edge of eye. 2150X
(D) Base of bristles on antenna 18,060X
(E) Base of bristle; higher magnification 43,000X

Plant tissue (Courtesy of Dr. M.M. Laane)

(F) Mitotic prophase nucleus of *Haemanthus katherinae.* 5250X
(G) Late mitotic anaphase (one daughter nucleus) of *Haemanthus katherinae.* 4200X

Chapter 15

HISTORICAL REVIEW

TWO MAJOR scientific contributions in this century that took advantage of J. J. Thomson's[174] confirmation of the existence of the electron in 1897, provided the foundation for the development of the transmission electron microscope. These were the de Broglie[1] hypothesis in 1924, on the wave nature of moving electrons, and the discovery by Busch[2] in 1926, that electric and magnetic fields possessing axial symmetry act upon electron beams as lenses (see p. 4). The same two contributions led to the development of the scanning electron microscope and thus, in essence, they served to open up the science of electron-optical instrumentation.

Milestones in the development and perfection of the scanning electron microscope can be summarized as follows:

1935	Knoll[31] suggested the possibility of developing an SEM.
1938	Von Ardenne[175] built the first SEM.
1942	Zworykin et al.[176] developed an SEM capable of producing a 500 Å probe.
1948	Oatley et al.[177] began work that eventually led to the construction of the first commercial microscope in 1965.
1953	McMullen[178] solved the noise problem in the electron collection system.
1955	Smith and Oatley[179] added gamma controls, stigmator, water vapor cell, and hot stage to the SEM.
1960	Everhart and Thornley[180] developed the scintillator photomultiplier system of electron detection.
1963	Pease[181] demonstrated a beam diameter of 50 Å and resolution of ~ 100 Å.

To date, well over 1,000 instruments have been sold by more than 15 manufacturers in the United States, United Kingdom, France, Germany, Holland, and Japan. Most current models have a resolution capability below the 100 Å level and a magnification extending beyond 100,000 ×.

Chapter 16

BASIC THEORY

THE SEM CAN be viewed as somewhat akin to the reflection ("metalographic") light microscope, where the specimen is illuminated and viewed from the same side. The SEM makes use of a finely focused electron beam formed by the demagnification of the image of the source by several electron lenses. Such a beam will have a small diameter (*circa* 10 nm or less), affording it the quality of a microprobe. The electron beam bombards specific points of the specimen as it scans its surface in a predetermined regular pattern. The interaction of these primary electrons with the specimen surface provides backscattered and induced emission of electrons with different properties (see p. 234), depending upon the physical characteristics of the object. The secondary electrons generated carry a variety of physical, chemical, and electrical information, but cannot be focused in the optical sense. Another method must be employed in the SEM in order to obtain the necessary one-to-one correspondence between points on the specimen and points on the image.

Image formation in the most common version of the SEM occurs in the following manner: the secondary electrons emitted from a point on the bombarded specimen are collected and converted into a minute current that is amplified to produce a signal voltage (Fig. 107)*; this signal is passed on to a cathode-ray tube (CRT) where it determines the potential of the regulating or modulating electrode which controls the current in the cathode-ray tube writing beam. Thus, a point on the screen of the CRT is formed whose brightness is controlled by the current reaching the collector.

If current from a generator passes through pairs of symmetrical coils positioned on both sides of the electron microprobe and cathode-ray beams, both the beam in the microscope and the spot on the face of the CRT will be deflected. Such a double system is used in order to produce deflections at

*In order to prevent specimen charging resulting from secondary emission which would otherwise distort the resulting images, it is usually helpful to coat the specimen with a thin conductive layer. See p. 274.

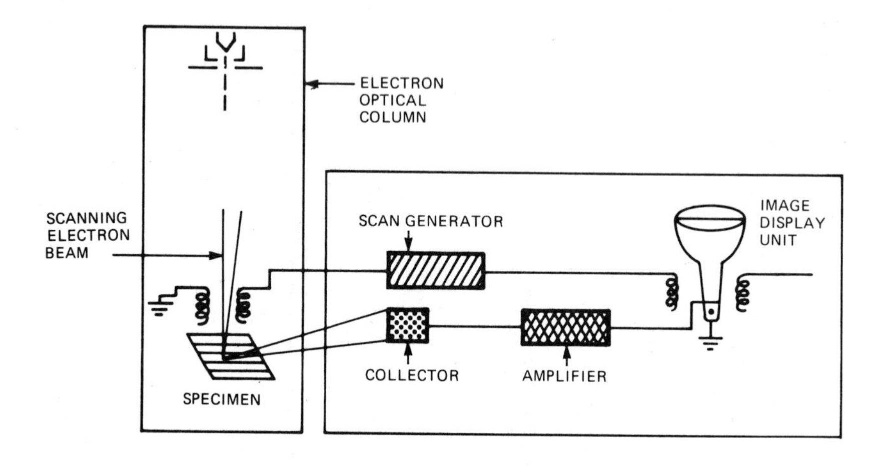

FIG. 107. Basic principles of scanning electron microscopy.

right angles to each other, so that the electron beam scanning the specimen and the CRT beam can describe a rectangular zig-zag array of lines or raster in synchrony. Thus, all points along a line in the CRT are in direct spacial relationship to the points on a line on the specimen. As the electron beam scans the surface of the specimen, differences in a particular property (such as texture, composition, or topography) at the point where the beam strikes the specimen, result in variations in the electron current that reaches the collector and, therefore, a change in the brightness of the CRT spot. Consequently, an image is formed on the screen of the CRT which may be viewed or photographed. The image represents a signal map of the scanned area with the signals displayed in their proper positions. The image of the specimen surface is built up point by point. The operating parameters are selected to optimize the quality of the resulting image and to assure complete coverage of the cathode-ray tube. The ratio between the large scanning width on the CRT surface and the width of the smaller raster on the specimen surface is equal to the image magnification.

Other kinds of signals (see p. 238) can be utilized in a manner analagous to that described above for secondary electrons. Thus, if x-ray signals are used to modulate the beam of the CRT, an x-ray image is obtained and the instrument can be used as an electron probe x-ray microanalyzer.

Chapter 17

DESIGN OF THE SEM

THE SEM IS made up of two basic systems with the specimen at their boundary (Fig. 108). The first system is the electron optical column which provides the beam of "illumination" that is directed at the specimen. The second system consists of the electron collection, signal amplification, and image display units which convert the electrons emitted from the specimen into a visible image of the specimen. The various components of an SEM will be considered individually.

A. ELECTRON OPTICAL COLUMN

The same elements of the column of a TEM—the electron gun and electron lenses—are present in an SEM.

1. Electron Gun

Three types of electron guns will be discussed. They vary in the intensity of beam brightness that they can produce.

a. TUNGSTEN FILAMENT CATHODE (Fig. 109A)

The electron source is most commonly the hairpin tungsten filament located in a triode electron gun (see p. 38). The electrons are emitted by the filament or cathode and accelerated by a field produced by the anode, usually at a positive potential on the order of 15 KV with respect to the cathode. A third electrode, the shield, lies between the anode and cathode and is negative with respect to the cathode.*

The electrons that leave at different points along the filament, cross over at a point beyond the anode. This crossover region can be regarded as a small electron source and its demagnified image, induced by the electron lenses, forms the *electron probe* that impinges on the specimen, which is almost identical to the case for the TEM (see p. 42).

*The shield, which is the control electrode in this type of electron gun, usually ha
concave surface facing the anode. If a convex surface is used, the crossover
is drastically reduced. Thus, at 25 KV a crossover diameter of 12 μm and b
A/cm^2 - ster was obtained (compare with the same characteristics in colum

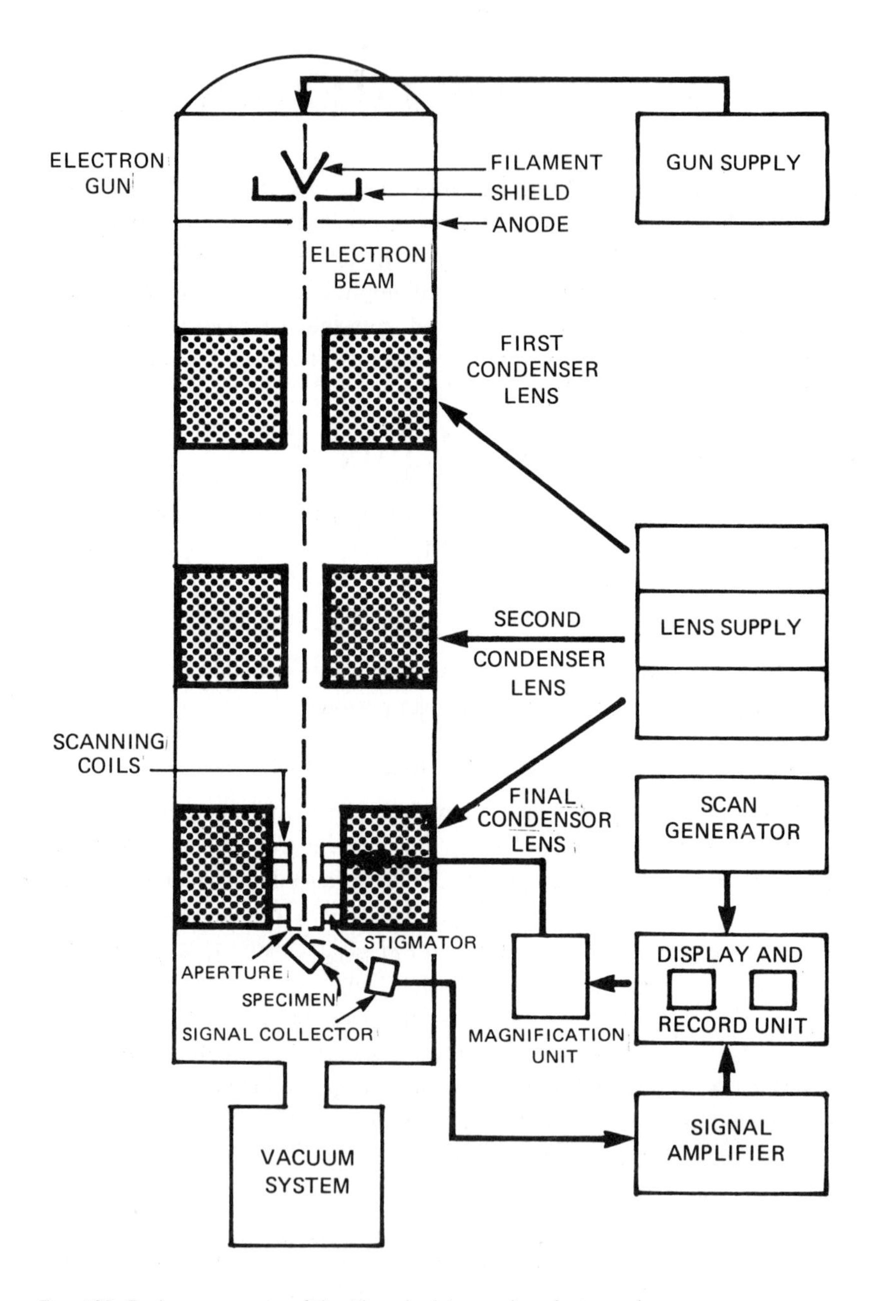

FIG. 108. Basic components of the (thermionic) scanning electron microscope.

The number of electrons emitted from a thermionic source, such as a tungsten filament, per unit of time increase with temperature. The *emission current density* or rate of electron emission, expressed in current per unit surface of the hot metal surface area (A/cm^2), was already noted (p. 18) to be:

$$I_o = AT^2_\epsilon -(We/kT) \quad (1)$$

where:

I_o = emission current density (amps per cm^2).
A = constant which is a function of the material.
T = absolute temperature (degrees C + 273).
ϵ = natural base of logarithms.
W = work function (volts).
e = charge of the electron (1.6×10^{-19} coulombs).
k = Boltzmann's constant (1.39×10^{-9} joules per degree absolute).

To determine the emission current density of a filament, one has only to substitute the appropriate values in equation (1), which is also known as *Richardson's equation*. The value of I_o for tungsten is equal to 1.75 A/cm^2 at a typical operating temperature of 2700°K and where A = 60.2 A/cm$^{2°}$K and W = 4.52 V. At this emission current, filament life should average around 40–80 hours in a vacuum of 10^{-5} torr.

The most important performance parameter of the electron gun is that of beam brightness. The brightness sets the upper limit on the current density in the focused spot. Langmuir[182] showed that maximum current density is determined by the equation*:

$$I_m = I_o \cdot \left(1 + \frac{eV}{kT}\right) \cdot \sin^2 \alpha \quad (2)$$

where:

I_m = maximum current density in focused electron spot.
I_o = emission current density at the cathode.
e = electronic charge.
V = accelerating voltage.
k = Boltzmann's constant.
T = absolute temperature of the cathode.
α = semiangle of the electron beam converging on the image to form a point at the focal spot.

*For the derivation of this equation, see Appendix M.

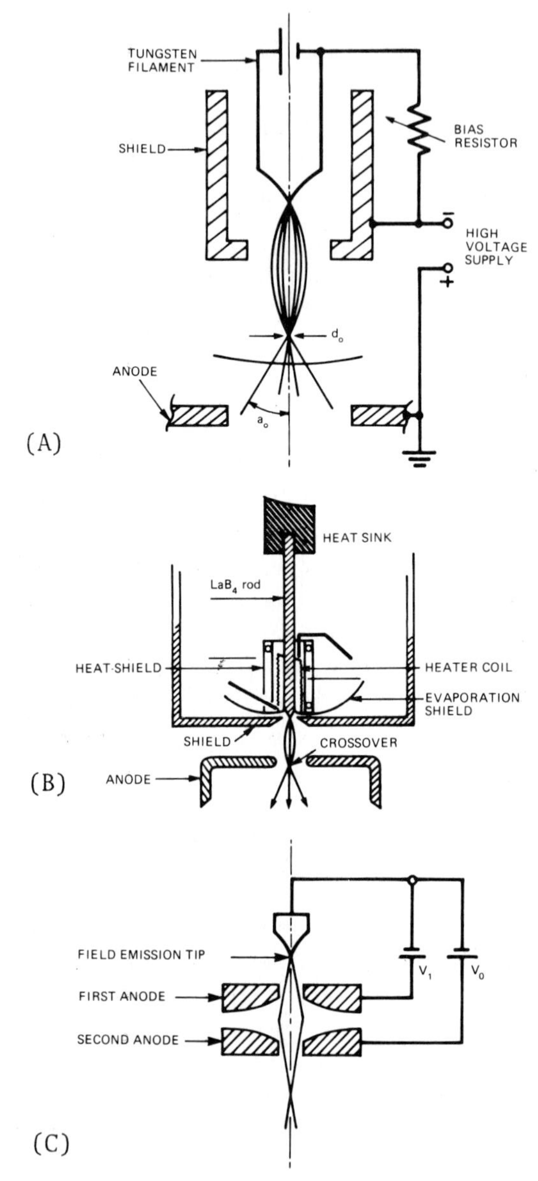

FIG. 109. Electron guns
(A) Tungsten filament cathode.
(B) Lanthanum hexaboride cathode.
(C) Field emission gun.

This equation indicates that I_m, the maximum current density that can be focused into an electron spot is limited by three factors: (1) the accelerating potential, V; (2) the cathode parameters I_0 and T; and (3) the solid angle subtended by the electron-optical system at the surface of the specimen, $\sin \alpha^2$.

The Langmuir equation can be considered as establishing that the maximum brightness β in a focused electron spot can be no greater than the brightness of the source. Brightness is the current density per solid angle, which is expressed in units of amps per square centimeter per steradian (A/cm²-ster). The solid angle in a cone of semiangle α is $2\pi(1\text{-}\cos \alpha)$. This is approximately, $\pi\alpha^2$ where α is small. Thus Langmuir's equation can be rewritten as:

$$\beta \simeq \frac{I_m}{\pi\, \alpha^2} \tag{3}$$

Substituting for I_m from equation (2) we get:

$$\beta = \frac{I_o}{\pi} \cdot \left(1 + \frac{eV}{kT}\right) \tag{4}$$

since for small α, $\sin \alpha = \alpha$. Since $\frac{eV}{kT}$ is $>>1$, equation (4) can be rewritten as:

$$\beta \simeq I_o \frac{eV}{\pi kT} \tag{5}$$

The typical value for a tungsten filament used in the SEM for the factors in equation (5) are: $I_o = 1.75$ A/cm², $V = 25$, $T = 2700$°K which gives a brightness of about 60,000 A/cm²-ster. Because voltages on the gun are held within relatively narrow limits, significant improvement in brightness for tungsten filaments will be made by increasing filament temperature in order to raise the emission current density. Thus if the temperature is increased 300°K, from 2700°K to 3000°K, I_o and β is increased more than seven fold (I_o from 1.75 A/cm² to 14.2 A/cm² and β from 60,000 A/cm²-ster to 440,000 A/cm²-ster.) Under these conditions, however, filament life is very drastically reduced from more than *circa* 60 hours to about one hour. Thus, in practice, elevating filament temperature is an unacceptable solution for increasing brightness. This situation has resulted in a search for cathodes that are more optimally suited to provide higher brightness.

b. LANTHANUM HEXABORIDE CATHODE (Fig. 109B)

In the SEM it is important to achieve the highest possible current density in the focused spot. Therefore, incorporation of an electron gun capable of producing a high brightness source is very desirable.

During the past decade there has been a renewal of interest in developing emitters superior to the tungsten hairpin filament. The possibility of using high-melting-point rare-earth borides, particularly lanthanium hexaboride (LaB_6), as a cathode emitter was noted by Lafferty[183]. He emphasized that since the work function of LaB_6 is 2.7 (as compared with tungsten, 4.5), it had higher ratios of electron emission density to evaporation rates than tungsten. These observations were confirmed by Broers[184] who developed a functional high-brightness gun using a LaB_6 rod cathode.

The LaB_6 cathode is a solid rod* (typically 1.6 cm long by 1 mm square) whose emitting tip has been milled to a very fine point ($\sim$ 20 μ diameter). The other end is brazed into an oil-cooled heat sink, thereby holding its temperature down. The cathode is heated to a high temperature by passing a current through a tungsten wire coil surrounding the lower third of the rod. A double walled shield which improves heating efficiency surrounds the coil.

LaB_6 has been shown to produce a very high, I_o, emission current density of 65 A/cm^2 at 1600°C as calculated from Richardson's equation (1); where A = 40 A/cm^2 °K^2 and $W = 2.4$ V. Such an I_m would, according to the Langmuir equation, produce a brightness of about 10^7 A/cm^2-ster which has been experimentally confirmed. While maximum brightness often drops off after 50 hours, satisfactory operation can still be maintained for several hundred hours.

The LaB_6 cathode, thus, has not only increased gun brightness but has also greatly extended cathode life. As a result, a number of manufacturers have made a LaB_6 gun available for use with their instruments. These can be utilized to full advantage where a vacuum of 5×10^{-7} torr can be maintained at the gun. Poorer vacuum causes rapid deterioration of the LaB_6 point.

c. FIELD EMISSION GUN (Fig. 109C)

The source capable of providing the highest brightness is the field emission gun. In its simplest form, this gun is a two electrode or diode gun. If a voltage (*circa* 1–5 KV) is applied to a very sharp metal point ($\sim$ 1000 Å radius), a high negative field is produced at the emitter surface and produces electron emission by "tunnelling" through the potential barrier (see p. 46). Electrolytically etched tungsten rods are usually used in the field emission gun and the emission current density produced in this way is extremely high

*Single crystal LaB_6 sources have recently been introduced and may be significantly better than the polycrystaline rods because they are more readily alignable and have qualities more similar to point sources.

(up to 10^6 A/cm^2) compared with that produced by thermionic emission. The brightness of such a gun has been estimated at 23 KV to be about 2×10^9 A/cm^2-ster (about two orders of magnitude higher than the LaB_6 gun). Because the virtual source is some 30 times smaller than the emitter point, field emission guns of this type can produce a focused spot of electrons smaller than 100 Å.

Two major factors have impeded commercial adoption of the field emission gun. The first is that it will operate reliably only under ultrahigh vacuum conditions (10^{-10} torr or better), which results in problems of microscope design and operation (see p. 245). To cope with this problem, some designs incorporate a special high vacuum pump for the gun chamber (e.g., a titanium sublimation pump, see p. 341). The second factor is the relative instability of the cathode tips. This is created by backscattering of ions from the anode, which causes etching of areas of the tip and can result in severe "migration" of the emission area, or by contamination of the tip, which is usually operated at ambient temperature. This can be corrected by periodic "cleaning" of the tip, usually by brief heating. Tips operated constantly at elevated temperature can produce beam currents of high stability, and currents several orders of magnitude higher than those operated at room temperature. However, there is some increase in spot size under these conditions.

The many advantages of the field emission gun have encouraged commercial manufacturers to pursue its development. Among these advantages is the fact that field emission tips, for an instrument in daily use, can have a life of six months or more.

d. DISCUSSION

Examination of table IX, which compares the three different types of electron sources, demonstrates that the field emission gun can provide a source about 1,000 times brighter than a thermionic gun. This permits the use of probes many times smaller at equal signal currents and therefore greatly increases attainable resolution. In addition, the reduced "energy spread" of the field emission gun results in less chromatic aberration, thus aiding further in achieving minimum spot size.

There is usually only one condenser lens in a SEM with a field emission gun because the very small virtual image of its source requires little demagnification (see p. 227) for other than maximum resolution. When maximum resolution is not required, the condenser lens may be de-energized, increasing the probe size to about that with a thermionic source, but with still much higher brightness and much smaller aperture angle.

TABLE IX

COMPARISON OF ELECTRON GUN CHARACTERISTICS.

Characteristic	Electric gun		
	Tungsten	Lanthanum Hexaboride	Field Emission
Diameter of tip	~ 20 μm	~ 1 μm	~ 10nm (≃ 1nm virtual)
Beam brightness (A/cm²-ster) for 10 nm diameter probe; $\alpha = 10^{-2}$ rad.	5×10^4	1×10^6	10^8
Operating temp. (°K)	2859	1850	ambient
Probe current (amp) for 10 nm diameter probe	5×10^{-12}	8×10^{-11}	$10^{-9} - 10^{-8}$
Required vacuum (torr)	$\sim 10^{-5}$	$\sim 10^{-6}$	$10^{-8} \sim 10^{-10}$
Estimated life (hrs. at 20 KV)	35	250	indefinite
Minimum probe diameter (nm)	~ 9	~ 5	1–2

This provides two additional operating advantages:

(1) The specimen can be scanned at standard television rates (30 frames/sec) rather than at slow-scan rates. As a result, flicker-free direct observation of the image is possible. This also permits the facile observation of transient phenomena. Because the time-constant for the bleeding off of induced specimen charge is considerably longer than 1/30 sec., induced charge-related image movement and specimen damage are greatly reduced.

(2) The very low aperture of the field emission source (with the condenser de-energized) increases the depth of focus about 10 X (to about 100 μm at 3,000 X magnification).

Field emission sources, thus, have significant potential and will probably become increasingly available as the source for SEM's.

2. Electron Lenses

In the SEM, all the lenses serve to demagnify the electron image formed at the crossover to produce the final probe spot that impinges on the -imen. The extent of demagnification can vary from zero to the order of -mes (for a tungsten filament the crossover diameter 20–50 μm; final -eter 50 A – 1 μm; under some conditions with a field emission -nification is needed). The several magnetic lenses used are of the conventional electron microscope type, with the focusing of the beam taking place as a result of the effect of the magnetic field upon the stream of moving electrons.

a. IMAGE DEMAGNIFICATION

Demagnification of the beam is essential because a heated tungsten filament may have a diameter at its source of 20 to 50 μm. The reduction in electron image diameter is clearly evident from a schematic of ray traces as they pass through the first and final condenser lenses in an SEM (Fig. 110). The demagnification induced by the condenser lens is given by the equation:

$$D_m = \frac{s_c}{s'_c} \quad (6)$$

where:

D_m = demagnification.
s_c = the (fixed) distance from the crossover point to the condenser lens gap.
s'_c = the (variable) distance from the condenser lens gap to the point of focus on the other side of this lens.

The diameter of the intermediate image (electron image after passing through the condenser lens) is given by the equation:

$$D_c = \frac{d_s}{D_m} \quad (7)$$

where:

D_c = diameter of the focused electron beam just beyond the condenser.
d_s = diameter of the crossover spot.
D_m = demagnification.

We can apply the thin lens equation from geometrical optics (see p. 305) to lens systems in the SEM. This equation states:

$$\frac{1}{f} = \frac{1}{s_c} + \frac{1}{s'_c} \quad (8)$$

where:

f = focal length of the lens

Earlier in the discussion of the condenser lens (p. 46), it was noted that focusing takes place by varying the focal length. This is accomplished by changing the intensity of the lens coil current which in turn alters the intensity of the magnetic field that is generated by the lens. As the lens current increases, the lens strength increases and the focal length decreases. As a result, according to equation (8) S_c will decrease and in turn according to equation (6) D_m will increase.

Since decreasing the first condenser focal length increases the divergence angle of the beam from the intermediate image, α_o, an aperture (100–300 μm diameter) is positioned below this image (in the final condenser lens). This aperture decreases the α_o entering or passing through the final condenser lens. The demagnification provided by the final condenser lens is given by an equation like (6) above (with the parameters being S_o and S'_o). The probe spot size which is the result of the total demagnification by both lenses of the initial crossover image is expressed by:

$$d_o = \frac{d_s}{D_{m_c} = D_{m_{c'}}} \tag{9}$$

where:

d_o = probe spot diameter.
d_s = crossover image diameter.
D_{m_c} = demagnification of the first condenser lens.
$D_{m_{c'}}$ = demagnification of the final condenser lens.

b. LENS ABERRATIONS

It was shown earlier (eq. 3) that the Langmuir equation can be rewritten as:

$$I_m \simeq \pi \cdot \beta \cdot \alpha^2 \tag{10}$$

This suggests that if there were no aberrations in the electron column, one could increase the current density at a constant probe diameter by simply increasing α. However, since aberrations are present in the system, α must be kept small and the current available for a given probe diameter is restricted.

It should be noted at the outset that even where spherical and chromatic aberrations are insignificant, resolution is limited by the phenomenon of *diffraction* (p. 8). As already shown (p. 15) in the case of the electron microscope. Abbe's equation, which defines the limit of resolution due to diffraction (p. 11), can be written as:

$$d = \frac{7.5}{a\sqrt{V}} \text{Å} \tag{11}$$

where:

d = diameter of Airy disk (angstroms).
α = angle between the converging ray and electron optical axis (radians).
V = accelerating potential (electron volts).

It is apparent that to minimize the effect of diffraction, thus keeping a minimum value for d, the value of α should be as high as is practical in view of the requirements imposed by the effects of spherical and chromatic aberration.

Three types of aberrations can be present (p. 53).

1. Spherical Aberration. *Spherical aberration* results from the fact that electrons moving along trajectories further away from the optical axis are focused more strongly than those near the axis. The limit of resolution in the presence of this aberration was shown (pp. 54 and 311) to be:

$$d_s = k_s \cdot f \cdot \alpha_{c'} \qquad (12)$$

where:

d_s = separation of two object points which are just resolved.
k_s = dimensionless proportionality constant characteristic of the final condenser lens.
f = focal length.
$\alpha_{c'}$ = final condenser aperture angle.

This equation indicates that aperture angle and focal length are the critical factors which determine the magnitude of spherical aberration. To reduce the effect of d_s on the electron probe diameter, the $\alpha_{c'}$ should be decreased. To do this the size of the objective aperture must be decreased, which in turn will reduce the current density in the probe spot. The other alternative is to place the specimen closer to the final condenser lens pole piece. This will necessitate an increase in lens excitation, thus decreasing f (and also k_s).

2. Chromatic Aberration. *Chromatic aberration* is the result of variations in energy and thus in velocity of the electrons passing through a lens or a variation in the magnetic field of the lens. Either will alter the point at which electrons emanating from a point are focused. Variations in energy and magnetic field may be due to imperfect stabilization of the various power supplies of the microscope, but this potential problem can be corrected so as to be negligible. Another factor inducing this aberration, namely variations in the energy of the electrons due to the spread of the initial velocities of the

emitted electrons may be higher than originally thought (p. 55). They have recently been shown to be on the order of 3 V for tungsten and LaB_6, and 0.5 V for field emission cathodes.

The equation for the limit of resolution in the presence of this abberration was shown to be (pp. 55 and 313):

$$d_c = k_c \cdot f \cdot \alpha_{c'} \cdot \frac{\Delta E}{E} \tag{13}$$

where:

d_c = separation of two object points which are just resolved.
k_c = chromatic aberration coefficient.
$\alpha_{c'}$ = final condenser aperture angle.
$\frac{\Delta E}{E}$ = fractional variation in electron beam energy.

As in the case of spherical aberration, the value of d_c can be minimized by decreasing the value of $\alpha_{c'}$.

3. Astigmatism. *Astigmatism* is due to the asymmetry of the magnetic lenses (p. 89). If uncorrected, it will act to enlarge the effective size of the final electron probe diameter. As with the TEM, a stigmator (p. 91) which produces a weak correcting field can be used to minimize this aberration.

The astigmatism of the SEM lens system varies with the state of cleanliness of the lens apertures and with the presence of any electrostatic charge on the specimen resulting from beam-specimen interaction. Thus, every time the position of the specimen is altered, astigmatism must be checked (and if necessary corrected with operator controls that alter the current to the stigmator) in order to ensure high resolution. Moreover, astigmatism should be checked before *each* micrograph is taken and it may be desirable to do so at a level of magnification somewhat higher than is to be recorded. If astigmatism cannot be compensated for by the controls, the splash and final apertures and the final aperture holders (and if necessary the entire column) should be cleaned.

The importance of lens aberrations insofar as SEM is concerned is summarized in Table X.

In general, all aberrations can occur concurrently and it is necessary to add together the quantitative estimates of the individual errors. The effective spot size d_{eff} is then regarded as the square root of the sum of the squares of the separate diameters of the discs of least confusion, namely:

$$d^{o}_{eff} = \left\{ d_s{}^2 + d_c{}^2 + d_a{}^2 + d_d{}^2 \right\}^{1/2} \tag{14}$$

TABLE X

SIGNIFICANCE OF LENS ABERRATIONS IN THE SEM

Aberration	Significance
Diffraction	Sets basic limit to resolution
Spherical aberration	Can be kept low using optimal final condenser aperture
Chromatic aberration	Can be kept low by using stable power supplies
Astigmatism	Can be compensated with a sigmator
Distortion	Correctable, as in the TEM
Coma	Correctable, as in the TEM

where:

d^0_{eff} = effective spot diameter.
d_s^2 = square of the diameter of the disc of least confusion due to spherical aberration.
d_c^2 = square of the disc of least confusion due to chromatic aberration.
d_a^2 = square of the disc of least confusion due to astigmatism.
d_d^2 = square of the Airy disk diameter.

In the best of cases, d^0_{eff} approaches 50 Å with beam currents of 10^{-12} amps at 20 KV with low values for the aberration constants. The range of α typically is 0.0024 to 0.010.

With a tungsten filament cathode, the diameter at crossover is about 100 μm. Since for high magnification work the diameter of the incident electron probe should be about 0.01 μm or less, a total demagnification of at least 10,000 times is needed (see Fig. 110). To attain this goal, while utilizing a final condenser lens whose focal length is no less than 1 cm and avoiding an excessively long column, a three lens system is required. However, when a pointed tungsten or LaB_6 filament is used, since it provides a smaller crossover and thus less demagnification is needed, a two lens system may be adequate. When a field-emission cathode is used, little or no demagnification of the virtual source is necessary since it is so small (0.001 μm). Thus, an electron gun can be designed that serves both to accelerate and to focus the electron beam on the specimen, thereby making it unnecessary to use special lenses, although in practice most field emission microscopes do incorporate one additional lens.

c. FINAL CONDENSER LENS

Most SEMs have three magnetic lenses in their column: the first, second, and final condenser lenses.

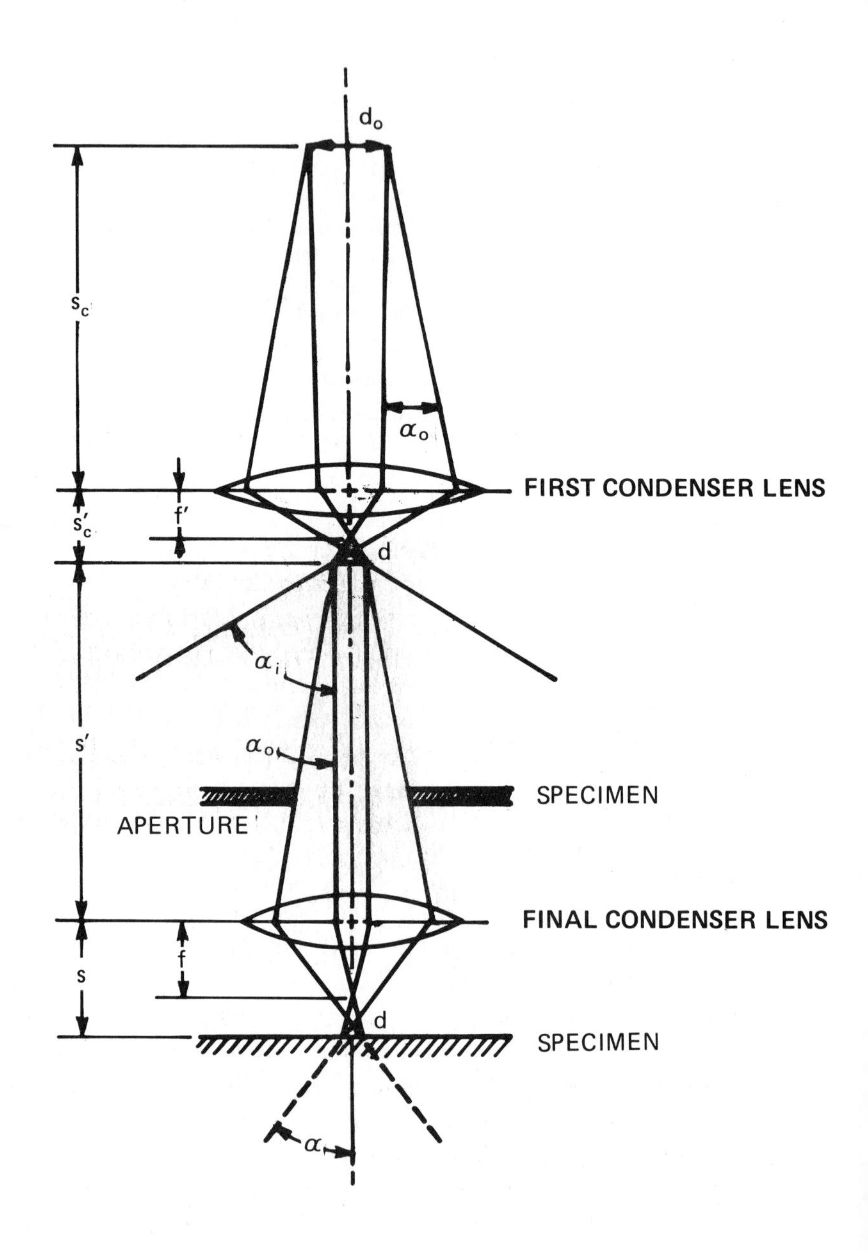

FIG. 110. Image demagnification by the first condenser and final condenser lenses.

In designing the final condenser lens, several performance characteristics must be considered:

(1) Aberrations. Since the intermediate images of the crossover produced by the condenser lenses have significantly larger diameters than the final spot size, the effects of aberrations on these lenses are relatively small. It is thus the effects of spherical and chromatic aberrations as well as astigmatism of the final condenser lens that is as critical in the design and performance of the objective lens of a TEM.

(2) Magnetic field. As a result of electron bombardment, secondary electrons in the SEM are emitted over a wide solid angle. These have energies of only a few electron volts, yet they must be able to reach the detector to produce the necessary signal. As a result the magnetic field at the specimen must be designed so that it will not restrict effective secondary electron collection.

(3) Bore. Enough room has to be allowed in the bore of the lens to place in it a stigmator, an aperture, and scanning coils.

(4) Focal length. Since the extent of lens aberrations is dependent upon the focal length, it is most desirable to keep the latter as short as possible in order to help minimize the effects of aberrations.

To meet the aforementioned performance characteristics, it is desirable to place the specimen just outside the lower bore opening of the lens (Fig. 111). A "pinhole" variety lens is usually used in order to keep the magnetic field low, and to minimize the focal length. The aperture is usually placed near the inner face of the outer pole piece. In this type of lens, the diameter of the outer pole piece is much smaller than the inner pole piece.

B. IMAGE TRANSLATING SYSTEM

1. Scanning System

A focused electron beam of the smallest possible diameter is scanned across the specimen surface. It is essential to avoid unnecessary components between the bottom of the final condenser lens and the specimen in order to keep the working distance short (~1 cm), and thus avoid the need for a relatively long focal length lens which would have a higher coefficient of spherical aberration. As a result, the scanning coils are usually incorporated into the final condenser lens housing (Fig. 111).

It has been recommended that two sets of magnetic coils be used for beam deflection. By this means, the beam is deflected twice away and back to the axis, through angles such that the ray which was originally coincident with the axis of the beam passes through the center of the final condenser

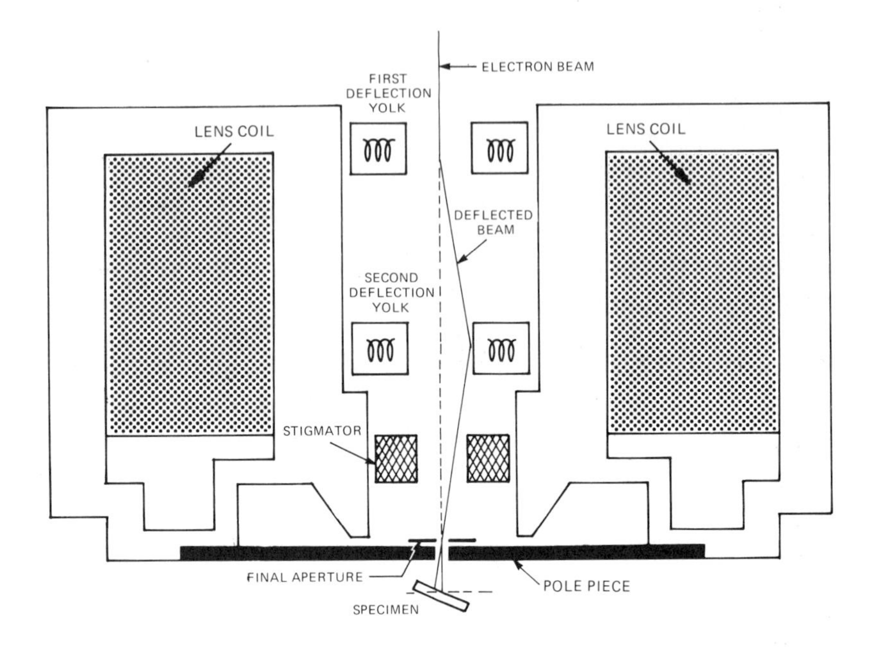

FIG. 111. The SEM final condenser lens and its components.

aperture (Fig. 111). As a result aberrations are kept to a minimum and maximal deflection is obtained. If two sets of coils are used, they should be spaced apart so that the direct interaction between their magnetic fields is small. Both air-cored and iron-cored coils have been used successfully in SEM's.

A single scan generator is used for the deflection coils of both the final condenser lens and cathode ray tube. It provides the needed a.c. voltages to energize the coils and cause the beam to be deflected in a raster-like pattern over the specimen surface (Fig. 112). Since the half-angle at which the beam converges is generally less than half a degree, the beam diameter does not increase significantly on either side of the plane of best focus. As a result, the imaging system has a great depth of focus so that all parts of an object appear to be essentially equally in focus.

2. Signal Processing

a. BEAM–SPECIMEN INTERACTION

As a prerequisite to a discussion of the mechanism involved in signal processing in the SEM, one must understand the nature of the interaction

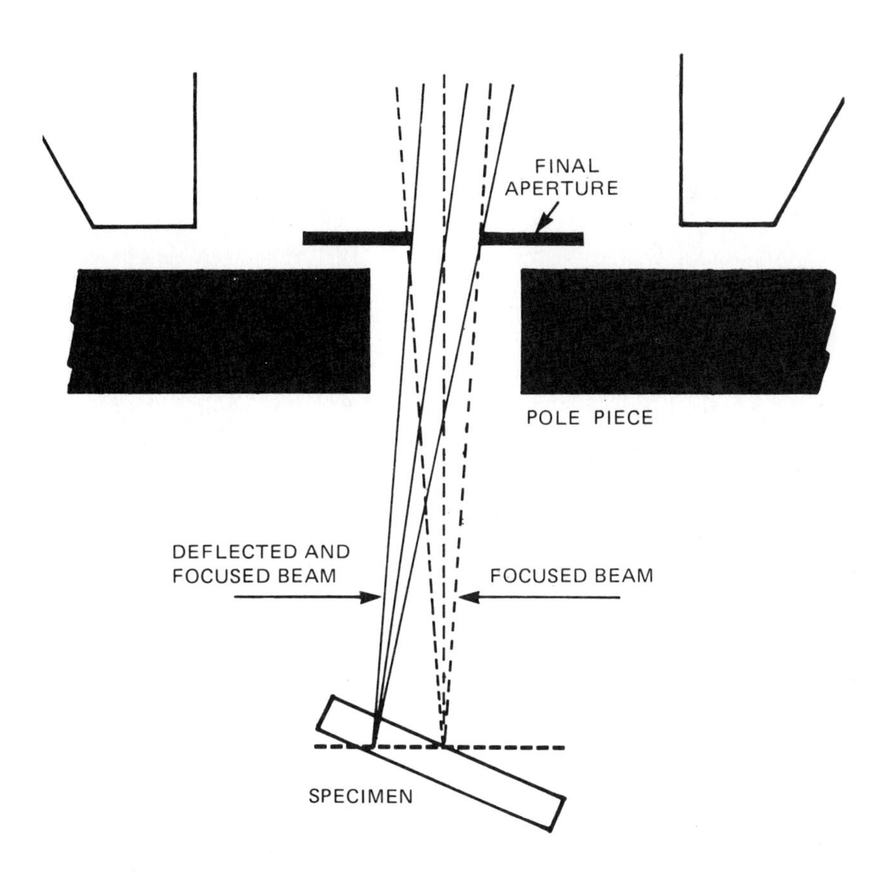

FIG. 112. Beam focusing and deflection on the specimen. (Enlargement of specimen area in Fig. 111)

between the electron probe and the specimen. The electron beams in question have been accelerated through a voltage of 1 to 50 KV.

Three basic possibilities exist as to the nature of the beam specimen interaction used to generate the image (Fig. 113A).

(1) Some primary electrons, depending upon accelerating voltage, penetrate the solid to depths as much as 10 μ. Their trajectories vary and they lose most of their energy as they penetrate the specimen. (If the specimen is sufficiently thin, some electrons pass through it with reduced energy and changed direction, while others may pass through with no interaction.)

(2) Some primary electrons interact with the uppermost atoms of the specimen so that there is a change in momentum (but no interchange of energy) such that the electron is scattered through a large angle and is

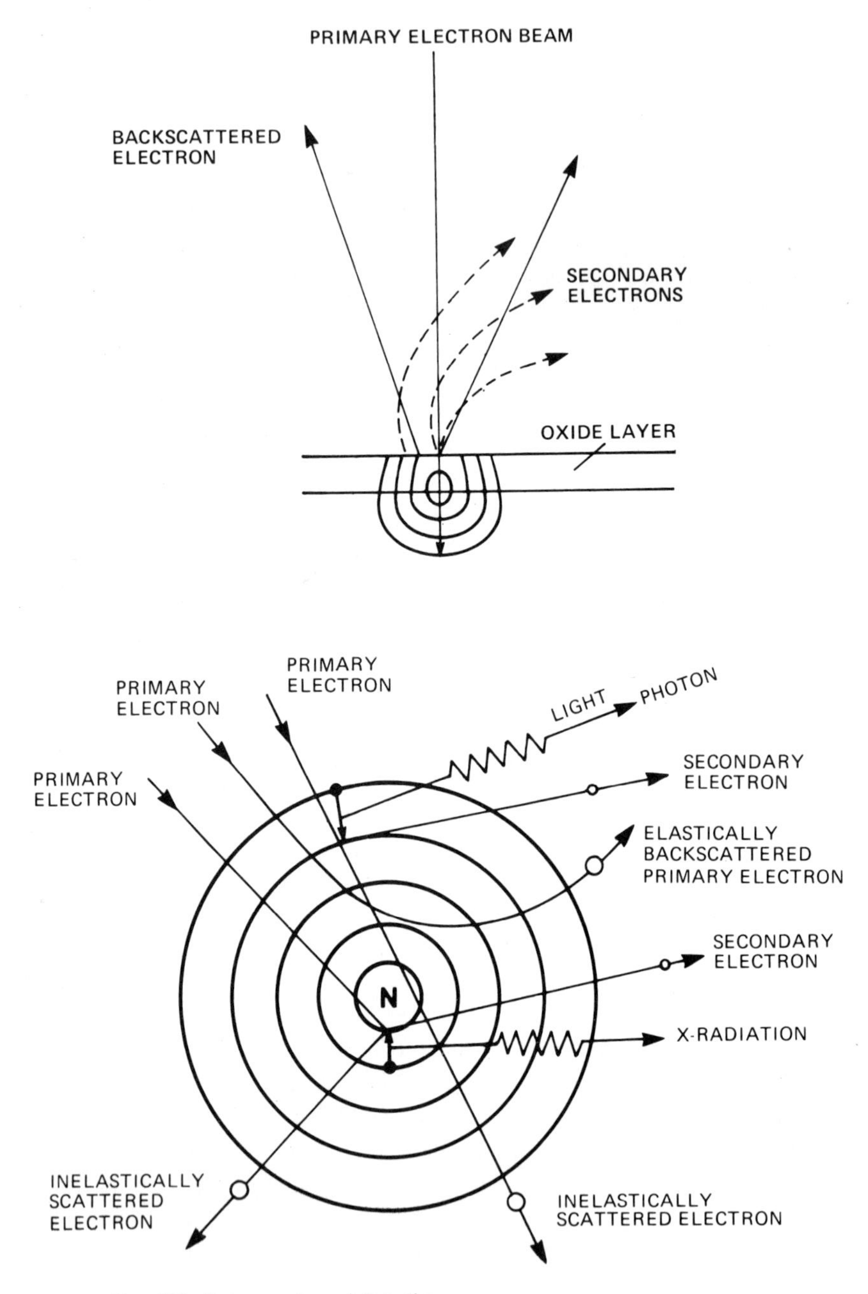

FIG. 113. Beam-specimen interaction
(A) The two major types of electrons emitted from the specimen surface.
(B) The variety of signals generated.

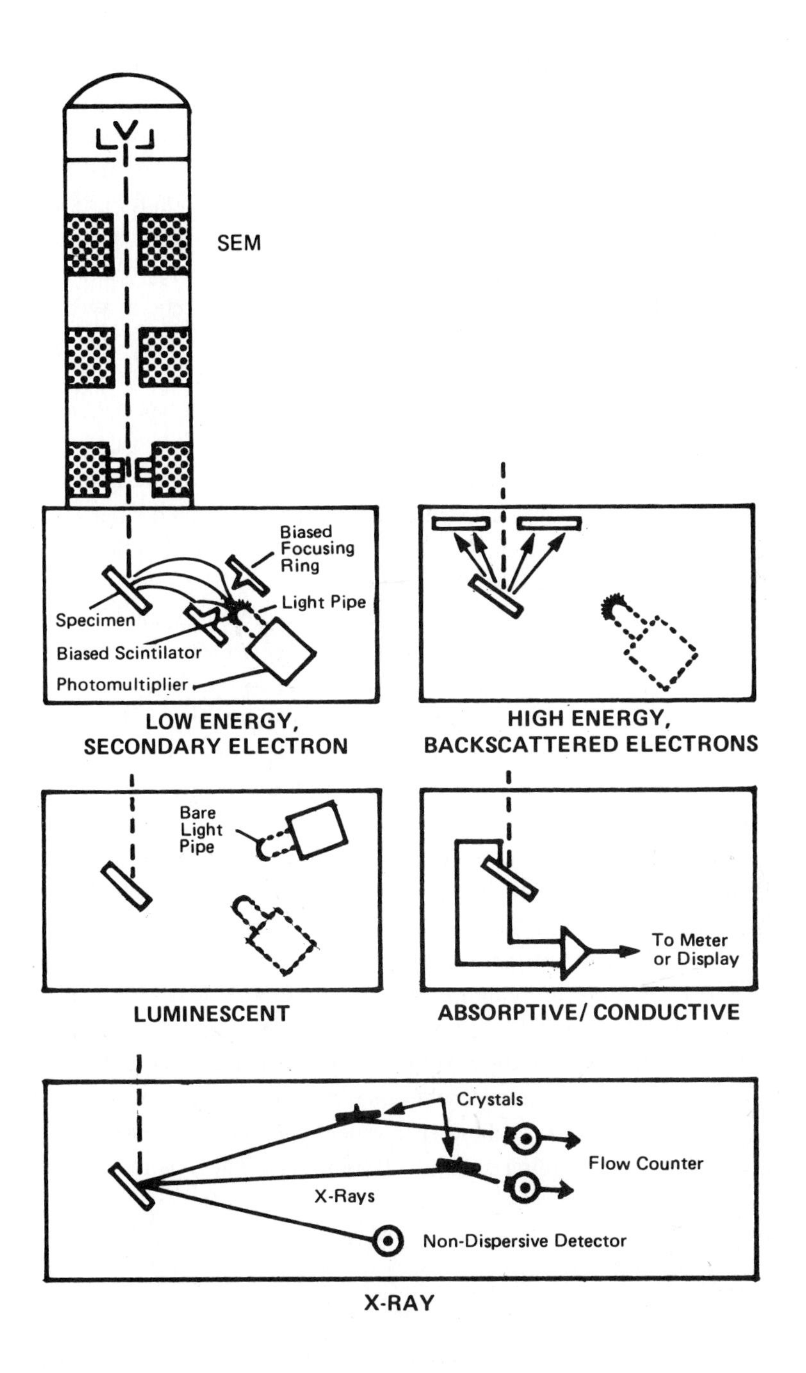

(C) Signal collection.

effectively reflected from the specimen. Such elastically reflected primary electrons are known as *backscattered electrons*.

(3) Some primary electrons interact with the host atoms so that as a result of collisions, a cascade of *secondary electrons* is formed along the penetration path. Some of these electrons located in the outermost stratum (5 to 50 Å) will diffuse towards the surface losing energy. But, if they still retain enough kinetic energy in excess of the surface barrier energy (2 to 6 V), they will often escape from the surface. This process is known as secondary electron emission.

The distribution of the signal intensity yielded in a specimen is a function of the incident electrons' energy *(E_0)* and the specimens atomic number *(Z)* and density *(P)*. Experimental studies[185] have shown that the larger the value of E_0 and the smaller the value of Z and P, the deeper electrons will penetrate the specimen. Conversely, the smaller the value of E_0 and the larger the value of Z and P, the more electrons will diffuse in a horizontal rather than a vertical direction.

Specimen-beam interaction, in addition to producing backscattered and secondary electrons, also produces photons, specimen currents, Auger electrons, and x-rays,* which are characteristic of the probed specimen (Fig. 113B). While in principle any signal generated can be utilized to produce an image, in practice it is the secondary electrons released from the sample that are most commonly used for this purpose in biomedical investigations with the SEM.

Using surface emitted electrons, rather than those passing though the specimen as in TEM, surface images revealing some three-dimensional quality can be obtained.

The impression of three-dimensional surface relief seen on the screen of the display tube or scanning electron micrograph is the result of the distribution of light and dark areas. This distribution is accounted for largely by the fact that the incident beam generates more collectable secondary electrons per unit area when it strikes a sharply curved edge or sloping surface than when it hits a flat surface. Biological specimens, being heavily contoured, facilitate inducing such a differential effect. Where the surface is

*When a high-energy incident electron beam hits an atom, excitation takes place and an electron from an inner orbit may be ejected. As a result, a "hole" is created which is filled by electron transition from an outer orbit. Electron "relaxation" then occurs and a photon is generated. This photon can be internally converted and an *Auger electron* emitted or a *characteristic x-ray* photon can be emitted. When some materials are subjected to electron bombardment, a part of the absorbed energy is emitted as light in the course of its relaxation to a state of equilibrium. This phenomenon is known as *cathodoluminescence*.

smooth, tilting it at an angle to the probe will enhance the desired variation in collected secondary electrons. More subtle effects involving the manner in which structures lying above a primary surface either deflect or absorb the probe electrons also come into play in the process. In any case, the "shadows" seen are a true representation of the three-dimensional character of the specimen surface under study.

b. ELECTRON COLLECTION

The released secondary electrons must be collected but need not be focused since those collected at any given instant come only from the area of the specimen upon which the probe is focused. To collect the secondary electrons, a suitable electrode is held at a positive potential and serves to attract them and produce an emission current. The strength of this signal is proportional to the number of electrons impinging on the collector. This signal is used after amplification to modulate the intensity of the cathode-ray tube beam as it moves across the tube face, synchronously with the path of the electron probe across the specimen surface. Typically, the secondary electron collector comprises a cylindrical metal box with a metal gauze window. Inside is an aluminum-coated plastic scintillator that is attached by a short connecting tube to a lucite rod (light pipe) whose other end rests against a photomultiplier window (Fig. 114). The secondary electrons are accelerated toward the scintillator by a potential difference of a few hundred to a few thousand volts. Upon hitting the scintillator, each electron produces many photons that are guided by the light pipe to the photomultiplier.

The backscattered electrons that have undergone relatively little energy loss, are not appreciably deflected by the small potential difference on the collector; thus, most of them do not enter the collector, and a minimum noise is introduced into the system. There are alternative specific detectors that can be used for collecting backscattered electrons or other signals that are generated by specimen-beam interaction. Most SEM systems are designed to permit switching from one signal to another.

c. SIGNAL AMPLIFICATION

The photons produced by the electrons hitting the scintillator serve to excite photoelectrons when they reach the window of the photomultiplier. Each photoelectron triggers a release of two or more secondary electrons at the first electrode (dynode) and the process cascades, yielding from 100,000 to 50 million additional electrons (Fig. 115). Thus, the photomultiplier reconverts the light to an electron current and provides a high degree of

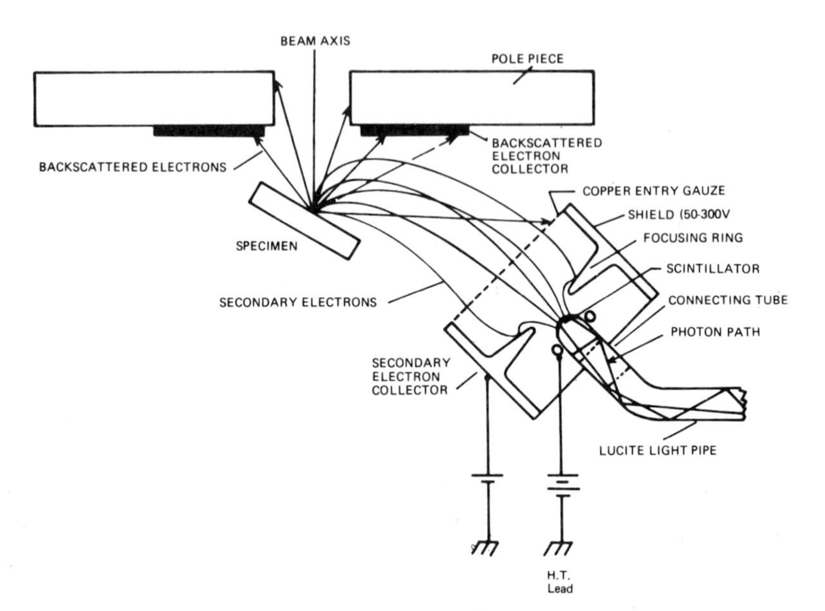

FIG. 114. Electron collection.

amplification which can be controlled by variation of the potential applied to the dynodes.

d. SIGNAL DISPLAY

The final magnified image in the SEM is formed on a cathode-ray tube. The usual commercial instrument has two separate display channels: one for visual observation, and one for photographic recording (see front-piece). The visual tube is scanned at rates varying from many times per second (TV rate) to once per several seconds, depending upon the signal strength. Using a screen coated with a phosphor having a relatively long afterglow greatly reduces flicker at the slower scan rates. It is the impact of electrons on the phosphor that excites phosphorescence (see p. 69). While the effective spot size is larger than the electron beam diameter, resolution is satisfactory for visual observation (e.g., 500 lines per 10 cm square screen).

For photographic recording a high brightness phosphor with a very short afterglow is chosen and a much higher resolution (up to 4200 lines per 10 cm square screen) is achieved. Relatively long recording times (8 to 100 sec or more) are required at high magnifications in order to permit enough

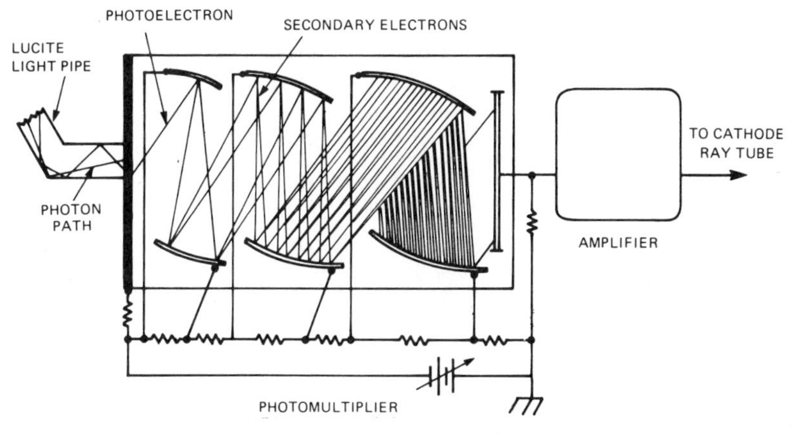

FIG. 115. Signal amplification.

electrons to probe each component of the specimen. A single long-interval scan (rather than multiple scans) is ordinarily used since the photograph will be sharp (though distorted), even if some specimen drift occurs. Stable accelerating and lens voltages are a prerequisite for good quality photography as in TEM.

The CRT used as the display unit for the SEM image reflects the signal intensity generated as differential areas of brightness on the tube surface. As noted earlier, the signal is commonly generated by secondary electrons emitted from the specimen; these are captured by a collector after deflection along curved trajectories more or less irrespective of the initial direction of their emission (see Fig. 114). The specimen appears as if it were broadly illuminated and the secondary electron image has no fuzzy shadows.

e. SIGNAL SELECTION

The scanning method was first employed as one of the functions of the electron probe x-ray microanalyzer. The scanning method was subsequently developed as a technique for the SEM. Since the fundamental principle is the same, information obtained from both is complementary and their functions can be combined in one unit, the *electron probe instrument*. This can serve as a measuring instrument by obtaining quantitative data about the various signals produced in the course of beam-specimen interaction, namely electrons (secondary, backscattered, transmitted, absorbed, and Auger), and electromagnetic waves (x-rays and cathodoluminescence). The nature of the information sought will essentially determine the signals selected, as can be seen in Table XI.

TABLE XI

SIGNAL SELECTION

Information Sought	Signal that can be used
Morphology	All signals (except x-rays and Auger electrons
Element Analysis	Backscattered and Auger electrons; x-rays; cathodoluminescence
Chemical bond	Auger electrons; x-rays
Crystallography	Backscattered, secondary and transmitted electrons; x-rays
Electromagnetic	Secondary electron

C. OTHER INSTRUMENT COMPONENTS

1. Apertures

The final condenser aperture serves to reduce spherical aberration in the lens to an acceptable value and also defines the all important angle subtended by the beam at the specimen. It is useful to be able to vary this angle by altering the aperture size used. Thus, larger apertures are used in initially lining up the column, while smaller ones are used in routine microscopy. To facilitate changing apertures, a holder carrying several apertures is usually incorporated into the microscope. The range of semiangles desirable to meet different operating beam current density needs makes it useful to have accessible apertures of several sizes in the range between 50 and 500 μ in diameter. These apertures are placed in series on a micrometer-type holder so that they can be positioned with accuracy.

The final aperture, requires the most care in terms of position and cleanliness. The other apertures in the lens are present to prevent stray electrons from hitting the bare walls and contaminating them (see Fig. 112).

They are less critical but nevertheless require cleaning or replacement when contaminated. In the most recent designs, such "splash" apertures are placed in a removable assembly. Such apertures serve to extend substantially the total operating hours between "down-time" intervals when the column is dismantled in order to remove contamination from the internal surface.

Scanning electron microscopes equipped with a field emission gun generally do not require apertures.

2. Specimen Stage

There are three basic requirements that a stage should meet to be suitable for general operational needs:

(1) *Movability*. The specimen movements should be smooth, finely controlable and readily repeatable. Ideally, translation in three mutually perpendicular directions, as well as rotation and tilt, should be possible.

(2) *Stability*. Movement of the stage should not introduce vibrations or drift which could lead to degradation of the resolution capability of the instrument; nor should it create beam distortion, or astigmatism, in the region between the objective and specimen.

(3) *Flexibility*. The design should be such as will permit the secondary electron collector (or other detectors) to be moved about or replaced.

Specialized stages have been developed that can create a desired environment around the specimen. Thus, temperature-controlled stages are available for use with the SEM which permit examination of specimens at levels ranging from +25° C to −180° C.*

3. Magnetic Shielding

Because, as with the TEM, stray variable electromagnetic fields from apparatus and power supplies can degrade microscope performance, the location of the microscope should be carefully selected. The aim is to limit the varying component of the external magnetic field to a level of 5-10 milligauss in the vicinity of the column. This can be facilitated by using Mu-metal shields around the lenses, especially the final condenser.

*Frozen tissue specimens are prepared by quench-freezing in liquid nitrogen or Freon 22. Studies have shown that the cold storage is useful in examining biological specimens (e.g., microalgae, protozoa, and plant material), and that they retain their natural three-dimensional configuration. A major advantage of the cold stage is that images can be obtained within a few minutes after the specimen is removed from its source.

4. Vacuum System*

It is essential to reduce the pressure in the column so that most of the electrons complete their trajectories before they collide with gas molecules. The majority of commercial SEM's use a tungsten filament cathode for which a vacuum of 10^{-5} torr is adequate to avoid excessive corrosion and/or ion bombardment of the emitting surface. The desirable vacuum level is attained by using an oil diffusion pump backed up by a mechanical rotary pump (see p. 335). Vibrations may be introduced by both these pumps. As with the TEM, to minimize such vibrations, the mechanical pump should be placed away from the microscope and the oil diffusion pump should be mounted on shock absorbers.

To prevent oil vapor from the pumps from reaching the specimen chamber, a trap may be inserted between both pumps and a thermoelectrically-cooled baffle between the diffusion pump and the specimen chamber. Oil vapor reaching the chamber is absorbed on the specimen surface, and is subsequently decomposed by the electron probe leading to contamination. As a result, it limits the time available for specimen examination. Internal contamination can also degrade the performance of the electron optics by creating random charging and astigmatism (see the discussion on p. 230).

The vacuum level required for the SEM varies according to the type of electron gun being used. While an oil diffusion pump that can easily provide a vacuum of 10^{-4} torr may be adequate when a tungsten filament is used, for a LaB_6 cathode a higher vacuum is needed ($\sim 10^{-6}$ torr). Where a field emission gun is used an ultrahigh vacuum (10^{-9} torr) is needed. This necessitates specialized vacuum equipment (see below).

The use of an oil diffusion pump in the SEM (and TEM) represents a trade-off: in return for a modest expenditure and the possibility of speedy specimen change, the risk of contamination is accepted. There are two classes of alternatives to the oil diffusion pump and these may be used in multiple locations in the instrument to facilitate optimum performance and minimum pumpdown time. They must be used with a field-emission gun, because a 10^{-9} torr vacuum level must then be attained.

(1) USE OF OIL-FREE SYSTEMS.

Oil-free systems significantly reduce contamination, and easily provide a moderately high vacuum (10^{-7} torr), but they have prolonged pump-

*For a more detailed discussion of the Vacuum System, see Appendix N.

down times (24 hours). The aforementioned disadvantage is overcome in instruments where the gun and specimen chambers can be differentially evacuated. There are four types of oil-free pumps.

(a) *Sputter-ion pumps*. These work by accelerating electrons and spiraling them in a strong magnetic field. The electrons collide with residual gas molecules and ionize them. The ions are in turn accelerated into a freshly evaporated titanium surface where they are captured. Such pumps have the advantage of eliminating the major source of hydrocarbon contamination.

(b) *Orbitron pumps*. These function like sputter-ion pumps but use electrostatic fields for spiraling the electrons.

(c) *Titanium sublimation pumps*. These function by the getter-like action, like sputter-ion pumps, but without ionization of the pumped gas.

(d) *Cryopumps*. These function by virtue of the fact that, at very low temperatures, gas is removed by condensation on a surface which is at a temperature below the condensation temperature of the gas.

(2) USE OF ULTRAHIGH VACUUM SYSTEMS.

An ultrahigh vacuum system aims to eliminate even the residual contamination present in the oil-free system. In order to maintain ultrahigh vacuum levels (*circa* 10^{-9} torr), the columns used with these systems are periodically baked to desorb residual contaminants. To do this, strip heaters are mounted around the column and fluorocarbon gaskets or metal seals are used.

In a compromise arrangement, an LaB_6 gun can be employed with a vacuum system consisting of a cold-trapped mechanical pump to attain a vacuum of 100×10^{-3} torr followed by a sorbtion pump to bring it down to 5×10^{-3} torr. An ion pump then pumps down to 10^{-7} torr. Under these conditions, the contamination rate can be reduced from 500 Å/min (of an oil-pumped system) to 1 Å/min.

Chapter 18

IMAGE CHARACTERISTICS

A. CONTRAST

1. General Considerations

ONE OF THE significant features of the SEM is the great variety of contrast mechanisms possible for forming images. Theoretically, any measurable short-term effect that arises in the course of beam-specimen interaction can be utilized to modulate the display cathode-ray tube, thereby providing information concerning the specimen. Thus, for example, x-rays emitted when the beam strikes the specimen can be monitored to provide a picture of the elemental distribution in the specimen. In biomedical research, the secondary electron emission, at present with the typical SEM, is the principle source of contrast and only collection of these particles was discussed (see p. 239). Most emphasis will be placed on this contrast mechanism.

2. Secondary Electron Emission Mode

Contrast generation in the SEM is a complex subject which is not as yet understood fully. The interplay of several independent factors produces the overall result.

The eye is able to detect fractional changes of intensity between adjacent areas of a picture. There are three groups of factors which cause variations in intensity of the SEM image.

a. FACTORS DEPENDING ON THE ANGLE OF INCIDENCE OF THE ELECTRON BEAM.

The factors in this group are largely determined by the topography of the specimen and are of major importance in determining the SEM image. This is due to the fact that the incident probe reaches the specimen from a sharply defined direction and the image which becomes displayed represents the picture of the specimen as viewed along this probe. Moreover, the factors in this group are responsible for providing the illumination that serves to reveal the structure of the walls of deep surface cavities.

Factor (1): To facilitate electron collection, the mean specimen surface is usually tilted at an angle (of 30–60°) to the probe. More collectable secondary electrons are generated per unit of projected specimen area when the probe strikes a curved or sloping surface than are collected with a flat surface. The tilting enhances this phenomenon and makes the distribution of light and dark areas, or contrast, more pronounced.

Factor (2): The local irregularities in the surface of the specimen are likely to cause the local angle of incidence θ to vary from point to point. The number of collectable secondaries generated is largely independent of the material of the specimen, but is roughly proportioned to the sec θ. Since brightness in the image is proportional to the number of secondaries leaving any area in the specimen, a small change in θ will produce a change in brightness that should be reflected in the image, in the form of different contrast between adjacent areas of a specimen.

b. FACTORS CAUSING VARIATION IN THE PORTION OF THE SECONDARY AND BACKSCATTERED ELECTRONS COLLECTED.

The factors in this group (like those in group a) are largely determined by surface topography. They are subordinate in importance to those in group (a) and serve to modify the contrast between one topographical feature and another.

Factor (1). The accelerating potential in the SEM is usually 2–30 KV. Under these operating conditions, the ratio of low-energy secondary to high-energy backscattered electrons will be greater at low probe energy values. Moreover, for specimens containing elements of low atomic number, the ratio will usually be greater than unity (over the entire energy range). With high atomic number elements, the ratio may fall a little below 1 at the upper incident energy levels. This ratio, however, is not important because the collection efficiency for secondaries is very much greater than for backscattered electrons for the secondary detectors which have a small solid angle. Only large angle detectors pick up enough backscatter to mean anything.

Factor (2). The ratio of secondary to backscattered electrons collected by the collector can be affected by its size, position, and the potential difference between it and the specimen (which is usually several hundred volts positive).

Factor (3). While the presence of a potential difference between the collector and specimen increases the number of secondaries collected, the extent of the increase depends on local specimen topographical conditions (in addition to the considerations enumerated in item 2). Thus, low-energy secondaries emitted from a concavity may not be significantly affected by the potential difference and many of them therefore may re-enter the surface before they can be accelerated towards the detector. Thus, at such sites, the gain in secondaries collected will be much less than at elevated sites. Considerable contrast may result from this effect, with the concavities appearing dark and the elevations bright.

c. FACTORS DEPENDING ON THE NATURE OF THE SPECIMEN.

The factors in this group are not of primary importance, but the contrast they produce does convey additional information about the nature of the specimen.

Factor (1). Specimens which are insulators often have secondary emission coefficients* that are much higher than unity. While this would be responsible for strong contrast, the need to coat the specimen with a conductive material so as to prevent it becoming charged (see p. 274) dampens this contrast stimulating potential, since the secondary emission coefficient would reflect, in part, the characteristic of the conducting layer rather than of the specimen.

Factor (2). When conductive coating of the specimen is required for high voltage scanning, the nature of the coating material may have a significant effect on both the secondary and backscattered electrons.

In summary, contrast results from the interaction of three groups of factors. The first and second can be considered as providing the "black and white" image which essentially provides a complete view of the topographical features of the specimen. The factors in the third group provide, so to speak, qualitative "coloration" to the image. Some of these factors can be controlled by the microscopist (e.g., energy of the probe, specimen tilt angle, detector specification, and especially specimen preparation), but the most useful controls of contrast are provided by operator adjustment of controls which effect the processing of the final image.

3. Enhancement in Final Image

With the SEM, the contrast can be further controlled even after beam-specimen interaction by increasing or decreasing the amplification of the electrical signal that is generated and on which the contrast is based, and by biasing the signal. Thus, low contrast means that the average current component of the signal is very large in comparison with the sample detail current

*Secondary emission coefficient is expressed as:

$$\theta = \frac{\text{number of secondary electrons}}{\text{number of primary electrons}}$$

(with the secondaries being electrons with energies of less than 50 electron volts).

component. At high gain, and with the subtraction of a current slightly smaller than the background, contrast can be enhanced enormously.

B. RESOLUTION

For the SEM, the resolution achieved in any given instance depends on the specimen as well as on the microscope. As far as the former factor is concerned, contrast between two close but separate topographical features of the specimen must be resolvable. For the microscope, it appears theoretically that distances between two points that are slightly smaller than the probe diameter could be resolved. However, in practice background noise and current density changes complicate the situation so that, in reality, even the beam diameter does not normally limit resolution.

More specifically, the factors affecting resolution can be placed into three catagories:

1. MICROSCOPE OPERATING CONDITIONS.

There are four microscope operating conditions: (a) accelerating potential; (b) beam brightness; (c) lens aberrations; and (d) lens operating parameters. These are all interrelated factors, as expressed in this equation for the total beam current:

$$i_b = \frac{\pi^2}{4} \cdot d_o^2 \cdot \beta \cdot \sin^2 \alpha \qquad (15)$$

where:

i_b = total beam current (amp).
d_o = aberrationless beam diameter.
β = brightness (A/cm^2-ster).
α = aperture angle.

This equation can be rearranged* to read:

$$d^2 = \frac{C_o^2}{\alpha^2} + C_s^2 \cdot \alpha^6 + \left(\frac{\Delta V}{V}\right)^2 \cdot C_o^2 \cdot \alpha^2 + \frac{1.5\lambda^2}{\alpha^2} \qquad (16)$$

*The mathematical steps involved in rearranging equation (15) into equation (16) are given in the footnote on the next page. By this rearrangement it is more readily possible to substitute known values for different electron guns and then calculate the minimum probe spot diameter (see p. 251).

where:

C_o^2 = the square of the effective diameter of the probe.
C_o^2 = $4_{i_b}/\pi^2\beta$ (see footnote).
α^2 = the square of the aperture angle.
λ^2 = the square of the electron wavelength.
C_s^2 = the square of the spherical aberration constant.
C_c^2 = the square of the chromatic aberration constant.
ΔV = instability in the accelerating potential.
V = accelerating potential.

---

*Equation (15) states that the total beam current, i_b, is equal to the product of the area of the probe, πr^2, its brightness, β, and its solid angle, $\pi \sin^2 \alpha$. That is:

$$i_b = \pi r^2 \cdot \beta \cdot \pi \sin^2 \alpha \tag{1}$$

Since $\pi r^2 = \pi\, d_o^2/4$, equation (1) can be rewritten as:

$$i_b = \frac{\pi^2}{4} \cdot d_o^2 \cdot \beta \cdot \sin^2 \alpha \tag{2}$$

Equation (2) can, for small α, be rearranged as:

$$d_o^2 \approx \frac{4i_b}{\pi^2\beta} \alpha^{-2} \tag{3}$$

Let $C_o = \sqrt{\dfrac{4i_b}{\pi^2\beta}}$

Since the total beam diameter is estimated by equation (13) as:

$$d^2 = d_o^2 + d_s^2 + d_c^2 + d_d^2 \tag{4}$$

where:

$$d_o = \frac{C_o}{\alpha}$$

$$d_s = ½\, C_s\, \alpha^3$$

$$d_c = \frac{\Delta V}{V}\, C_c\, \alpha$$

$$d_d = \frac{1.5\lambda}{\alpha^2}$$

Substituting these values in equation (3) gives:

$$d_o^2 = \left(\frac{C_o}{\alpha}\right)^2 + \left(½\, C_s\, \alpha^3\right)^2 + \left(\frac{\Delta V}{V}\, C_c\, \alpha\right)^2 + \left(\frac{1.5\lambda}{\alpha}\right)^2 \tag{5}$$

This equation can be rewritten as:

$$d_o^2 = \frac{C_o^2}{\alpha^2} + C_s\, \alpha^6 + \left(\frac{\Delta V}{V}\right)^2 \cdot C_c^2\, \alpha^2 + \frac{1.5\lambda^2}{\alpha^2} \tag{6}$$

The aperture angle α is of special significance, as it is present in all four sets of variables. Where the operating values are known, the effective probe spot size d_o can be graphically plotted as a function of the angular aperture and a value for α at which d_o is minimal can be determined. Thus, it has been shown that for a tungsten filament cathode (see Table IX) where $C_s = 1$ cm; $C_c = 1$ cm; $V = 20$ KV; $\Delta V = 10^{-4}$; $i_b = 10^{-11}$ amp; $\beta = 10^5$ amp/cm^2-ster, the angular aperture $\alpha = 0.005$ radians, and minimum probe spot comes out to be 70 Å. Moreover, by using an LaB_6 gun to increase brightness, β, there will be a decrease in the aberrationless beam diameter at the specimen, d_o, such that resolution will almost double to *circa* 35 Å, where other factors such as beam penetration (see below) come into consideration as limiting agents.

2. BEAM-SPECIMEN INTERACTIONS.

This catagory consists of three factors: (a) specimen material; (b) specimen geometry; and (c) beam penetration. Specimens of differing materials and geometry have varied beam-specimen interactions which are not predictable in advance.

To obtain high resolution, relatively low accelerating voltages (*circa* 10 to 40 KV are used. Under these conditions, the incident electrons will penetrate the specimen to depths of about 1.0 μm. Some of the escaping secondary electrons, especially when the probe strikes the specimen obliquely, may escape at a distance up to possibly 1.0 μm from the point of beam impact, instead of the usual 0.005 μm. Some of these widely deflected secondaries that are collected may be influenced by the surface features that are remote from the probe penetration site. If this occurs, effective resolution will be reduced and micrograph interpretation hindered. In addition backscattered electrons leaving the surface at considerable distances from the impact site will add to the background noise, and reduce the signal/noise ratio. The magnitude of the adverse effect of penetration of primary electrons will vary greatly from one specimen to another. Electron penetration, rather than electron optics, is the major factor limiting resolution. The highest resolution will be achieved only with very thin specimens, as in the TEM.

3. INDUCED INTERFERENCE PROBLEMS.

This catagory consists of three factors: (a) vibrations; (b) stray magnetic fields; and (c) contamination. Since they are discussed in another section (p. 93), suffice it to say that they can be reduced or eliminated by taking the necessary preventive measures.

Chapter 19

CURRENT SEM DEVELOPMENTS

WHEN THE SEM became commercially available, it was initially used only to obtain attractive electron micrographs of the surface appearance of a wide variety of interesting specimens. Subsequently, the SEM began to be used in a systematic scientific manner to collect significant experimental data. Moreover, the use of this instrument not merely to record topographical details but also to examine specimens by a variety of specialized techniques was proposed.

Concomitant with the above evolution in SEM usage was the development of the instrument itself. As a result, two groups of instruments have come into being over the past decade. One group, which is relatively less expensive, is designed for routine inspection and testing of specimens as well as for teaching purposes. A second group consists of research instruments which are characterized by more versatility and complexity, and thus are more expensive. Continued development along these two lines can be expected with improvements in engineering, especially by the use of solid-state circuits simplifying the electronics in both groups.

Future developments of the SEM are proceeding along three lines: low voltage microscopy; scanning transmission electron microscopy; and scanning electron microprobe analysis. Each of these areas will be discussed briefly.

A. LOW VOLTAGE MICROSCOPY

When electron voltages of 10 KV or higher are used in the examination of all thick specimens, resolution is limited by electron penetration rather than by electron optics (see p. 53). Resolution can obviously be improved by using low voltage acceleration since electron penetration is also lowered, but aberrations and diffraction limit increase and bar the way to very significant improvements in resolution.

Additional advantages of low voltage operation are:

(1) Insulating specimen surfaces will not become excessively charged under these conditions; thus, conduction coating (see p. 274) is

unnecessary and specimens can be examined directly, especially if scanning takes place rapidly. This is due to the fact that at about 1 KV the secondary emission coefficient is about 1 for many substances.

(2) With low energy primary electrons (using 5 KV or less), the total energy dissipated in the specimen per electron is lowered and, as a result, specimen damage due to beam heating is smaller. This is especially useful when examining thick heat-sensitive specimens.

(3) Reduced penetration, which is a consequence of low voltage acceleration, serves to improve the visibility of the surface.

B. SCANNING TRANSMISSION ELECTRON MICROSCOPY

During the past decade, advances have been made in developing a TEM-SEM hybrid instrument, appropriately called a scanning transmission electron microscope (STEM). In this system, a specimen is sufficiently thin, electrons can pass through it and be collected by a collector located directly behind the specimen. The image is derived from such transmitted and collected electrons.

A principle advantage of the STEM is that it permits convenient energy analysis of the transmitted electrons. In the TEM, the electron beam traverses simultaneously through all parts of the specimen, making energy analysis difficult. In the STEM, an image can be formed by selecting electrons whose loss of energy after passing through the specimen lies in any specific range. Moreover, signals from the electrons in any specific range can be correlated with specific points in the image of the specimen. An STEM can simultaneously record the transmitted electrons which have been separated into different energy-loss groups, and the signals from the groups can be mixed, stored, and modulated to produce a variety of final images. It is also interesting to note that by recording an image, by using electrons of only a specific energy (e.g., those suffering a loss of $\Delta V \pm 1$ volt), a high-contrast image virtually free of chromatic abberation can be obtained from specimens as thick as a few tenths of a micrometer at 100 KV and a micrometer at 1000 KV. This offers a way to overcome the major deficiency of High Voltage Electron Microscopy (see page 107). However, to collect enough electrons to provide such an image with a reasonably high signal-to-noise ratio, requires that it be exposed to especially large numbers of illuminating electrons and, therefore, the probability of beam damage increases (see pp. 97 and 191).

Pioneering work with the STEM has been done by Crewe[186] and his associates.

C. SCANNING ELECTRON MICROPROBE ANALYSIS

It was noted earlier (see p. 242) that elements of matter (except hydrogen and helium) emit a variety of particles and photons, including x-ray photons when bombarded by an electron beam. In 1913, Moseley[187] observed that the atomic numbers of the target elements can be determined by observing the wavelengths of their characteristic x-ray emission lines. To a first approximation, the intensities of these lines are related to the local concentrations of the components. For high spatial resolution, a well-focused electron beam or microprobe must be employed to induce x-ray excitation, in order to limit the dimensions of the volume which emits radiation. The focusing of the beam can be attained by standard electron optical principles.

The brightness of a finely focussed electron beam, however, is also limited; thus, maximum x-ray intensity is restricted and an efficient x-ray collection system is desirable.

Castaing and Guinier[188] reported in 1950 the development of the first electron probe microanalyzer for the spectrochemical analysis of microscopic regions of a specimen. This instrument had an optical system similar to that in Von Ardenne's SEM.[175] But it was Cosslett and Duncumb,[189] who first demonstrated that the basic features of the SEM can be incorporated into the design of an electron probe microanalyzer to produce a scanning electron probe microanalyzer (SEMA), which opened up the field of scanning x-ray microscopy.

In general, an SEMA in operation focusses an electron beam of 0.1 μm diameter on the sample. The number of x-ray photons and their wavelengths (or energies) are measured to give a quantitative analysis in favorable samples of picogram (or smaller) amounts of an element. The usual sensitivity of the method is in the 10^{-14} to 10^{-16} grams (10^6–10^8 atoms) range. It can detect all but the lightest elements in the concentration range of 10–1000 ppm in large specimen areas, but elements above atomic number 20 (calcium) can be identified only if present as major constituents (>10%) in particles as small as 0.1 μm (10^{-15} – 10^{-16} grams).

Chapter 20

OPERATIONAL CONSIDERATIONS

A. INSTRUMENT TESTING

OCCASIONAL examination of a suitable standard specimen to test the operational status of the instrument is desirable. Standard biological specimens that have been used for testing are serum lipoprotein macromolecules and tobacco mosaic virus crystals. A strongly recommended nonbiological standard is precious opal etched with hydrofluoric acid and coated with a conducting layer.

B. INSTRUMENT SETTINGS

There are several instrument settings of special importance. These are considered individually.

(1) Accelerating voltage. The range of accelerating potentials available with the standard SEM is usually 1 to 50 KV. For observation of biological material, low voltage, i.e., no higher than 10 KV, has been recommended when adequate resolution is attainable at these levels and it is seldom necessary to exceed 20 KV. It may be desirable to try several potentials to determine the optimal one for visualizing the specimen while incurring the least damage.

(2) Specimen current. This should be as low as possible; for surveying the specimen, currents in the 10^{-10} amp range are satisfactory, while for high resolution studies, about 10^{-11} amps is recommended. The readings are made on a picoammeter.

(3) Final condenser aperture size. This is dependent on the resolution, magnification, and depth of focus desired. At low magnification a large depth of focus is desirable and thus a small aperture (*circa* 50 μ) is used. For high resolution investigations, a large final condenser aperture is desirable (*circa* 200 μ). As in all electron microscopes, the aperture should be centered with reference to the optical axis of the final condenser lens and should be very clean.

(4) Astigmatism correction. This factor will be critical in determining the final image. This defect should be kept to an absolute minimum by

maintaining a clean column and cleaning the final condenser aperture regularly. The first step is to determine the magnitude and direction of the astigmatism. The next is to introduce into the final condenser lens an artificial astigmatism of equal magnitude but opposite direction to that inherent in the lens as with other electron microscope objectives (see p. 89). Astigmatism also results from local specimen charging and must be continually re-corrected as one moves the specimen stage from place to place. Astigmatism azimuth and amplitude are corrected before taking each high resolution micrograph.

(5) Contrast level. This factor is often controlled by adjusting photomultiplier tube gain to set visually the desired contrast within the range allowed by the film on the cathode-ray tube. Alternately, the proper setting can be determined on some instruments by measuring the peak height of the signal monitor display (see section 3 below). When the CRT adjustment is used, the contrast level selected should be a little below that which appears to be visually optimum.

(6) Magnification. This is the ratio of an adjustable current applied to the deflection coils in the scope to the current applied to the deflection coils in the CRT. The deflection coils control the size of the area of the specimen surface scanned in relation to the constant size of the face of the CRT. Since a change in magnification alters only the size of the specimen area scanned, magnification can often be varied over a wide range without refocusing.

(7) Scan rates. These have a wide range (from 1/30th of a second per field to hundreds of seconds). They are related to the level of magnification used. As magnification increases, a smaller probe diameter should be used to scan the smaller area. Since the total electron current in the focussed probe decreases as the probe diameter gets smaller, slower scan rates (*circa* 10–40 seconds per frame) are desirable in order to provide adequate exposure times during image recording.

Slow scan speeds are standard in the SEM to ensure that a high enough signal-to-noise ratio is achieved. However, with sacrifice in signal-to-noise ratios, the introduction of high speed scan system permits the image to be seen on the screen of an ordinary TV monitor which permits viewing of dynamic events and thus is of use in special circumstances.

One important advantage of field emission sources in SEMs is that fast scan rates (many per second) and small probe diameters may be used at all magnifications. The disorienting flickering screen seen with slow scan speeds and the inability to follow rapid specimen movements are completely obviated.

C. SIGNAL MONITOR

To appraise the SEM quantitatively, most commercial instruments are capable of monitoring the intensity of signals along a given scan line across the specimen separately. The signal monitor can help in analyzing signal distortions and acts as an exposure meter to facilitate efficient photographic recording. Automatic photographic systems incorporated in the SEM have also become available in recent years.

Chapter 21

COMPARATIVE VALUE OF THE SEM

UPON EXAMINATION of SEM micrographs, one is immediately impressed by their vivid, three-dimensional character. More specifically this advantage of the SEM stems from: (a) an ability to resolve topographical details of the external surface of specimens (obviating the need to make sections or replicas for examination by TEM); and (b) a great depth of focus which facilitates three-dimensional visualization of specimen surfaces.

The possibility of reducing total beam damage to the specimen is another important advantage. The total current falling on the specimen in the TEM is about $10^{-7} - 10^{-8}$ amps and about $10^{-10} - 10^{-12}$ amps in the SEM. Although the SEM current density at the specimen is often greater (especially with slow-scan), it is much easier to prevent cumulative damage *outside of the area of immediate interest* at high magnification with the SEM.

Further comparison of a typical SEM with a typical TEM (see Table X) summarizes some differences.

TABLE X

COMPARISON OF SEM AND TEM

Scanning Electron Microscope (including STEM)	Transmission Electron Microscope
An image can be obtained by any signal generated by the incident electrons and the signalling specimen particles need not be focused.	An image can be obtained only from the focusable transmitted electrons.
Only one minute spot of the specimen is irradiated at a time.	A relatively large area of the specimen is continuously irradiated.
Signal can be processed in a variety of ways.	Image is not processed prior to photography (Image is "processed" by the fluorescent screen).
Is usually designed with fewer lenses.	Design is usually more complex.
Specimen observation at progressively higher magnification is usually possible without any change in focus or image brightness.	Adjustment of focus and image brightness are usually necessary as magnification is progressively increased. (However these are automated in several current models.)
Specimen area center at low magnification remains centered at higher magnification.	Recentration of specimen area selected may be necessary as magnification increases. (However, this problem is eliminated by "perfect" alignment.)
Same specimen area identifiable on light and SEM micrographs.	Specimen area identified on TEM micrograph of thin section requires adjacent thick section for light microscopic comparison.
The specimen need not be cut into thin sections to be examined.	Primarily only ultrathin sections can be examined.
Specimen preparation is usually simpler.	Specimen preparation is usually more complex.

Part Two

METHODOLOGY

Chapter 22

INTRODUCTION

TO CARRY OUT scanning electron microscopy successfully, it is necessary not only that the instrument be in good working order, but also that the specimen be as undistorted as possible. Fortunately, preparation of specimens for examination is generally not a complex procedure. Moreover, some materials with natural surfaces (e.g., leaf, blood cells, protozoas) require only minimum handling.

It should be realized that while a very wide variety of biological material have been successfully examined with the SEM, discrimination—as in the selection of tissue for study—is nevertheless essential as only certain kinds of specimens can provide useful information. Among the factors that need to be considered in specimen selection are the nature of the material, the region of the specimen to be studied, the kind of information sought, and the effect of preparation and beam exposure on the specimen.

Specialized SEM texts and papers (p. 364) provide detailed technical descriptions of specimen preparation methods. The following chapters are therefore devoted to outlining in general terms the basic procedures—prefixation handling, fixation, and postfixation handling—used in SEM specimen preparation.

While the SEM is a superb instrument for providing information concerning the surface topography and shape of a specimen, success is dependent upon proper preparation of the material to be examined. The goal is to preserve the material in a state that is as close to life as possible. When the specimen is wet and soft (Fig. 116), it usually requires fixation, dehydration, and drying before it can be mounted. On the other hand, for a specimen that is dry and hard (Fig. 117), it need only be mounted on a specimen holder and coated (see below). In both cases adequate care needs to be exercised to insure attainment of high quality electron micrographs of undisrupted material. Careless preparation can cause changes in the structural integrity of the surface of the tissue block or section and readily introduce artifacts.

Specimens examined with the SEM must be able to tolerate very low pressures. In addition, the surface of the specimen should not become charged when subjected to the electron beam. To reduce specimen charging, specimens are generally coated with a conductive material so that excess incident electrons are bled off.

Chapter 23

HISTORICAL REVIEW

BECAUSE OF serious technical problems associated with the development of preparation methods for use with the TEM, application of this microscope in biological research initially lagged behind its commercial development. This has not been the case of the SEM, although full appreciation of the usefulness of this instrument did not occur until the early 1970s.

The highlights of the methodological advances that made preparation and study of biological material with the SEM profitable are summarized below and will serve as a brief historical review.

1962	Boyde and Stewart[190,191] first used the SEM in the study of teeth and other hard, cornified tissues.
1965	Jaques *et al.*[192] first applied the SEM to the study of soft tissues.
1966	Pease and Hayes[193] published the first SEM study of free cells.
——	Pease, Hayes and Camp[194] demonstrated the possibility of using SEM with living material.
1968	Barber and Boyde[195] introduced an effective method for specimen dehydration and drying.
1969	Boyde and Wood[196] introduced the use of enzymes for removing contaminants from various types of tissue.
——	Boyde and Wood[196] introduced the use of critical point drying and freeze-drying methods for SEM specimen preparation.
1970	Echlen, Paden, Dronzek and Wayte[197] examined frozen specimens in the SEM using a secondary electron detector.
1971	Scheid and Traut[198] introduced cytological techniques for isolating cell organelles for scanning electron microscopy.
1972	Haggis[199] introduced the freeze-fracture technique for SEM specimen preparation.
——	LoBuglio, Rinehart, and Balcerzak[200] introduced the use of an immunologic-marker for the scanning electron microscope.
1975	Flood[201] introduced the dry fracture technique for studying the internal surfaces of soft tissues.

Over the past few years there has been considerable refinement, improvement and expansion of a variety of the techniques of specimen preparation that have increased both the usefulness and applicability of SEM studies. This is especially true in so far as critical point drying, freeze-drying, ion beam etching and cell surface labeling is concerned.

Chapter 24

PREFIXATION HANDLING

THERE ARE A variety of methods by which tissues may be treated prior to fixation in order to prepare them for satisfactory fixation. The choice of approach depends on the organization and chemical composition of the native specimen. The approaches discussed will be grouped into three categories: mild pretreatment, dissection, and chemical pretreatment.

A. MILD PRETREATMENT

The SEM has been used successfully in many histological studies of tissues in which their naturally existing surfaces were examined and care was taken that distortion did not occur during processing. Specimens in this category include epithelia, endothelia, cornea and lens of the eye, insect cuticle, plant leaf, as well as individual cells such as blood cells, protozoa, and tissue cultured cells. The tissue need not be pretreated prior to fixation for such studies.

When blood covers the tissue surface, it can be removed by exposing the surface briefly to a gentle stream of isotonic saline. With other adhering contaminants such as mucous, wax, or a gelatinous membrane, it may be necessary to use a digestant, detergent or solvent, or mild acid, respectively, as a prefixation cleanser. With plant material, a jet of clean air (clean Freon gas) will often remove contaminants.

B. DISSECTION

In most biological material, all of the interesting surfaces are not naturally exposed. The simplest approach for revealing the internal organization of soft tissue is to tear fixed bulk pieces (Fig. 116)[202]. To minimize tissue damage, it may be advisable to carry out the dissection only after aldehyde perfusion (p. 121), and fixation. Parducz's fixative (p. 269) was also found to be a good hardening agent for reducing tissue distortion. Interpretation of SEM images of such torn surfaces must be made with care because bulk tissues may not be fixed well, and the tearing process can grossly displace structures.

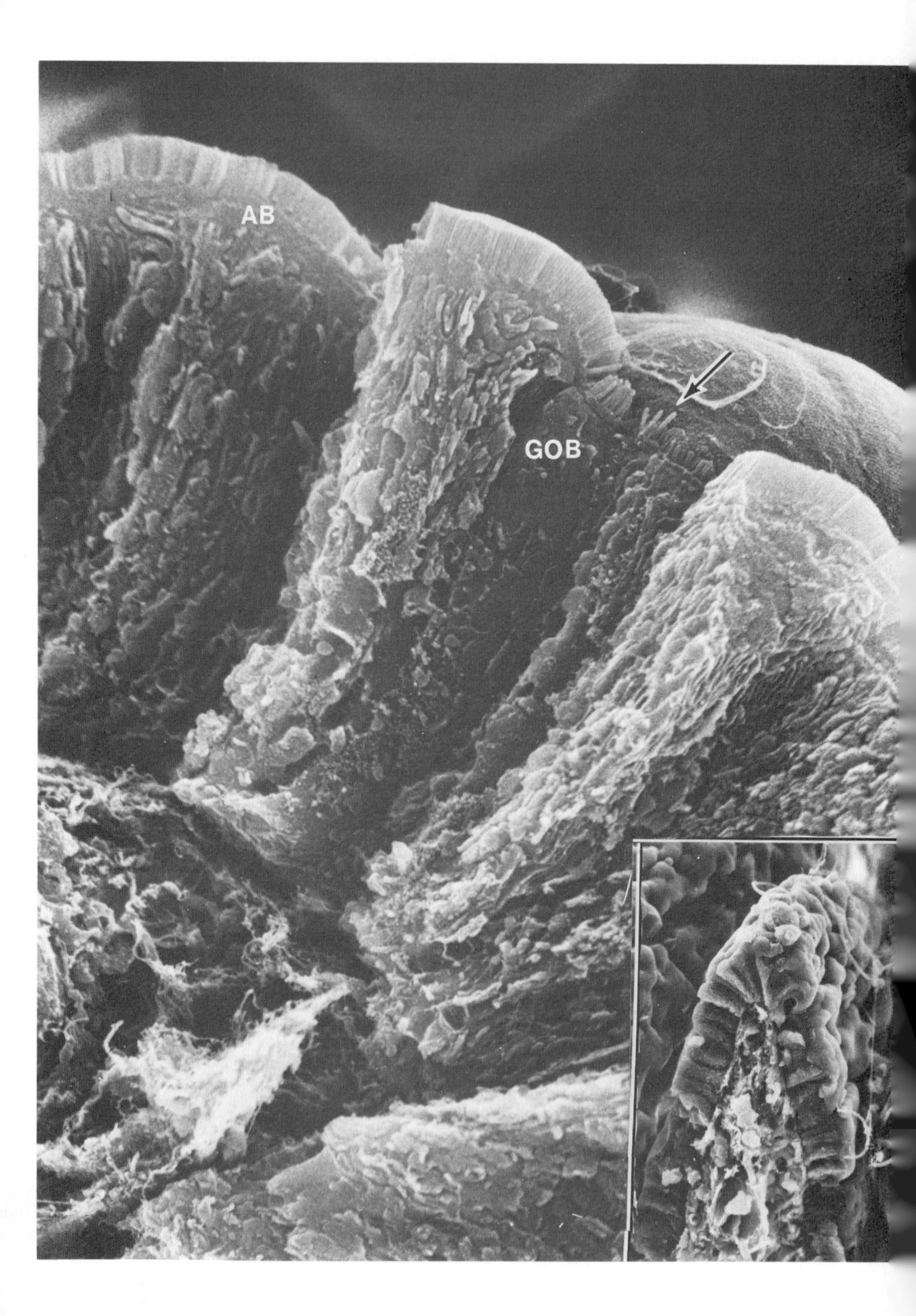
AB
GOB

C. CHEMICAL PRETREATMENT

Selective enzymatic digestion of fresh (unfixed) soft tissues can help to reveal a desired surface or components[203] (e.g., collagenase to remove collagen, or hyaluronidase to remove intercellular ground substance). After pretreatment, the tissue should be thoroughly washed with isotonic solution to remove the enzyme before fixation.

In studying calcified tissues (Fig. 117), removal of the mineral component may be useful. There are two alternate methods for doing this. One is by using a chelating agent (e.g., EDTA). Due to its relative gentleness, this approach is especially beneficial if one desires not to disturb the organic matrix. Acid decalcification is another possible method, but if carried out rapidly the matrix is distorted when some of the carbonate is given off as CO_2 gas.

In seeking to remove the organic matrix so as to expose the mineral component, treatment with strong alkalies has been used. However, this requires washing out of a large concentration of the mineral ions to avoid contamination and obscuring the surface. An effective alternative that therefore has been advanced is the cold ashing technique.[204] This involves placing the specimen in a tube and then passing a stream of oxygen at low pressure over the specimen and exciting a glow discharge. The highly mobile oxygen ions of the glow discharge will unite with the carbon and hydrogren of the specimen, removing them as CO, CO_2 and H_2O without causing distortion.

FIG. 116. Demonstration of the application of simple dissection as a prefixation method. This scanning electron micrograph shows a view of the entire thickness of the epithelial lining of the mouse jejunum. A prominent apical band (*AB*) marks the boundary between the microvilli and the elaborations of the cell membrane into the lateral intercellular space, which in this case forms a series of folds or ridges running the length of the lateral surface. More loosely organized microvilli (arrow) may indicate the position of a goblet cell (*GOB*). The cells rest on the connective tissue of the lamina propria. 7,000X. Insert: An entire villus can be dissected in such a manner as to be able to determine the position of cells on it easily. 400X. (Courtesy of Drs. M. M. Miller and J. P. Revel.)

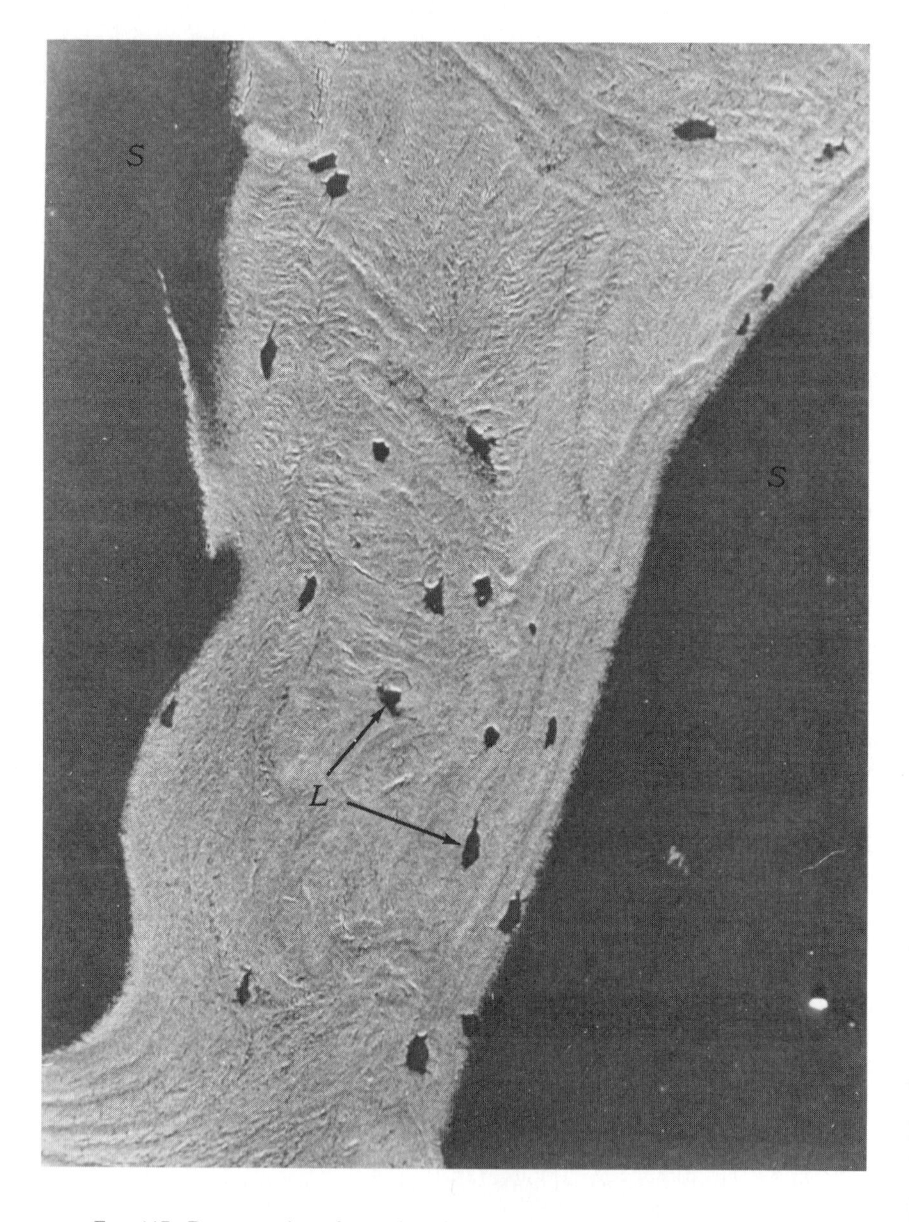

FIG. 117. Demonstration of scanning electron microscopy applied to a hard tissue. (Courtesy of A. Boyde.)

Backscattered electron image of a high polished piece of bone. It contains numerous osteocyte lacune (L). The black space around the bone fragment represents the bone marrow space (S). This type of image can be used for automatic image analysis to measure the volume of bone in bone and the proportion of bone cell spaces and osteocyte lacunae inside the bone matrix. 375X.

Chapter 25

FIXATION

THE COMMONLY used TEM fixative, glutaraldehyde, is also highly recommended for SEM studies.[205,206] This is because it cross-links free amino groups and toughens the tissue, making it more resistant to collapse during the dehydration process. Some investigators have recommended Parducz's[207] fixative, an osmium tetroxide-mercuric chloride mixture, to harden the tissue, as well as Karnovsky's[208] formaldehyde-glutaraldehyde mixture.

Since only surface layers of the specimen are usually viewed with the SEM, the fixatives used should be isotonic so as not to expose the cells to osmotic shock. In addition to considering the final osmolarity of the fixative and especially of its vehicle (buffer), the concentration of the fixative is of importance.

A. FIXATION PARAMETERS

A number of other factors are also of importance in the fixation process. These are briefly discussed.

1. pH. This variable should be as close to the pH of the specimen as possible (typically 7.4 for animal tissues, 7.1 for plant tissues). Both collidine and cacodylate have been used routinely as buffers with considerable success.

2. Temperature. Generally, the temperature of the fixative can be higher for SEM specimens than those prepared for TEM (where penetration is a critical factor so that temperatures near 0° C are used). Thus, the damage produced by room temperature fixation at some depth in the tissue is usually inconsequential and may be absent on the tissue surface.

3. Duration. With an average size specimen, the length of immersion in the fixative is between two and eight hours. For larger specimens, it can be extended up to 15 hours. Gentle agitation of the fixative from time to time, especially during the first hour, is desirable.

4. Size. The size should be kept as small as possible since this will ensure quicker fixation, dehydration, and better coating. The maximum size should be about 1 mm^3.

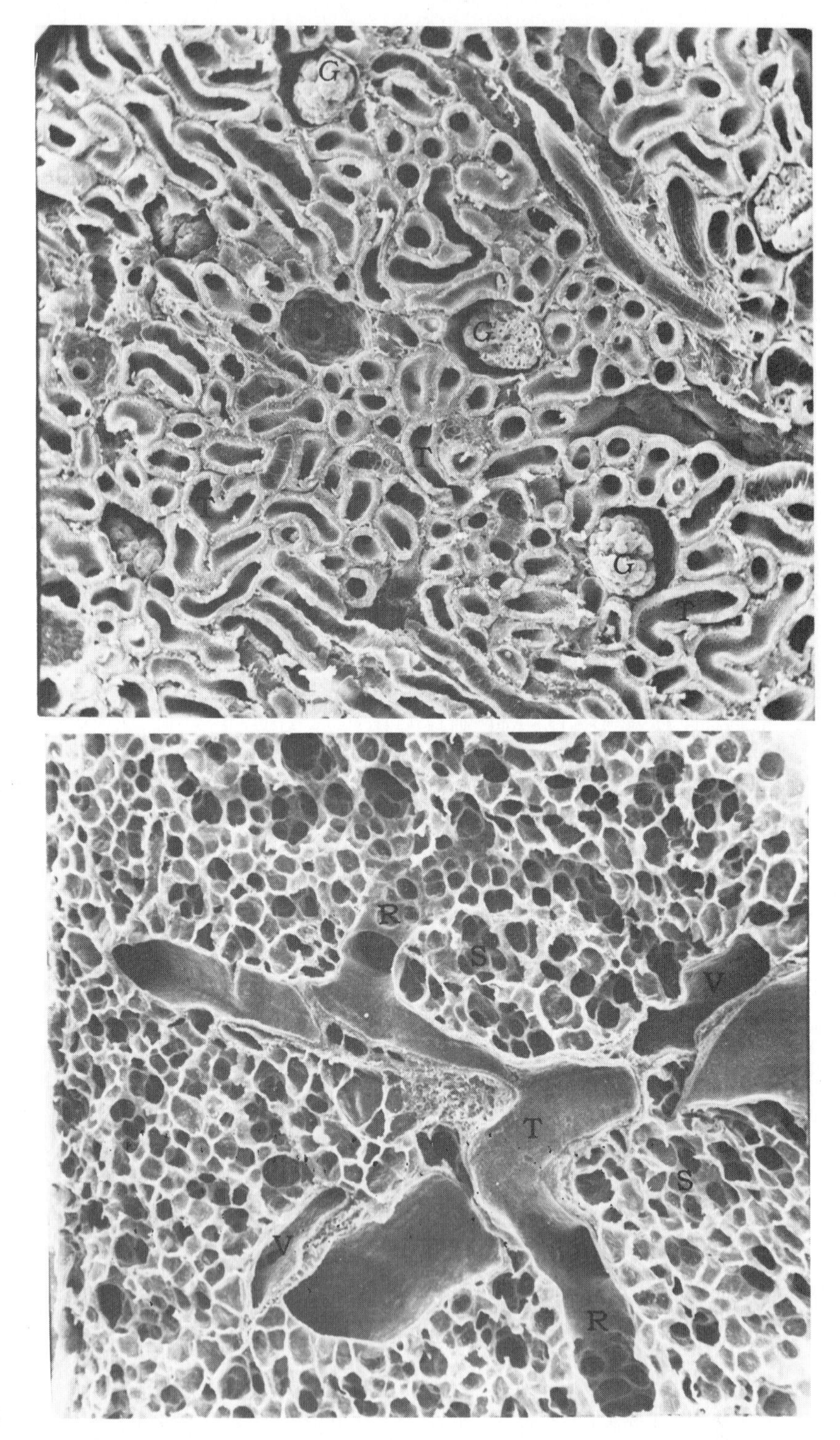
G
G
T
G
T
A
R
S
V
T
S
V
R
B

B. FIXATION METHODS

There are four major classes of fixation methods, each of which has its own advantages and disadvantages. These methods are discussed briefly here since the subject has already been treated in detail in Chapter 8.

1. Immersion. This method involves *excision* of the desired tissue and its submersion in the fixation medium. This approach is attractive because of its simplicity. Fixation artifacts, however, are introduced by immersion, since (for example, with animal tissue) the blood supply has been interrupted and internal anoxia can develop due to slow and irregular penetration. For the study of interior parts of tissues such as nerve, kidney, and lung, immersion fixation is clearly not ideal.

2. Dripping. The approach involves exposing the tissue surface to be studied by partial dissection of an anesthetized animal and flooding it with a large volume of the fixative. New fixative is added at decreasing intervals (from every few seconds to every few minutes). After about 20 minutes, the tissue is excised and immersed in fresh fixative. This method of *in situ* fixation reduces the extent of autolytic changes since both blood pressure and oxygenation are maintained until the moment of fixation. For SEM studies, the dripping method is particularly useful.

3. Injection. This is also an *in situ* method used for preserving the surface structure of internal organs. The fixative is injected directly into the cavity of an exposed living organ with a hypodermic needle or a micropipette. The specimen is then excised and further fixed by immersion. This method has the same advantages as dripping, as far as SEM is concerned.

4. Vascular Perfusion (Fig. 118). This is the most useful approach, it overcomes the major objections associated with other methods and, when effectively carried out, can yield excellent results. There are, however, limitations and difficulties associated with this method, such as the need for elaborate equipment and skill in carrying out the procedures. If carefully carried out through the left ventricle and aortic arch, fixation can be attained with a succession of organs—liver, gastrointestinal tract, spleen, kidneys, gonads, caudally, and the brain cranially—in the course of the same procedure.

FIG. 118. Demonstration of SEM of soft tissues fixed by vascular perfusion. ɔurtesy of Dr. P. Andrews)

(A) A survey scanning electron micrograph of a rat kidney cortex fixed by vascular perfusion. A number of glomeruli (*G*) and the lumina of many uriniferous tubules (*T*) can be seen on the cut surface. (Vascular perfusion fixation has preserved the proximal convoluted tubules in open states and prevented the blood elements from obscuring the cortical surface.) 120X.

(B) A survey scanning electron micrograph of the lungs. Terminal bronchioles (*T*) can be seen branching into respiratory bronchioles (*R*), alveolar sacs (*S*) and numerous alveolae. The lumen of a blood vessel (*V*) is also exposed. 50X.

Chapter 26

POSTFIXATION HANDLING

A. DEHYDRATION

THE STANDARD use of a graded series of alcohols for dehydration during the preparation of tissues for TEM (see p. 137), is a prerequisite to embedding in the non-aqueous media that is generally used. In SEM, the specimens examined are not usually sections and do not therefore require enclosure in a supporting matrix. The need is to eliminate the water entirely from the specimen rather than to substitute an embeddment for it. One of three drying techniques is used. Even so, it is necessary to perform solvent dehydration prior to air-drying in order to reduce shrinkage, prior to critical point drying so that the critical point solvent is conveniently exchanged, or prior to freeze-drying when one wishes to carry out this process from a non-polar solvent.

B. DRYING

If a specimen has been subjected to one of the prefixation approaches described in Chapter 20 it will be left in a wet condition. Three methods will be outlined that can be used for drying specimens for study by scanning electron microscopy.

1. Air Drying

If soft tissue water is evaporated by exposing the tissue to the air, severe shrinkage will occur. It may be reduced by drying from a liquid phase, i.e., from volatile low-surface tension solvents which have replaced the water (e.g., absolute alcohol, amyl acetate, Freon 113, etc.). The specimen must pass through a graded series of solvents prior to drying, as in the classical dehydration schedules. Some shrinkage can be helpful, for example, by inducing cell separation or by revealing components near the surface. The degree of shrinkage may be reduced by using a fixative that hardens the tissue (e.g., Parducz's) or by partially substituting, after solvent dehydration, a porous but solid substance (e.g., nitro-cellulose).

While the air drying technique is useful when special information, such as comparative form, is sought, it is not a substitute for the more reliable methods of critical-point drying and freeze-drying described below.

2. Critical Point Drying

The basic aim of this procedure is to remove the specimen's water without damaging it by the surface tension forces which act to collapse structures as the receding liquid evaporates. In a two-phase system, namely the liquid water of the specimen and the gaseous water over it, there is a "critical point," in terms of the environmental temperature and pressure applied to the two phases where they become indistinguishable. At this point the liquid phase can be converted to the gas phase, which in turn is removable from the system without the effects of surface tension coming into play.

The critical point process cannot be applied directly to water because it occurs at very high temperature and pressure. The tissue water must first be replaced by a liquid with a low critical point before drying. In the most widely used method, the sequence is first ethanol as a dehydrating agent, then pressurized liquid CO_2 at room temperature as the transitional liquid. The final step takes place in a drying bomb, that preferably has a transparent window so that the exchange can be followed visually. The temperature of the sealed bomb containing liquid CO_2 is raised above 31°C, the critical point of CO_2, at which temperature the liquid CO_2 and its gas are an indistinguishable single phase. Next, the pressure is slowly released, holding the specimen above 31°C, and allowing the gaseous CO_2 to leave the bomb; the specimen is then dry.

3. Freeze-Drying

This frequently used drying technique involves the total removal of the specimen water or solvent by freezing it and subliming it off in a high vacuum system. The latter process causes the least disturbance and thus leaves the solid elements of the specimen in their original configuration. The freezing process itself causes disturbances due to ice crystal formation. Ice crystals can puncture the cell surface, causing the formation of artificial holes. This, however, can be reduced in a number of ways, one of the most effective of which is to fix and first dehydrate with ethanol and then substitute amyl acetate for the ethanol prior to freeze-drying. The use of this non-polar solvent serves both to shorten the freeze-drying time and to eliminate ice crystal formation (since this solvent freezes into an amorphous glass). Freeze-drying is carried out at temperatures of about -70°C in a vacuum of 10^{-1} to 10^{-3} torr. The most suitable of the commercial freeze-drying units is one in which the specimen temperature can be continuously monitored.

When freeze-drying is completed, the specimen is warmed to a little above room temperature to prevent surface condensation of water prior to breaking the vacuum. The specimen can then be stabilized by carbon coating.

The specimen water can be replaced with other solvents which do not tend to form crystals near their freezing points. For example, water is substituted by ethanol, which is replaced by propylene oxide and then replaced by warm liquid camphene, which is quickly frozen at room temperature. The specimens are then placed in a vacuum chamber where rapid sublimation, i.e., freeze-drying, takes place. This procedure also has the advantage of permitting transfer of the specimen into the vacuum chamber at room temperature as well as preserving the interior of the freeze-dried specimen (not only its surface) so that it may be divided to expose other surfaces suitable for SEM study.

One may use both critical-point and freeze-drying techniques on the same fixed material so as identify artifacts induced by chemicals or freezing.

C. SPECIMEN MOUNTING

Large specimens are best mounted after drying. They may be glued (e.g., with conductive television tube coating paint, Apiezon wax* or epoxy resin) or attached by other means (e.g., double-sided adhesive) to a standard SEM specimen holder. For cells or subcellular particles in suspension, a drop is pipetted onto an abraided aluminum support surface before freeze-drying. Specimens that are already attached to a substrate (e.g., cells being cultured on coverslips) are quenched (e.g., in liquid N_2) as such prior to freeze-drying. The substrate is then attached to a specimen stub after drying.

D. SPECIMEN COATING

A specimen that is a conductor when examined in an SEM can be maintained at a desired potential, since any charge imposed on it by the incident electron beam can be led off. However, an insulated specimen exposed to the electron beam will become charged in the beam. This obstacle can be overcome by vacuum coating the specimen surface with a very thin layer of conducting material (0.01 to 0.1 μm thick) so as not to obscure the specimen surface. Electrical connection is made between the film and the specimen stage so that the charge is carried away.

*Apiezon W100 wax with a vapor pressure of about 6×10^{-9} torr at 20°C avoids drying and degassing problems occasionally encountered with other adhesives.

1. Coating Methods

Two basic methods are used for coating specimens: vacuum evaporation, and sputtering.

a. VACUUM EVAPORATION

In a evaporation unit (see Fig. L-1) where the pressure of its vacuum chamber is reduced to $\sim 10^{-5}$ torr, the heated coating metal rapidly evaporates into a monoatomic state. The vapor molecules are essentially unimpeded by air molecules, leave the source, and move in a straight line until they impinge on the surface of the specimen. There they condense to form a thin film. Using specimen holders mounted on a rotary device which facilitates presenting all the surfaces to the evaporation source, coating of complex contoured surfaces can be carried out. (The goal of vacuum evaporation is even coating and thus this process differs from shadowing [p. 329], where thickness variations are intentionally introduced to secure an artificial three-dimensional effect in the replicas examined by TEM.)

b. SPUTTER-COATING

Sputtering is a process that occurs when an appropriate target metal is bombarded by energetic particles.* Typically, the atoms from the surface of a target material are physically ejected with a glow-discharge instrument which bombards it with positive ions. A ''gas'' of metallic atoms is created in the poor vacuum by prolonged erosion of the target surface and they deposit uniformly as a coat on all specimen surfaces, regardless of orientation.

The simplest (diode) form of sputter-coating unit consists of two electrods: an anode on which the specimen is placed (and held at positive potential), and a cathode, the target metal. More recently triode sputter-coating units have been developed which virtually eliminate specimen heating. To achieve good results with sputter-coating, it is advisable to optimize the operating conditions.

2. Coating Materials

Materials used for coating fall into three catagories.

(1) Metal coating materials. Ideally, the material should be structureless (at the needed resolution level), possess contouring capability, and have a desirable electron emission. A wide variety of metals have been used singely

*This process is also used in some high vacuum pumps (see sputter-ion pumps, p. 337).

or in combination. Of these, gold, gold-palladium, or palladium are the most popular metals. Gold is especially useful because it is easily evaporated, has a reasonably small granularity, and has a high secondary electron emission coefficient, thus giving a high signal-to-noise ratio in the image.

(2) Carbon coating. Carbon is advantageous because it is easily evaporated, causes little specimen heating, is easily scattered to form a continuous layer, and forms a tough substrate which stabilizes the specimen surface; it is the choice coating material for x-ray microanalysis because it does not produce interfering x-rays.

(3) Antistatic agents. In some cases, such as low magnification studies, a conductive antistatic spray (e.g., Duron) can be used as an alternative approach to metal evaporation. Although convenient to use, the spray tends to cover small surface details, and so it is not applicable for high magnification studies.

Chapter 27

SPECIAL PROCESSING

A. SECTIONING

EXAMINING sectioned material prepared by standard electron microscope methods with an SEM reveals very little significant information. This is the result of the homogeneous nature of the embedded specimen, which usually does not provide meaningful topographical differences. There are several possible approaches which have been used with some success to overcome this problem.

(a) Fixed, unembedded, hand-cut (thick) tissue slices, sectioned across internal spaces, can provide information of interest.

(b) Fixed, unembedded, (Smith-Farquar) microtome cut tissue slices have also been examined.[209]

(c) Fixed, paraffin embedded, microtome cut sections have been used to reveal the internal architecture of both plant[210] and animal[211] tissues after removal of the paraffin and drying (preferably by the critical point method).

(d) Fixed, double-embedded, sectioned material can be profitably used after one of the embedments is removed; this serves to limit shrinkage while at the same time separating the layers.

(e) Fixed, Epon-embedded, thick sections have been examined after the plastic had been partially removed with an iodine-acetone solution.[212]

B. FREEZE-FRACTURING (Fig. 119)

A theoretically attractive procedure is to freeze the liquid component of soft (watery) tissues to make it brittle, and then fracture it. Unfortunately, ice crystal growth can cause displacement of cell components, and when the ice is sublimed in vacuum, holes are left making interpretation of SEM images difficult.

There are two basic methods of limiting ice crystal growth. One method involves the use of a protective agent (e.g., sucrose, gelatin, or DMSO, and/or glycerol[213]). These agents are used in high concentration prior to freezing and fracturing and are removed after the tissue has thawed, or by freeze-substitution. The tissue must then either be freeze-dried or critical-point-dried. A second method involves the substitution of the water

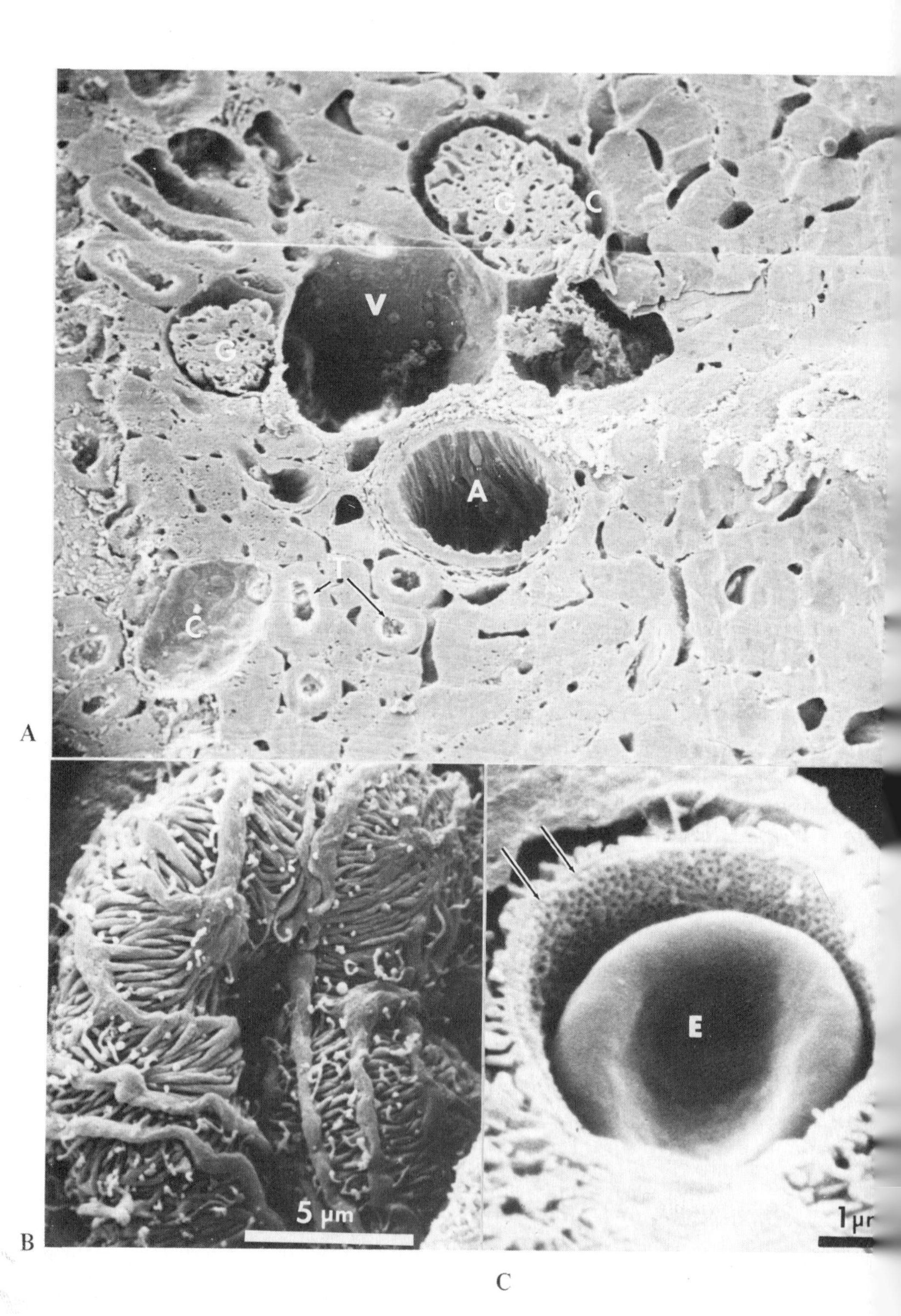
G
C
V
G
A
T
C
A
5 μm
B
E
1 μr
C

by nonaequeous solvents (e.g., ethanol,[214] amylacetate, or Freon 113) before fracturing. The fixed and dehydrated tissue is submerged in supercooled Freon 12 and then fractured. It is then either freeze-dried, air-dried or critical-point-dried (see Chap. 22).

C. FREEZE-ETCHING

For surface investigations (especially of membranes) the freeze-etching technique (see p. 201), in conjunction with TEM, presents a generally attractive alternative to SEM. An advantage in interpreting surface characteristics becomes even more striking by the use of stereo-pair micrographs, as well as in examination of opposing surfaces of a fracture, especially at levels of resolution beyond those routinely attainable with the SEM.

While it is nevertheless possible to study the shape and depth characteristics of replicas of freeze-etched specimens with the SEM, difficulties associated with the freezing process can result in images of the replicated frozen surface that may not easily be interpreted. The freeze-etching technique is therefore used more profitably in conjunction with TEM and it can possibly be useful in the correlation of TEM and SEM images of the same areas.

Mineralized tissues (e.g., enamel, shells) obviously lend themselves better to the fracturing procedure than does soft tissue. This technique is especially useful if the internal repetitive structure of the specimen affects the direction of the fracture. Such preparations will also be of value when cells or other components are intrinsic elements embedded in the mineralized matrix, since their spaces will be opened up by fracturing. The more densely mineralized the tissue, the cleaner the fracture will normally be.

D. ION BEAM ETCHING (Fig. 120)

This technique is an additional method useful in revealing underlying

---

FIG. 119. Demonstration of the application of the (ethanol) freeze-fracturing technique. (Compare with Fig. 118A) (Courtesy of Dr. W. H. Humphreys.)

(A) A fractograph of the mouse kidney cortex. Note the flatness of the plane of fracture. (*G*), glomerulus; (*A*) artery; (*V*), vein; (*T*), convoluted tubules; and (*C*), Bowman's capsule. 340X.

(B) A fractograph of Bowman's space exposing the exterior surface of a glomerular capillary loop. The fracture plane was above the loop and therefore exposed only the outer surface of the glomerulus. 5,200X.

(C) A fractograph of a portion of a capillary loop of a glomerulus. The basement lamina and associated structures are exposed. Arrows point to endothelial pores that were bisected by the fracture. (*E*), erythrocyte. 12,500X.

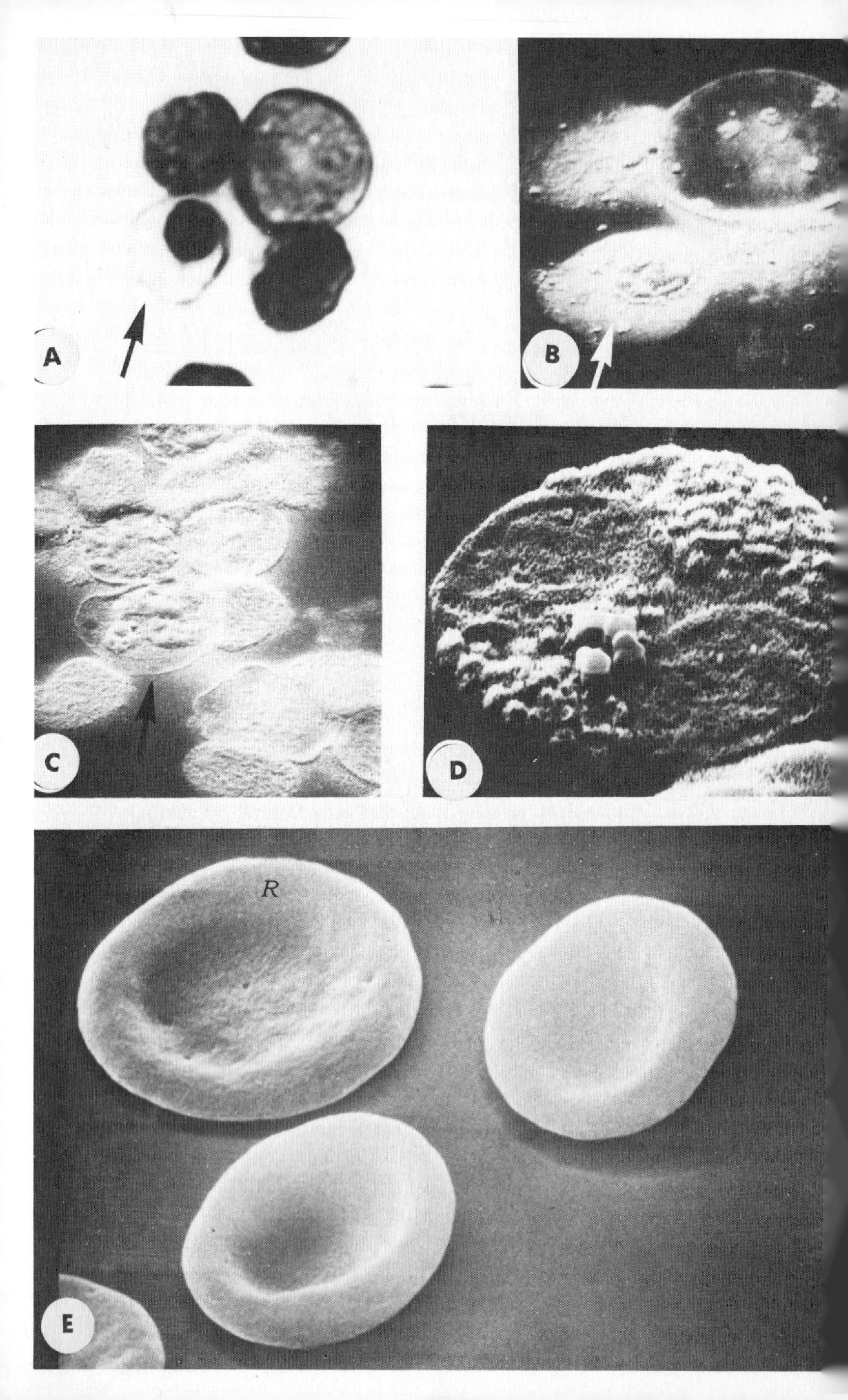
A
B
C
D
R
E

surfaces[215]. By using an ion beam source that can be operated in the specimen chamber of the SEM, it is possible to remove parts of specimens selectively. The ion beam source is a cold cathode argon discharge unit that can be focused on the specimen stub. The location of the ion source within the microscope permits the carrying out of a controlled etching process followed immediately by examining the uncoated specimen at low accelerating voltages. If the specimen is given a metal coat, it can be studied at high resolution levels. The coating can be removed if further etching is necessary.

It is not certain if the ion source removes material by simple vaporization or by carbonization. In any case, soft tissue will etch more rapidly than hard tissue; as a result, some problems in interpretation of the image may occur when both types of tissue are present together. In general, it is necessary to distinguish among differences arising from variations in the substructure of homogenious materials, those due to selective etching because of tissue resistance, and those due to the uneven charging effects of the ion beam.

E. REPLICATION

For certain types of studies, examination of replicas has special advantages. These are, for example, studies of sequences of events such as etching or digestion, or a series of growth states or deep surface depressions or casts of internal cavities. Replicas may be examined directly (after being given a conducting coat of carbon and/or metal), or may be replicated themselves to give a positive of the original. (This topic is also considered on pp. 204 and 329.)

F. CELL SURFACE LABELING

The SEM has the ability to be used not only for examining surface morphology, but can also be used for immunological studies. This is because

---

FIG. 120. Demonstration of ion beam etching.
(Courtesy of Drs. S.M. Lewis and B. Frisch)
(A) A Romanowsky stained blood preparation from a case of acute leukaemia.
(B) The same group of cells have been etched in H_2 gas. The normoblast (arrow) has a compact nucleus. The leukaemia is extremely etch sensitive.
(C) Other leukaemic cells with variable etch sensitivity. The arrow points to a mature neutrophil leucocyte.
(D) An eosinophilic leucocyte after etching.
(E) Erythrocytes lightly etched in argon. The reticulocyte (*R*) has lost part of its surface layer revealing a fibrillary network and some pitting. The mature cells are more etch resistant.

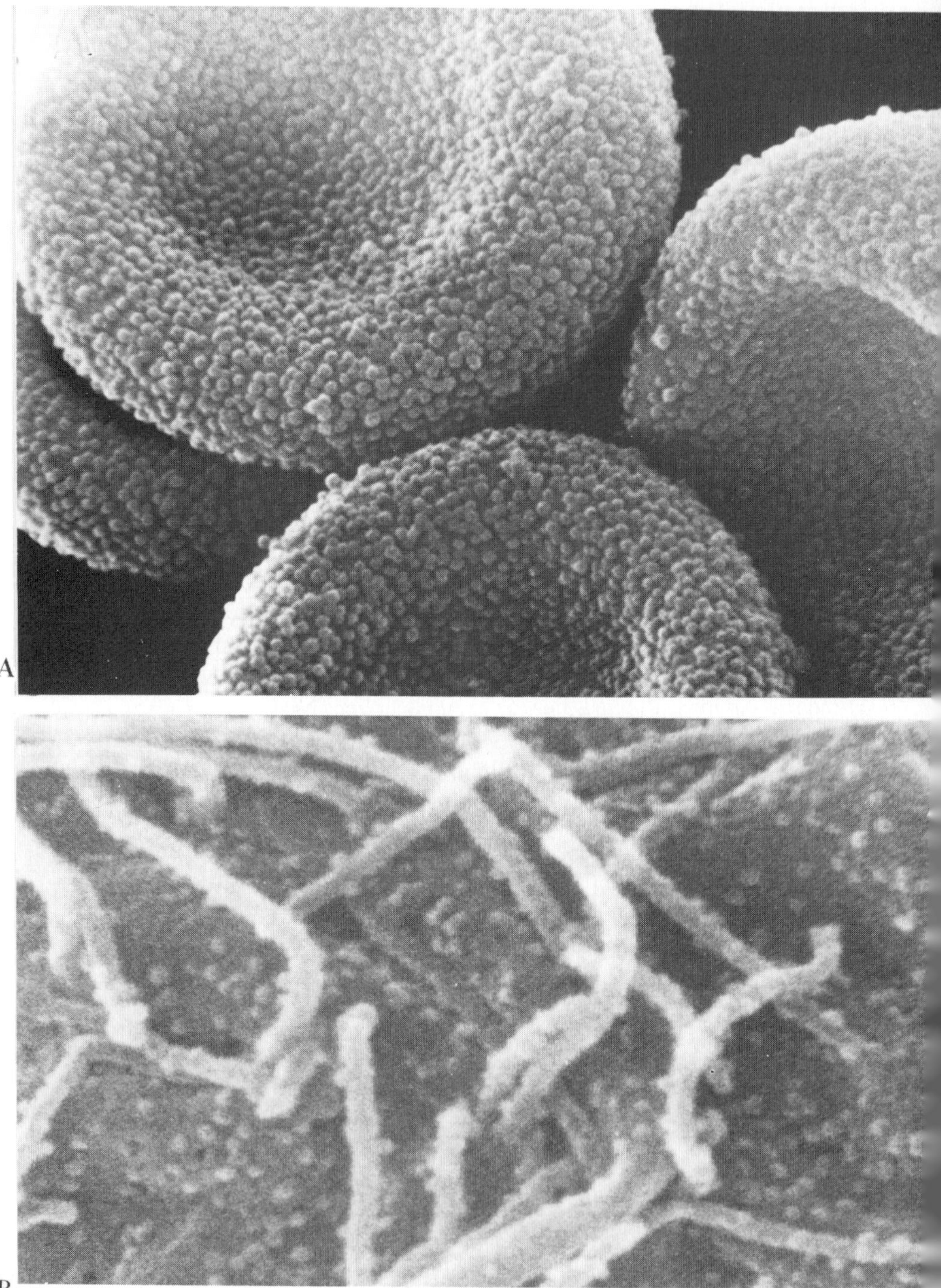
A
B

the relevant techniques that were originally developed for fluorescent and transmission electron microscopy have been adopted for use with the SEM. The nature, distribution and mobility of antigens and receptors on the surface of cells can consequently be studied. Both biological entities and synthetic particles can be used as markers. Reviews of marker techniques have been published by Nemanic[216] and by Molday.[217]

A wide variety of marker techniques have been developed. Each has its own advantages and limitations; the most common are summarized briefly.

1. Antibody-Latex Sphere Coupling (Fig. 121)

Latex spheres can be coupled to antibodies in order to localize specific antigens on a cell surface. Two basic approaches have been developed.

a. ABSORTION METHOD

Latex spheres can be noncovalently coated with adsorbed antibody.[218] The spheres attached to the cell surface can then be stabilized by postfixation with glutaraldehyde[219] preventing disassociation during dehydration and critical-point drying. This is a valuable method, but artifacts can be induced and specificity problems can arise (Fig. 121).

b. COVALENT LINKAGE METHOD

It is possible to couple an antibody covalently to prepared acrylic latex spheres.[220] The surface properties of the antibody ensure specific binding and its linkage to the polymeic matrix helps keep it intact even when subject to stressful processing.

2. Silica Sphere Labeling

Silica spheres of known size produced by polymerization of silicic acid have been used for labeling virus antigens at the cell membrane.[221] This marker may be particularly useful in high resolution studies because of its small size, uniformity, and stability.

FIG. 121. Demonstration of the cell surface labeling.
(Courtesy of Dr. R.S. Molday.)
(A) Scanning electron micrograph of human red blood cells treated with rabbit antired blood cell antiserum and subsequently labeled with goat anti-rabbit immunoglobin-latex microspheres, approximately 100 nm in diameter. 16,500X.
(B) Scanning electron micrograph of a mouse 3T3 fibroblast directly labeled with wheat germ agglutinin-latex microspheres, approximately 400 nm in diameter. 45,000X.

3. Antigen-Sandwich Labeling

A hapten is a small antigenic chemical that can elicit antibody production. The antigen-sandwich technique uses a hapten modified anti-cell-surface antibody, an anti-hapten antibody, and hapten-modified-markers (e.g., hemocyanin). The anti-hapten antibody links a hapten-modified marker to the hapten-modified anti-cell surface antibody that is bound to antigens on the cell surface.[216]

4. Ferritin Labeled Antibody

Ferritin, an electron dense, protein containing iron, has been discussed as a useful marker in immuno-transmission electron microscopy (see p. 181). It has been applied in SEM studies as well (see ref. in Hattori *et al.*[222]), and thus the possibility of correlative investigations exists. Ferritin has the advantage of a relatively small size (11 nm diameter), allowing resolution of fine details of the cell surface. It is, however, not a desirable marker for rough-surfaced cells, or where heavy metal coating is essential.

5. Other Antibody Markers

In addition to the relatively more common markers, several others can be noted.

(a) Plant viruses. Plant viruses, such as the tobacco mosaic virus, have distinct shapes and dimensions. Such viruses can be used by coupling them with an antibody as markers.[223]

(b) Bacteriophage T4. This marker has been used for detecting surface antigens on RBC's.[224] It can be used because antibodies that are specific for immunoglobins can be conjugated to T4 by glutaraldehyde, preserving its shape and the antibody activity.

6. Other Markers

(a) Concanavalin A–Hemocyanin. This complex can be used for visualizing Con A binding sites on the surface of blood and other cells.[225] Con A is a plant lectin that specifically binds to certain sugar residues associated with the protein that is bound with the cell surface. The blood protein, hemocyanin, binds to Con A and the complex is stabilized by glutaraldehyde fixation. The hemocyanin molecules in rectangular or circular profile can be seen on the cell surface.

(b) Colloidal gold. This marker can be readily prepared as homogeneous particles of various dimensions. Because these particles carry a negative charge, they bind to a protein such as Con A and can be used, for example, to visualize mannose on the surface of yeast.[226]

G. LIVING MATERIAL

The SEM has two advantages insofar as the examination of living specimens is concerned. It permits the examination of their entire external surface,[227,228] and it also exposes them to only small amounts of energy (because it uses low beam currents). However, only specimens that are not severely dehydrated by the specimen-chamber vacuum or that can tolerate a dehydration/hydration cycle can be examined ''alive.'' This severely limits the nature of the specimens that can be examined. Proposals have been made for using special specimen chambers, or for using a high voltage SEM to enable a broader spectrum of living specimens to be viewed.

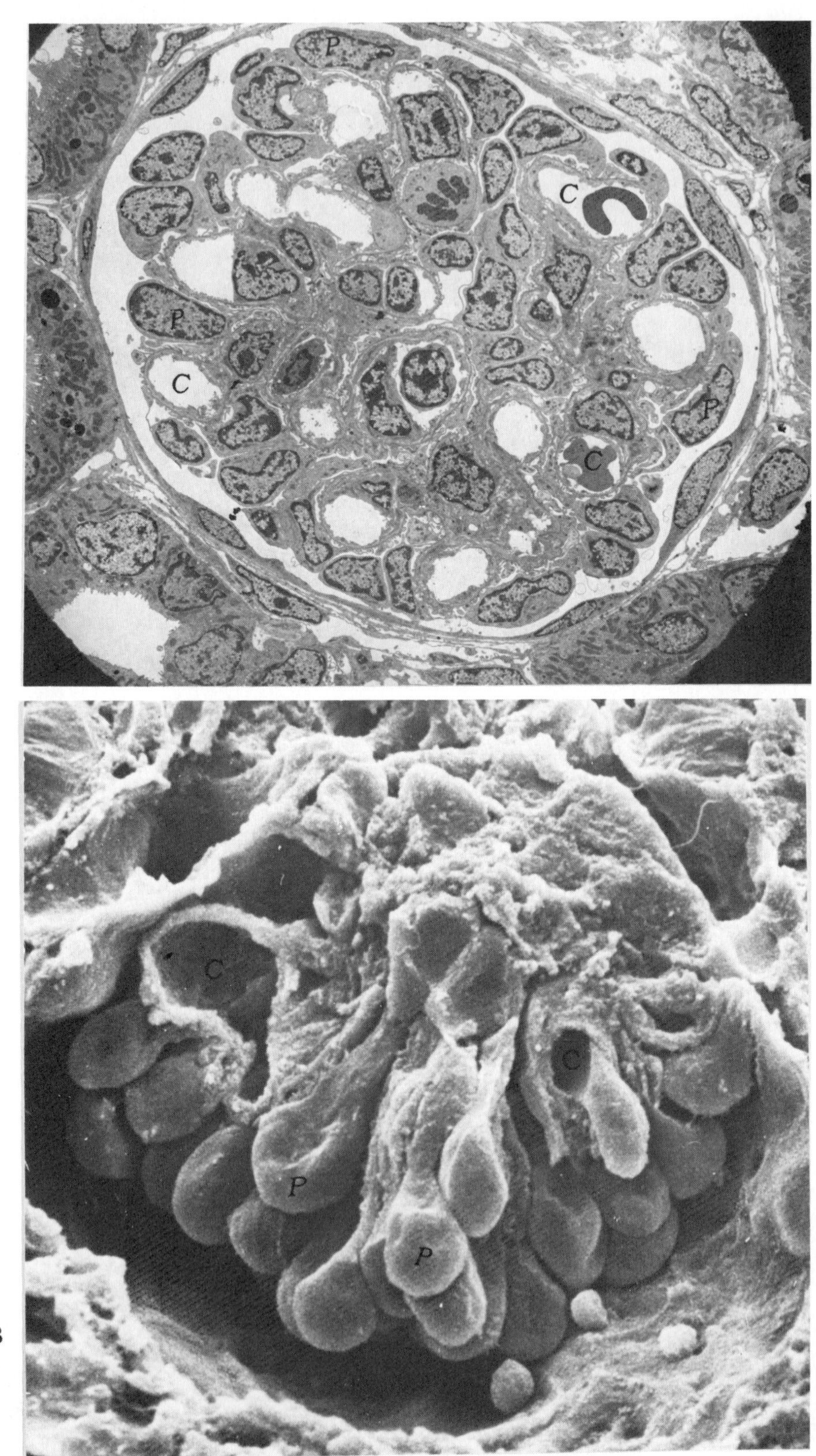

A

B

Chapter 28

COMPLEMENTARY TEM AND SEM STUDIES

TRANSMISSION and scanning electron microscopic studies of the same material can, *at times,* be correlated to provide useful data (Fig. 122). Correlative studies can be used in a variety of ways such as the following:

(a) To interpret new data obtained by the SEM.

(b) To verify data already obtained by the TEM.

(c) To evaluate the usefulness of preparatory procedures developed for use with the SEM or TEM.

(d) To correlate data about the same specimen examined in different orientations (e.g., surface view and cross-section).[229]

(e) To study isolated intact single cells (especially when a high voltage TEM is used).[230]

These advantages have led to an increase in the number of correlative studies being published.[231,232,233]

FIG. 122. Demonstration of the complementary value of TEM and SEM studies of the same type of meterial.

(Courtesy of Drs. M. Miyoshi, T. Fujita, and J. Tokunoga)

(A) A transmission electron micrograph of a glomerulus from the kidney of a fetal rat (10 days old). Numerous capillaries (*C*), two of which contain red blood cells, can be seen. The visceral epithelial cells, or podocytes (*P*) of the glomerular capsule are also evident. 2,100X.

(B) A scanning electron micrograph also of a glomerulus from the kidney of a fetal rat (2 days old). The glomerulus was broken during specimen preparation. A few blood capillaries (*C*) are cut open and glomerular epithelial cells or podocytes (*P*) are attached to the capillary walls—are seen from the side projecting into Bowman's space. 3,600X.

APPENDICES

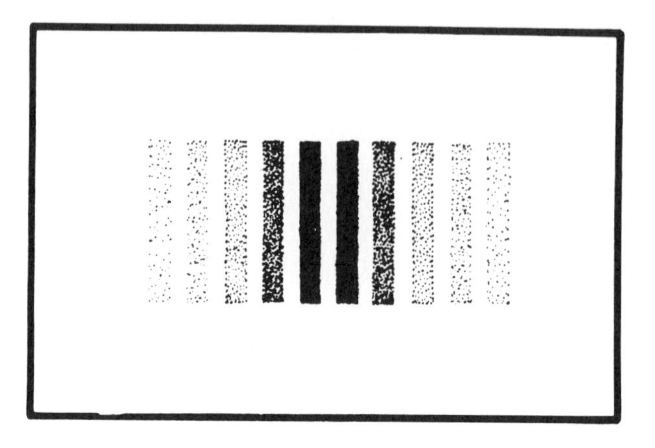

FIG. A-1. Young's screen.

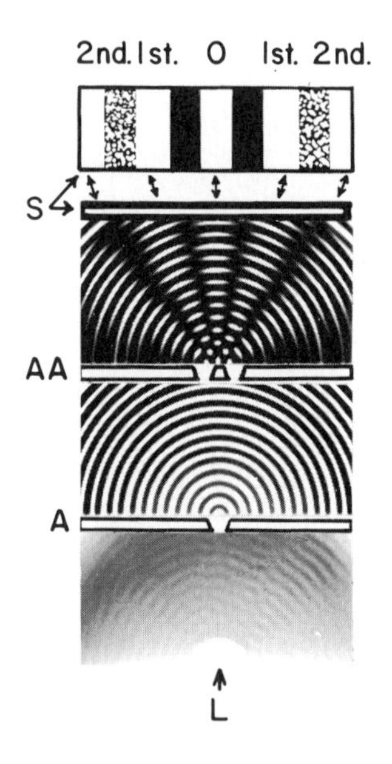

FIG. A-2. Interference phenomenon. A monochromatic incoherent light source, *L*, is placed in front of a pinhole aperture, *A*, which then becomes a second coherent source from which Huygensian wavelets issue. Behind this is a double slit aperture, *AA*. Each slit becomes a secondary source of light. At some distance away from *AA*, is a screen, *S*. On this screen appears an alternating series of bright and dark fringes. The former are produced by the wave trains arriving at the screen in-phase (constructive interference) and the latter are due to the wave trains arriving out-of-phase (destructive interference). The orders of the spectra are indicated.

Appendix A

THE NATURE OF LIGHT

During the latter part of the seventeenth century, Sir Isaac Newton was the most distinguished proponent of the corpuscular theory of light. He was able to explain most of the phenomena of light observed at that time, namely, rectilinear propagation, reflection and refraction, by making "adjustments" in the corpuscular theory or by making assumptions which later were proven to be incorrect. During this period, Christian Huygens, in a short pamphlet, set forth his wave theory of light. For more than a century Huygens' revolutionary theory was ignored. This was due partially to Newton's prestige. In addition, a wave concept of light implied that, like other waves, light waves should bend around corners and experiments which demonstrated diffraction of light were not made till later. Finally, there was no special medium which would account for the transmission of light waves; yet, at that time, a medium was considered a prerequisite for wave motion.

In 1801, Thomas Young demonstrated interference, a new phenomenon associated with light. If a source of light is placed a short distance away from a double-slit aperture, the image on a screen placed behind the aperture consists of a series of alternately dark and light narrow bands (Fig. A–1). To explain this phenomenon, the light from the source which passes through the slits is, at each slit, considered as a new light source producing bright bands of reinforced light where the waves reach the screen in step (in-phase) with each other (constructive interference) and producing dark bands where the waves are out-of-phase (destructive interference) (Fig. A–2). Young's experiment, which appeared to be explicable *only* in terms of Huygens' wave theory, served to revive and firmly establish the wave theory.

About the beginning of the twentieth century, it was discovered that light falling on the surface of certain metals causes them to emit electrons. These are known as *photoelectrons* and this phenomenon is known as the *photoelectric effect*. When it was found that the number of photoelectrons emitted was directly proportional to the amount of the light striking the sensitive metal, the wave theory of the nature of light was seriously questioned. According to this theory, an increase in the amount of the light used in producing the photoelectric effect should result in an increase in the velocities of the photoelectrons emitted. This, however, is not the case. Although the *rate* of electron emission (i.e., number emitted per second) increases at higher intensity, these electrons have the same velocities as

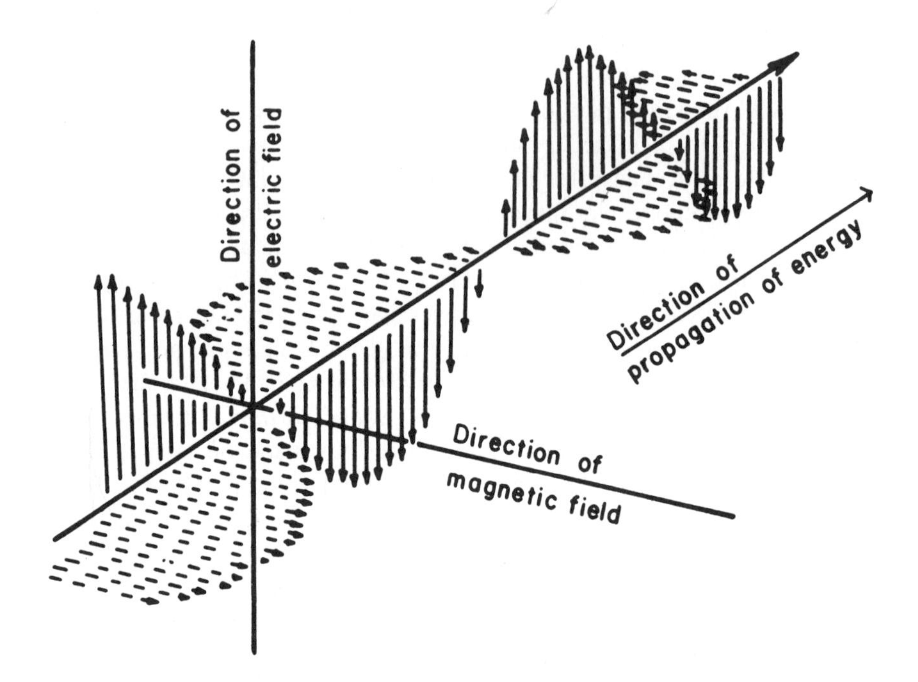

FIG. A–3. An electromagnetic wave. It consists of electric and magnetic fields at right angles to each other and both perpendicular to the direction in which the wave advances.

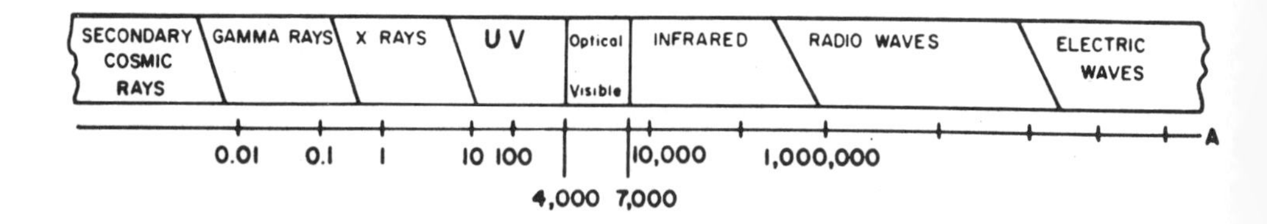

FIG. A–4. The electromagnetic spectrum.

those emitted at lower light intensities. On the other hand, Newton's corpuscular theory could not satisfactorily explain the discrete velocities of the emitted photoelectrons, even though it could explain the increased rate of emission. Thus both of the classical theories were found to be inadequate to explain the photoelectric (as well as other) effects.

The quantum theory, a product of the work of Planck and Einstein, was introduced in the early 1900's. It began to provide a basis for explaining such widely divergent phenomena as interference, diffraction, and the photoelectric effect, and thus offers a common basis for explaining the nature of the action of light. The quantum theory postulates that the transfer of energy between light and matter occurs only in discrete quantities proportional to the frequency of the light wave. Thus light with all its wave-like properties is looked upon as existing as energy packets or bundles, *quanta*, each of which is known as a *photon*. The energy of one photon is determined by the equation

$$E = hf \tag{1}$$

where:

E = energy of a photon (joules).
h = Planck's constant (6.624×10^{-34} joule-sec).
f* = frequency (cycles per second).

The current concept characterizes the nature of light in terms of corpuscular wave-like properties which find their quantitative expression in quantum mechanics.

A theory of the relationship of light to other electromagnetic phenomena was developed mathematically about 1865 by Maxwell. He theorized that the energy of electromagnetic waves is equally divided between an electric field and a magnetic field, one perpendicular to the other and both perpendicular to the direction in which the waves are propagated (Fig. A–3).

In 1885, Hertz experimentally confirmed the *electromagnetic theory* by showing that light waves and electrically generated (radio) waves are of the same essential nature. Naturally, many of their properties are manifested in qualitatively different ways due to the pronounced difference in their respective wavelengths. We now know that the electromagnetic spectrum has a very broad range which can be arbitrarily subdivided into 8 regions (Fig. A–4). Visible light, which belongs to the optical region, is the form of electromagnetic radiant energy which can be perceived by the human eye.

* Where: $f = c/\lambda$; λ = wavelength of light; c = velocity of light.

Appendix B

WAVELENGTH OF AN ELECTRON

In 1924, Louis de Broglie postulated that in every mechanical system wave-like properties can be associated with particles. His studies established the field of *wave mechanics* which encompasses the area dealing with the wave properties of masses of all sizes. This theory implies that a wavelength can be assigned to a particle and it is an extension of the quantum theory (see p. 209). The roots of de Broglie's formulation can be appreciated by taking the quantum equation [see (1), p. 209.]

$$E = hf \tag{1}$$

and substituting for E from Einstein's equation for the equivalence of mass and energy:

$$E = mc^2 \tag{2}$$

where:

m = mass.
c = velocity of light.

Thus we get for a photon:

$$mc^2 = hf \tag{3}$$

or

$$m = \frac{hf}{c^2} \tag{4}$$

but since (see footnote on p. 209):

$$c = f\lambda \tag{5}$$

and, therefore,

$$m = \frac{hf}{(f\lambda)^2} = \frac{h}{(f\lambda)(\lambda)} = \frac{h}{c\lambda}$$

or

$$mc = \frac{h}{\lambda} \tag{6}$$

Studies on the pressure exerted by light had led to the identification of h/λ with the momentum of a photon. Thus the wavelength of a photon may be expressed in terms of mc, another equivalent measure of its momentum.

$$\lambda = \frac{h}{mc} \tag{7}$$

In 1924, de Broglie proposed that similarly a wavelength for any particle of mass, m, having a velocity, v, can be expressed in terms of its momentum, mv, as:

$$\lambda = \frac{h}{mv} \tag{8}$$

As v approaches the velocity of light, m must be replaced by $\dfrac{m_0}{\sqrt{1 - \left(\frac{v}{c}\right)^2}}$

and

$$\lambda = \frac{\dfrac{h}{m_0 c}}{\sqrt{1 - \left(\dfrac{v}{c}\right)^2}} \tag{9}$$

The validity of this equation has been confirmed by all observations since that time and provides part of the basis for calculation of the resolving power of the electron microscope (see p. 14).

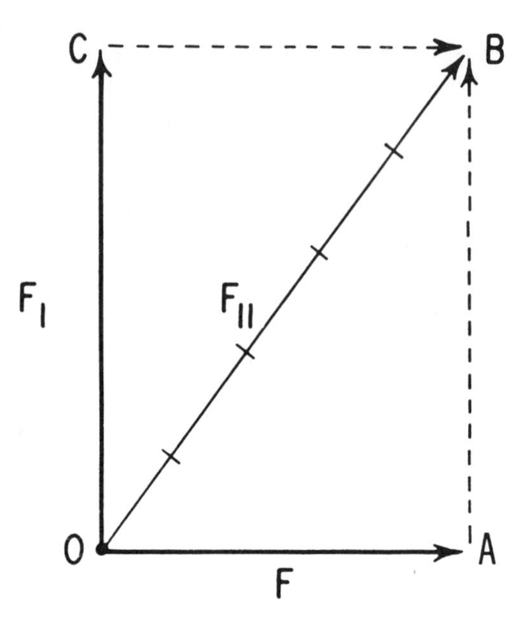

FIG. C–1. Addition of vectors.

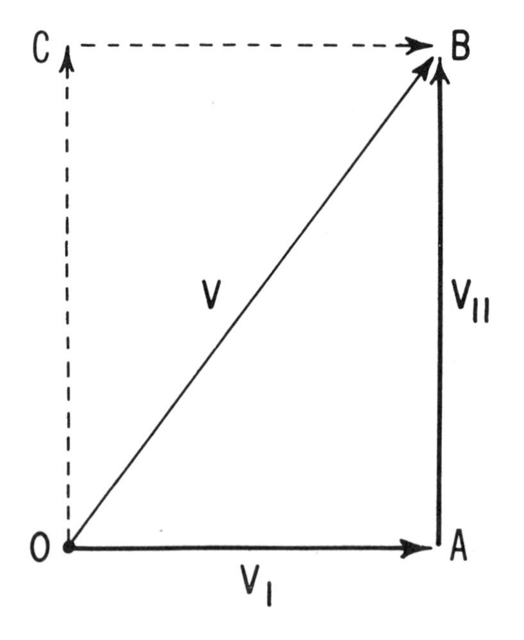

FIG. C–2. Resolution of vectors.

Appendix C

SUMMATION AND RESOLUTION OF VECTORS

Physical quantities generally fall into two categories. There are those that have magnitude but do not involve direction, such as density and temperature. These have only a numerical value and are called *scalar quantities.* In the second category are those, such as velocity, force and acceleration, which have both magnitude and an associated direction. These are called *vector quantities* and are conveniently described in a coordinate system by an arrow whose length represents its magnitude and whose head indicates its direction.

Graphically represented vectors can be added, or a single vector can be resolved into components. Both of these operations will be exemplified.

1. Summation of Vectors

In this case (Fig. C–1), the vector quantity, F, represents a force of the magnitude 3x (x = units) acting upon a body located at point O in the direction O-A. If a second force, F_1, of a magnitude 4x is applied to the body at point O in the direction O-C, the resulting force is the same as that which would be obtained by using a force, F_{11}, equal to 5x acting on the body at O in the direction O-B. Since this force F_{11} has a magnitude and direction which is equal to the combined effect of the forces F and F_1 acting together it, therefore, can be considered to be the sum or *resultant* of the individual vectors.

If we draw F_1 parallel to its original direction proceeding from point A, we can construct a right triangle with F_{11} as the hypotenuse. Conversely, drawing F from point C, parallel to its original direction, F_{11} is again found to be the hypotenuse of the resulting triangle as well as the diagonal of the rectangle formed by both triangles. In the general case where the forces are not at right angles to each other, we obtain a parallelogram. In general, a diagonal, such as F_{11}, represents the magnitude and direction of a force which is equal to the sum of two other forces, F and F_1, at given angles to one another.

2. Resolution of Vectors

In this case (Fig. C–2), the vectors also represent another physical quantity, namely, velocity.

Thus, if the vector, V, represents the velocity and direction of the movement of a body from O to B, the same motion can be described in terms of the velocities in the directions OA and AB. These are found to be equal to V_1 and V_{11}, respectively. We see that the original vector is the diagonal of a rectangle or parallelogram and it can be "resolved" into and represented by two appropriate components. This procedure is applicable to the situation under discussion, namely, analysis of the influence of magnetic and electrostatic fields on trajectories of electrons that are moving through the fields.

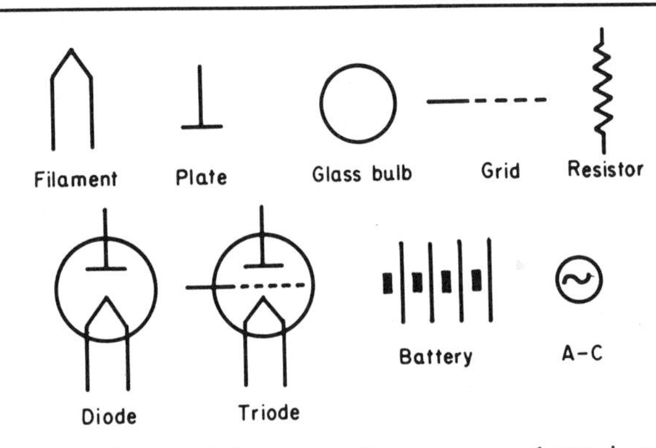

FIG. D–1. Symbols used in representing common electronic components.

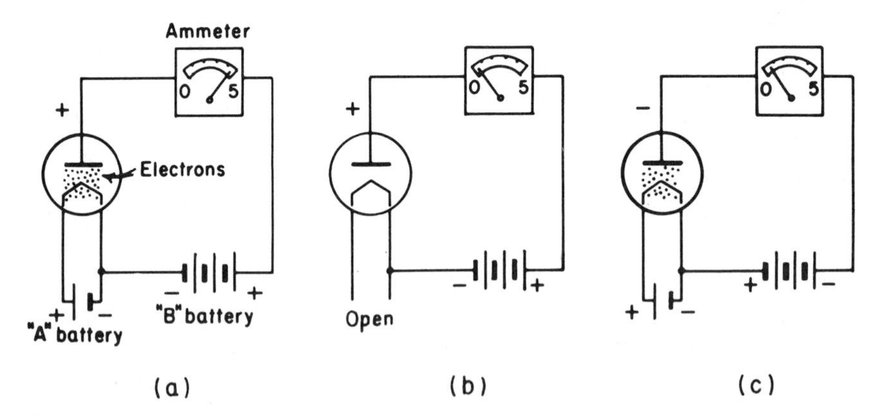

FIG. D–2. A diode circuit. (a) Filament is hot resulting in current flow. (b) Filament is cold thus no current flow. (c) Filament is hot but polarity of the "B" battery is reversed thus no current flow.

Appendix D

FUNDAMENTALS OF ELECTRONICS

Since the electron gun is in essence a type of vacuum tube, the basic information concerning the latter provides an introduction to an understanding of this part of the electron microscope.

The simplest vacuum tube is the *diode*. It usually consists of two electrodes which are sealed in a glass cylinder from which the air has been evacuated. The two electrodes are the cathode (which in its simplest form usually is a single filament with two leads to provide the heating current) and an anode (or plate, another electrode separated by some distance from the cathode). A diode's elements are commonly represented schematically with symbols that have a 1:1 relationship to its physical components (Fig. D–1).

A simple diode circuit is set up by connecting the two leads of the filament to an "A" battery (or ac source) and a plate to the positive terminal of another "B" battery, with an ammeter interposed between the plate and "B" battery and with the negative terminal of the "B" battery returned to one of the filament leads (Fig. D–2). No meter deflection will take place, i.e., no current flow will register, unless one additional condition is satisfied, namely, the flow of current from the "A" battery through the filament has heated it to a high temperature. Under these conditions (Fig. D–2a), the filament will emit electrons (thermionic emission) and these electrons will be attracted to the plate which has a positive potential with respect to the filament. As a result, there is a measurable flow of electrons that is recorded by the meter. When the filament is not heated (Fig. D–2b), no electrons are emitted by the filament. If the plate is made negative by a reversal of its connection to the "B" battery, the emitted electrons are not attracted by the plate (Fig. D–2c). In both of the latter cases no current flow takes place in the *plate circuit*.

A somewhat more complex vacuum tube, the *triode*,* consisting of three elements, was introduced by Lee de Forest. He placed a wire mesh or

* A triode circuit can be subdivided into three distinct parts: (a) *filament circuit*, which consists of the filament and the "A" battery; (b) *plate circuit*, which consists of the cathode, the electron flow from the cathode to the plate within the tube, the plate, the load resistor and "B" battery; and (c) *grid circuit*, which consists of the cathode, grid and source of input voltage and source of bias voltage ("C" battery).

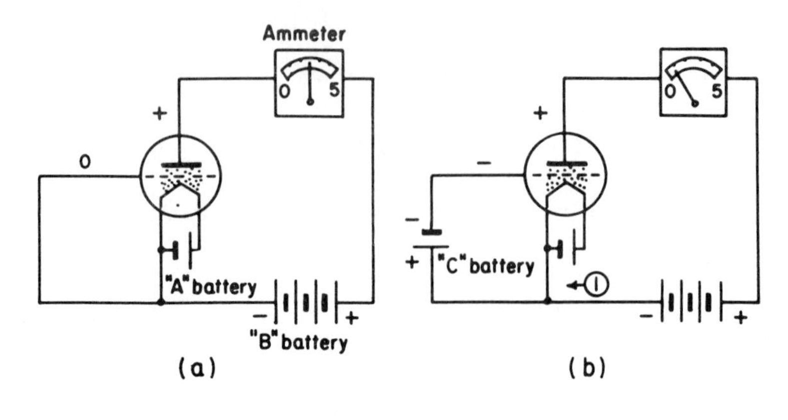

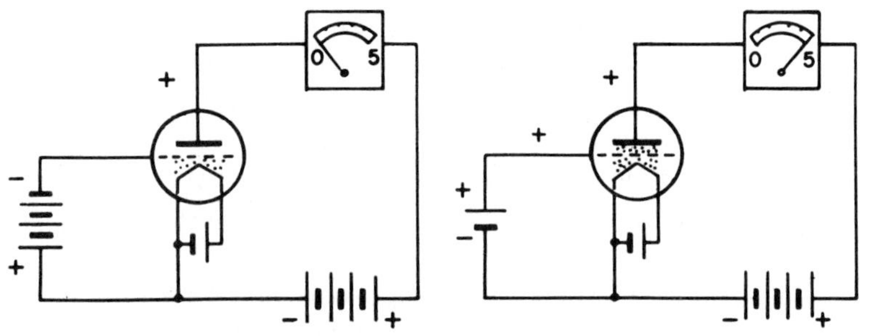

FIG. D–3. A triode circuit showing the effects of grid potential on the plate current. (a) Grid potential is zero and current flow occurs. (b) Grid potential is negative and current flow is diminished. (c) Negativity of grid potential is increased to the point of suppressing current flow. (d) Grid potential is positive and current flow is increased.

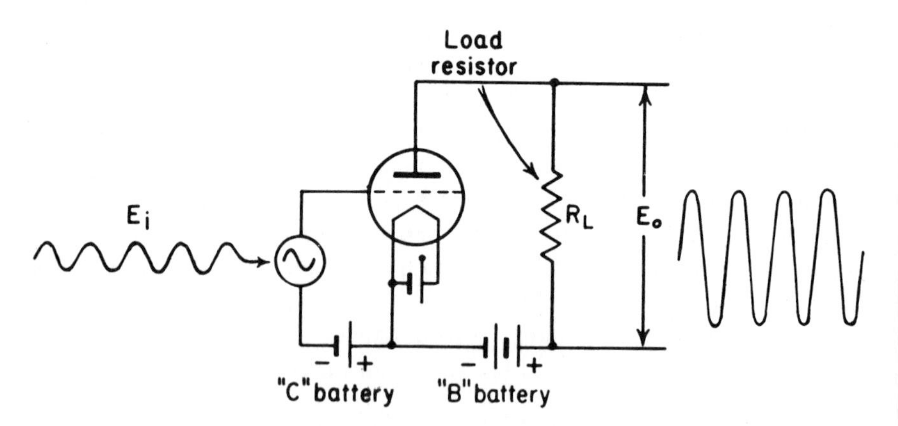

FIG. D–4. Amplification. A small alternating input voltage, E_i, plus dc voltage of the "c" battery, provides the grid potential in this circuit. A much larger output voltage, E_o, of the same wave form results.

grid between the filament and plate to control the flow of electrons between these two electrodes. If there is no voltage between the grid and the cathode in a triode circuit, the electrons will flow through relatively unimpeded and a current flow will be recorded (Fig. D–3a). For while some electrons may strike the wire of the grid, most will flow through since it is largely open space and the tube acts essentially as a diode. If the grid is kept at negative potential relative to the cathode, it will tend to repel electrons and the current flow will diminish (Fig. D–3b). If the grid potential is made sufficiently negative, the flow of electrons can be completely suppressed (Fig. D–3c). On the other hand, as the charge on the grid becomes positive the current flow will tend to increase (Fig. D–3d). Thus, by adjusting the grid voltage (or position), the plate current can be controlled.

The grid is closer to the source of electrons than the plate; therefore, a smaller variation in the charge on the grid will have the same effect on the plate current as a larger variation in the charge on the plate. Since a small change in grid voltage will produce a large change in plate current, the voltage change between the grid and cathode is amplified in the plate current—thus the triode is an *amplifier*.

The amplifying effect of a simple triode circuit on a weak alternating signal is shown in Figure D–4. In this circuit there is no meter, but a load resistor* is interposed in the plate circuit and the grid potential is provided by applying the small alternating *input* voltage, E_i, and an additional "bias voltage" is supplied with a "C" battery (see below). As a result, a much larger *output* voltage, E_o, of the same wave form and frequency, across the load resistor, is obtained. The triode takes electric energy from the power supply and uses part of this energy to increase the strength of the signal, i.e., amplify the signal. By proper operation of a vacuum tube, any voltage variations impressed upon its (control) grid can be faithfully reproduced as amplified voltage variations across an electrical resistance or impedance in its plate circuit.

As noted previously, the high plate current is obtained when the grid-to-cathode voltage or *bias* is positive (see Fig. D–3d). Under these conditions, some electrons are attracted to the grid itself and thus may produce an appreciable current in the grid circuit. Thus the vacuum tube may draw excessive power from the signal source (i.e., E_i) under these conditions. It is to avoid this that in normal use the grid bias is kept slightly negative, for example, with a small "C" battery.

* The load resistor is a device which extracts an appreciable portion of the electrical power which flows from the "B" battery and dissipates as heat and/or utilizes that energy in the performance of some kind of work. Very often, in voltage amplifiers, the load resistor is a simple ohmic resistor.

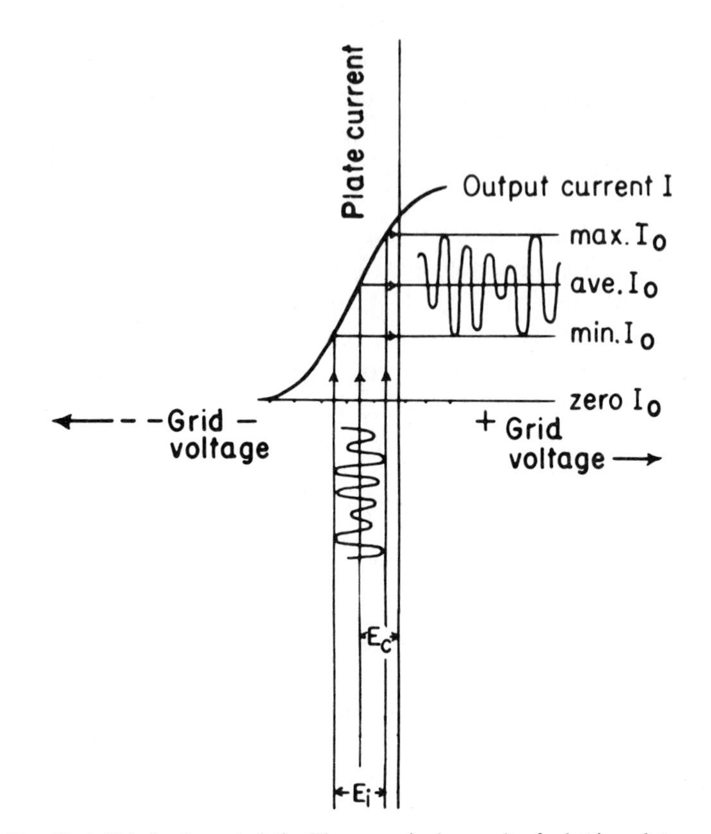

FIG. D–5. Triode characteristic. The curve is the result of plotting plate current against grid voltage changes. The voltage output $E_o = I_o \cdot R$ (where R_l is the load resistor).

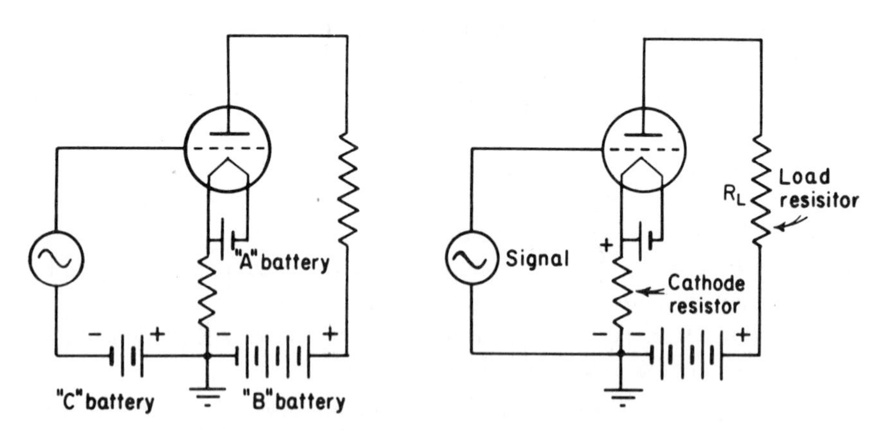

FIG. D–6. Two methods of obtaining a biasing voltage to make the grid negative with respect to the cathode. (a) Fixed bias. (b) Self-bias.

By plotting the changes in plate current (I_o) against changes in grid voltage, a curve will be obtained that summarizes the operating characteristics of a vacuum tube with a particular load resistance, R_l and "B" battery voltage (Fig. D–5). This curve is known as one of the *characteristics* of the triode. The bias applied to the grid is often set so that the operating point is near the middle of the straight line portion of the characteristic in order to provide the most faithful reproduction of the input signal. In this illustration it is assumed that the input voltage produces a grid swing (E_i) of equal magnitude in either direction from the operating point. As a result the plate current will move up and down between maximum (max I_o) and minimum (min I_o) values. Thus I_o through the resistor reproduces E_i but at an increased amplitude, $I_o \cdot R_l = E_o$.

There are a number of methods used to obtain a biasing voltage so as to set the grid at a potential, with respect to the cathode, which will allow the triode to operate on a straight line portion of the characteristic. Two methods are shown in Figure D–6. The first, known as *fixed bias* and illustrated above, involves the use of a small "C" battery or other dc voltage source to provide the grid bias. No current flows in the grid circuit as long as the grid is held negative with respect to the cathode. Therefore, C batteries may be made very small since they function under a no-load condition, i.e., no current is drawn from the batteries.

The second and most common method is known as *self-bias* or cathode bias. It involves putting a resistor in the circuit between the cathode and "B" battery. A small potential difference, i.e., voltage drop, is developed across this cathode resistor when current flows across the triode, making the cathode positive with respect to the bottom of the cathode resistor. Since the grid circuit is completed through a connection to the bottom of this resistor, it is, in effect, negative with respect to the cathode. Thus the constant component of the plate current of the tube itself is used to develop the biasing voltage in the grid circuit and, therefore, this method is known as self-bias. With self-bias, the varying portion of the plate current creates a varying potential across the cathode resistor which opposes the input voltage and thus reduces the gain of the amplifier. This is an example of *negative-feedback* and plays an important role in the operation of the electron gun of the electron microscope (see p. 41).

In early electron microscopes, the electron gun, which contained the same elements as a triode, had a grid that was not biased but was at cathode potential. In more recent times, one of several methods of supplying negative grid bias voltage has been used. Most are now self-biased guns. (See discussion, p. 42.)

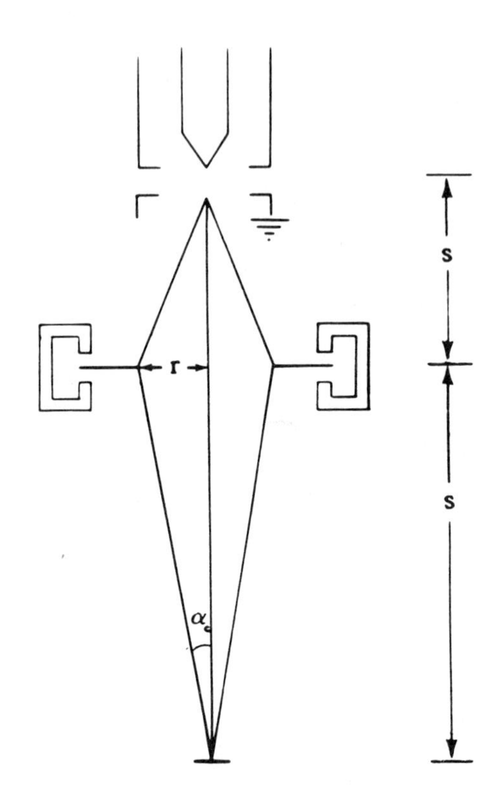

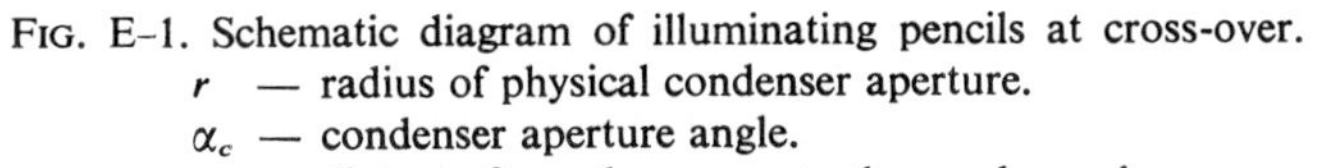

Fig. E–1. Schematic diagram of illuminating pencils at cross-over.
r — radius of physical condenser aperture.
α_c — condenser aperture angle.
s — distance from the source to the condenser lens.
s' — distance from the condenser lens to the specimen.

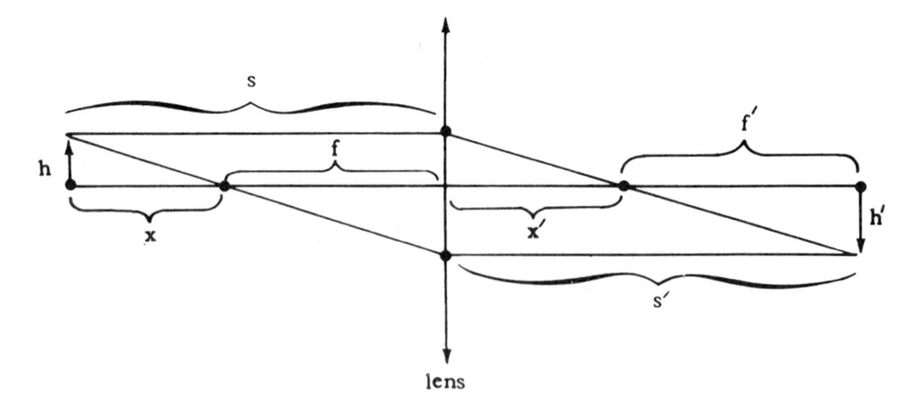

Fig. E–2. Image formation by a thin lens.

Appendix E

DEPENDENCE OF CONDENSER APERTURE ANGLE ON FOCAL LENGTH

From Figure E–1, which shows the illuminating pencils being focused by the condenser lens on the specimen (i.e., being at cross-over), the aperture angle can be defined under these conditions as:

$$\tan \alpha_c = \frac{r}{s'} \tag{1}$$

Since for small angles $\tan \alpha \simeq \alpha$, (eq. 1) can be be written as:

$$\alpha_c = \frac{r}{s'} \tag{2}$$

where:

α_c = condenser aperture angle.
r = radius of physical condenser aperture.
s' = distance from condenser lens to cross-over.

It is now necessary to express s' in terms of focal length. In order to do this, use is made of Newton's equation:

$$xx' = ff' \tag{3}$$

which mathematically expresses an imaging relationship in the case of thin lenses (Fig. E–2),
where:

f = focal length on object side.
f' = focal length on image side.

and where:

$$x = s - f \qquad (\text{since } s = x + f) \tag{4}$$

$$x' = s' - f' \qquad (\text{since } s' = x + f') \tag{5}$$

(s = distance from source to condenser lens.)
Since for thin lenses the following approximately holds true:

$$f' = f \tag{6}$$

eq. (3) can be written as:

$$xx' = f^2 \tag{7}$$

or

$$x' = \frac{f^2}{x} \tag{8}$$

on substitution of eq. (6), eq. (5) becomes:

$$x' = s' - f \tag{9}$$

Substituting eq. (9) and eq. (4) in eq. (8) we get:

$$s' - f = \frac{f^2}{s - f} \tag{10}$$

or

$$s' = f + \frac{f^2}{s - f} \tag{11}$$

which can be further simplified as follows:

$$s' = f\left(1 + \frac{f}{s - f}\right) = f\left(\frac{s - f}{s - f} + \frac{f}{s - f}\right) = f\left(\frac{s - f + f}{s - f}\right)$$

or

$$s' = f\left(\frac{s}{s - f}\right) \tag{12}$$

Substituting eq. (10) in eq. (2) we get:

$$\alpha_c = \frac{r}{f\left(\frac{s}{s - f}\right)} \tag{13}$$

When eq. (13) refers to a condenser, it is written as:

$$\alpha_c = \frac{r}{f_c\left(\frac{s}{s - f_c}\right)} \tag{14}$$

Appendix F

DEPTH OF FIELD

An objective lens forms an image on an intermediate screen, *sc*, of point P on the specimen (Fig. F–1,2). However, because of diffraction and lens aberrations the image point on the screen will appear as a disc of confusion whose radius referred to in *object* space is r. If there are objects at either point Q or R on either side of P, though they are out-of-focus on geometrical grounds they will, nevertheless, be in-focus on the screen.

Analysis of Figure F–2 (see ref. 22), which is an enlargement of the ray paths, provides an expression for the depth of field, D_{f_i} (i.e., the distance between Q and R), such that for small angles (where $\alpha \simeq \tan \alpha$):

$$D_{f_i} = \frac{2r_i}{\alpha_0} \qquad (1)$$

where:

D_{f_i} = depth of field.
r_i = radius of the disc of confusion.
α_o = objective aperture angle (to a good approximation, for thin specimens).

DEPTH OF FOCUS

A pencil of rays having an aperture angle, α_0, originates from a point P (Fig. F–3). If P is in geometric focus then it will be imaged on a screen, *sc*, as a disc of confusion whose radius is r_i. Upon defocusing, a satisfactory image will be obtained so long as the radius of the disc of confusion is not greater than r. The difference in image distance, D_{f_o}, for which this holds true is known as the depth of focus which, for small angles, is expressed as:

$$D_{f_o} = \frac{2r}{\alpha_i} \qquad (1)$$

where:

D_{f_o} = depth of focus.
r_i = radius of disc of confusion.
α_i = aperture angle in image space.
α_o = objective aperture angle.

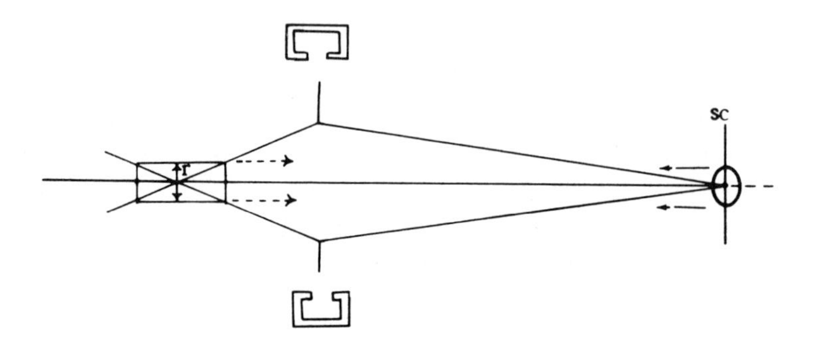

FIG. F–1. Schematic diagram of image formation of a specimen point by an objective lens in relation to depth of field.
r — radius of disc of confusion referred to object space.
sc — screen (the oval represents the disc of confusion formed on the screen).

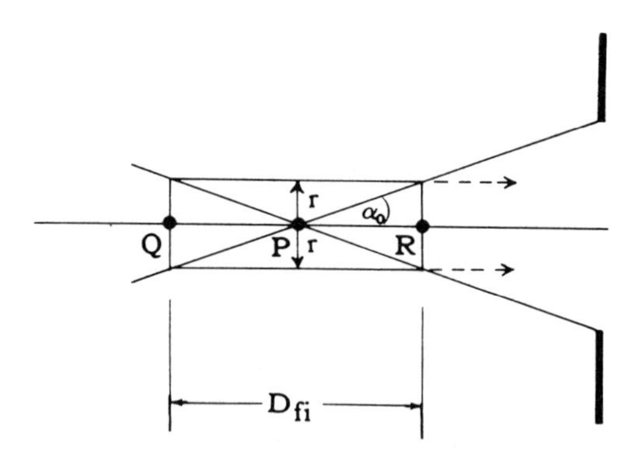

FIG. F–2. Detail of the ray paths at the specimen level.
P — point on a specimen.
Q & R — points on either side of P.
α_o — objective aperture angle.
D_{fi} — depth of field.

The relationship between r_i and r_0 is that:

$$r_i = r_0 \cdot M \tag{2}$$

where:

r_0 = radius of the smallest resolvable area in the specimen plane.
M = total magnification of the lens system.

And since for small angles:

$$M = \frac{\sin \alpha_0}{\sin \alpha_i} \simeq \frac{\alpha_0}{\alpha_i} \text{ (Abbe's sin law)} \tag{3}$$

Substituting eq. (2) and (3) in eq. (1):

$$D_{f_o} = \frac{2r_0 \cdot M}{\dfrac{\alpha_0}{M}} = \frac{2r_0 \cdot M^2}{\alpha_0} \tag{4}$$

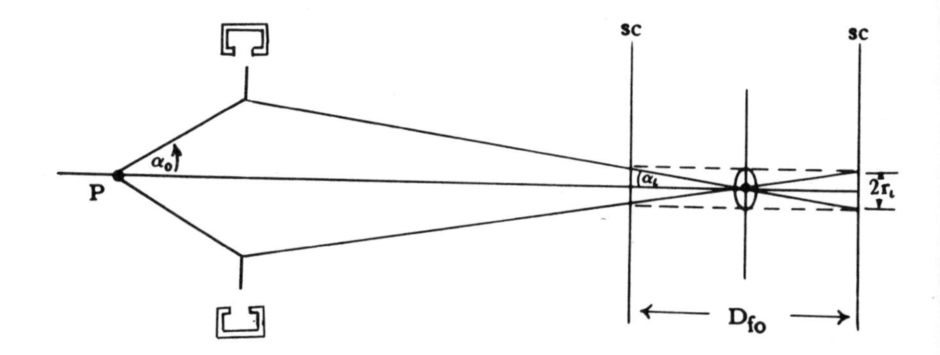

FIG. F–3. Schematic diagram of image formation of a specimen point by an objective lens in relation to depth of field.
α_o — objective aperture angle.
α_i — aperture angle in image space.
r_i — radius of the disc of confusion.
D_{f_o} — depth of focus.
sc — screen.

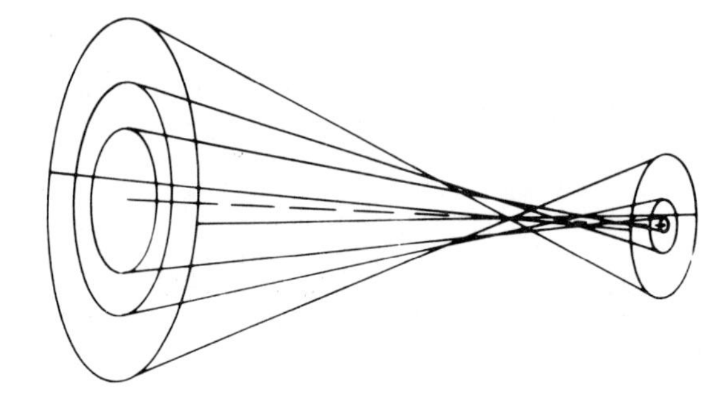

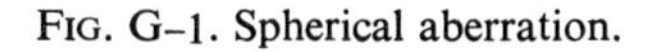

FIG. G–1. Spherical aberration.

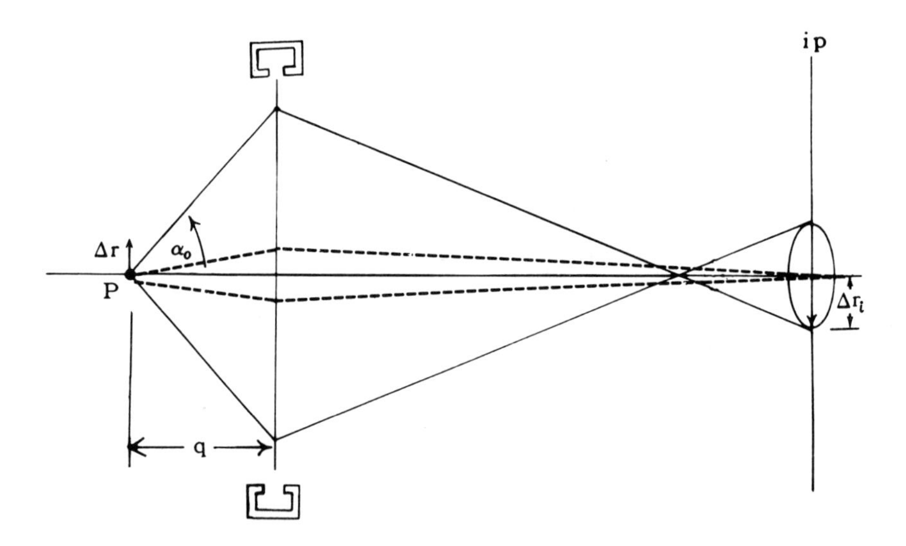

FIG. G–2. Ray paths in spherical aberration (compare with Fig. 48B).
p — axial point.
Δr — apparent radius of the disc of confusion in object space.
α_o — objective aperture angle.
q — screen distance (at high magnifications $q = f$).
ip — image plane.
Δr_i — radius of disc of confusion at image plane.

Appendix G

SPHERICAL ABERRATION

For a magnetic objective lens the spread of focal points due to spherical aberration will extend over a measurable portion of the optical axis. The magnitude of spherical aberration can be determined by reference to the size of the disc produced by the rays where they intersect the paraxial focal plane*. The result can be represented by a series of discs whose radii are used to measure the spherical aberration (Fig. G–1).

The effect of spherical aberration on the image of an axial point P is shown in Figure G-2. The rays contained within the aperture angle α_0 will spread over a disc of confusion of radius Δr_i at the image plane, *ip*.

Analysis of the distribution of focal lengths for the different zones of magnetic lenses, provides an expression for Δr_i:

$$\Delta r_i = k_s \cdot f \cdot \alpha_0^3 \tag{1}$$

where:

Δr_i = radius of the disc of confusion at the image plane.
k_s = dimensionless proportionality constant characteristic of the objective lens.
f = focal length.
α_o = objective aperture angle.

Since spherical aberration in electron microscopy can only be restricted but not eliminated, the magnitude of this aberration in any system, Δr_i, may become equivalent to the limit of resolution, d_s, which this factor imposes. When all other factors are controlled, spherical aberration imposes the ultimate limit on resolution of magnetic lenses.

Equation (1) can be rewritten as:

$$d_s = k_s \cdot f \cdot \alpha_0^3 \tag{2}$$

---

* The paraxial focal plane for magnetic lenses intersects the optical axis at right angles at the point where the rays for which the lens has the largest focal length cross the axis.

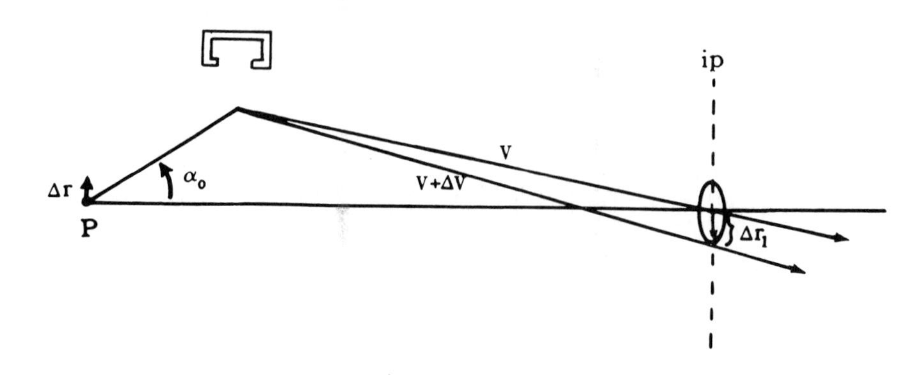

FIG. H–1. Ray path in chromatic aberration.
p — axial point.
Δr — radius of disc of confusion referred to object space.
α_o — objective aperture angle.
Δr_i — radius of disc of confusion referred to image space.
V — path of electron accelerated through potential V.
$V + \Delta V$ — path of electron accelerated through potential is negative.
ip — image plane.

Appendix H

CHROMATIC ABERRATION

The magnitude of chromatic aberration can be determined by considering the paths of two electrons with velocities equal to that due to acceleration by potential V and $V + \Delta V$ respectively (Fig. H–1). The two paths intersect the image plane (for electrons of voltage V) at different points separated by the distance Δr_i. If ΔV represents the maximum departure from V, then Δr_i is the radius of the disc of confusion containing all the electrons emanating from P. This disc of confusion may be considered as the image of a disc of radius Δr situated at P. Analysis of Figure H–1 (see ref. 23) provides an expression for Δr:

$$\Delta r = k_c \cdot f \cdot \alpha_0 \cdot \frac{\Delta V}{V} \tag{1}$$

where:

Δr = radius of the disc of confusion referred to object space.
k_c = a constant determined by the shape of the field and the object position (the value of which falls between 0.75 and 1.0 for magnetic lenses).
f = focal length.
α_o = objective aperture angle.
V = accelerating potential.
ΔV = maximum departure from V.

Since chromatic aberration in electron microscopy can only be restricted but not eliminated, the magnitude of this aberration in any system becomes equivalent to the limit of resolution which this factor, d_{c_v} (see eq. 13, p. 55), imposes. Therefore, eq. (1) can be written as:

$$d_{c_v} = k_c \cdot f \cdot \alpha_0 \cdot \frac{\Delta V}{V} \tag{2}$$

A similar discussion is applicable to variations in coil current intensity and results in eq. (14) (see p. 55), which is written as:

$$d_{c_i} = 2k_c \cdot f \cdot \alpha_0 \cdot \frac{\Delta I}{I} \tag{3}$$

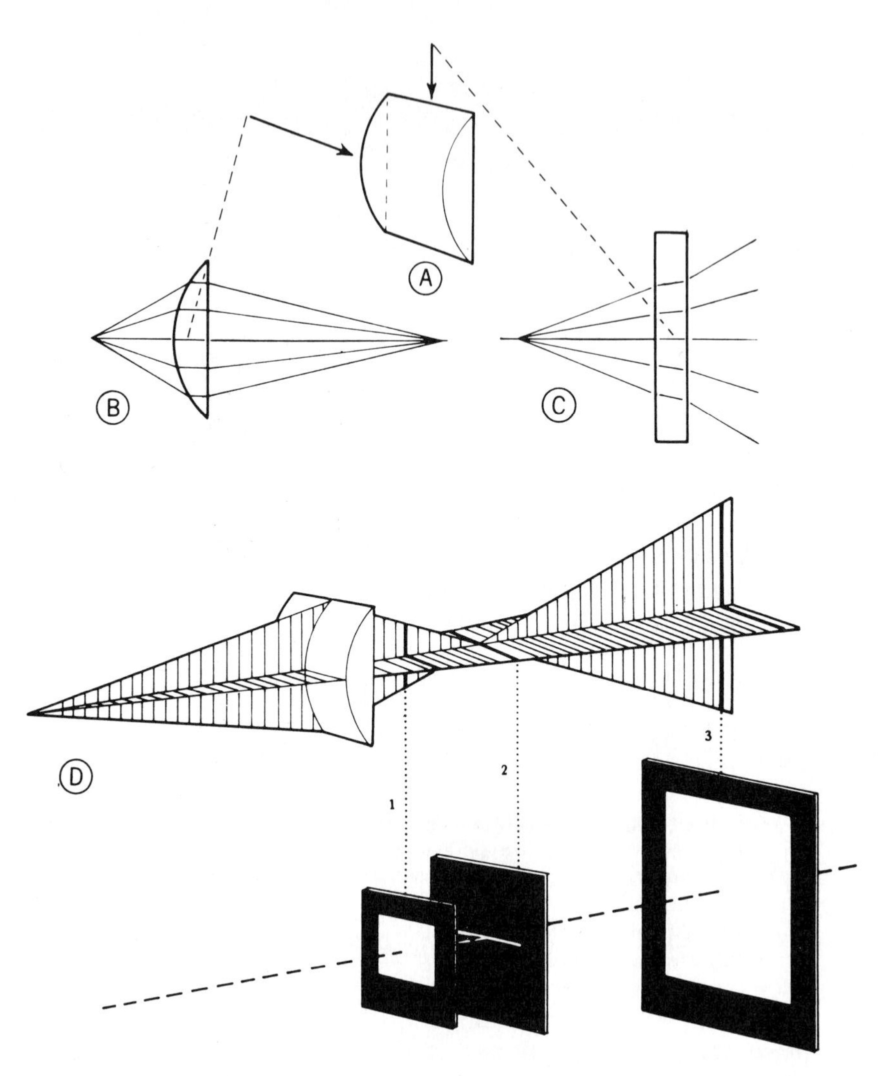

FIG. I–1. Optical theory of cylindrical lenses.
(A) A cylindrical lens.
(B) Rays passing through cylindrical lens as seen along axis of cylinder.
(C) Rays passing through cylindrical lens as seen at right angles to cylinder.
(D) Ray diagram, in perspective, of the focusing action of a cylindrical lens. Projections at the bottom of the figure show the appearance of the cross section of the beam as seen by interposing screens at points 1, 2 and 3 along the optic axis. Note that point source is imaged as a line at 2, the "focal plane" of this cylindrical lens.

Appendix I

OPTICAL THEORY OF CYLINDRICAL LENSES

The simplest cylindrical lens is illustrated in Figure I–1A. Such a lens can be visualized as resulting from cutting a cylinder of glass along its long axis. If the cylindrical lens is cut by a plane perpendicular to the axis and is viewed down that axis, a curved and a straight side are seen and, therefore, it will act as a plano-convex lens by bringing the rays traveling in this plane to a common focal point (Fig. I–1B). On the other hand, if viewed perpendicular to the cylinder's axis, straight and parallel sides are seen. Rays striking the lens in this plane will not be brought to a focal point (Fig. I–1C). By combining the separate optical actions in a three dimensional image, the effect of a cylindrical lens on rays from a point source can be visualized (Fig. I–1D).

A point source sends out rays in all directions, but only two pairs of perpendicular rays are diagrammed. The vertical pair comes to a focal point on the optical axis. All other rays in planes perpendicular to the plane which contains the point and the axis of the cylinder would come into focus on the latter plane at the same distance beyond the lens. Thus the lens acts to form a line of image points from a single object point. This can be seen (Fig. I–1D when a screen is interposed at level 2. If a screen is interposed closer to the lens (level 1), a small rectangular image will be formed. If it is placed at a distance further from the lens (level 3), another rectangular image will be formed. The long axis of such rectangles will be at right angles to one another.

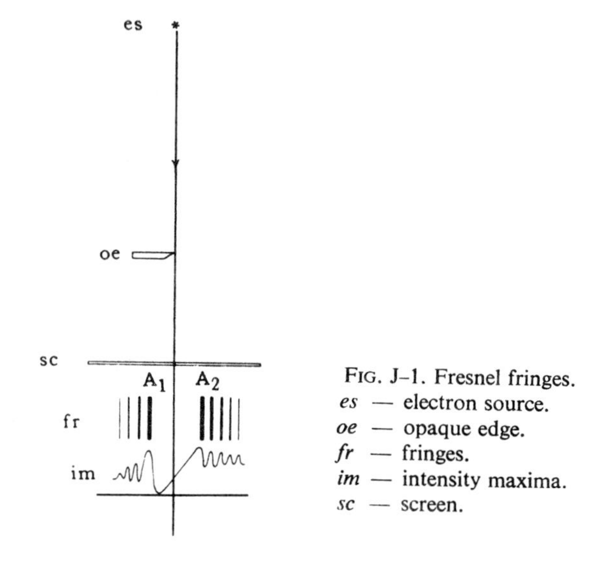

FIG. J–1. Fresnel fringes.
es — electron source.
oe — opaque edge.
fr — fringes.
im — intensity maxima.
sc — screen.

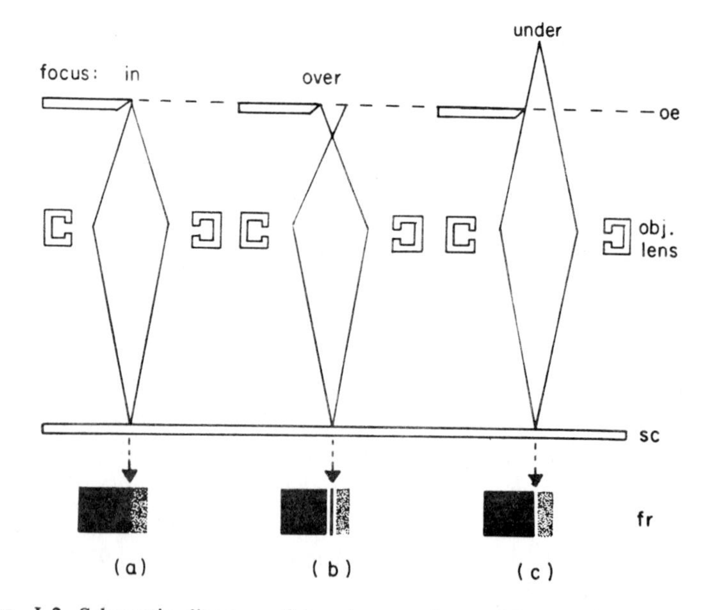

FIG. J–2. Schematic diagram of imaging pencils at different focal levels in relationship to an opaque edge.
(A) Lens current adjusted so that focal length corresponds to the position of the object.
(B) Lens current increased, thus focal length is decreased, resulting in the imaging pencil from the focal point originating from below the specimen.
(C) Lens current reduced, thus focal length is increased, resulting in the imaging pencil from the focal point originating from above the specimen.

Appendix J

THEORY OF FRESNEL DIFFRACTION FRINGES

Fresnel fringes are of practical importance in electron microscopy for they assist in evaluating the operating condition of the instrument. Specifically, they are useful in determining the extent of astigmatism due to asymmetry of the objective lens (see p. 90) and as a measure of the limit of resolution of the instrument. For this reason, some understanding of the theory of Fresnel diffraction fringes, even if only in a simplified form, is desirable.

The basic factor responsible for the presence of the fringes is the diffraction of an illuminating beam by an edge of an object. Thus, if an opaque object is interposed between a screen and a point light source, or between a screen and an electron source, a series of intensity maxima (in the form of fringes) of decreasing amplitude are distributed both outside and inside the geometric shadow of the edge (Fig. J–1). A similar phenomenon is noted when a perfect lens is introduced to image the edge on the screen. When the strength of a magnetic lens is slightly greater than that required for focus (i.e., therefore, forms an image with pronounced spherical aberration) and it is used with *semitransparent* objects, the first maxima of the *inside* set of fringes (fringe A_1) will be more contrasty than the corresponding one of those on the outside of the geometric shadow (A_2). With these considerations in mind we can now discuss the origin of the fringes.

Imaging pencils of electrons whose paths are determined by the objective lens current can be related to the edge of an opaque object in three ways (Fig. J–2). At exact focus (a), the edge appears on the screen as a sharp line with the fringe widths reduced to the limit of resolution set by the aperture of the optical system and the wavelength of the radiation. If the lens current is increased, the focal length is decreased (b) and, conversely, if the current is decreased, the focal length is increased (c). In both of these out-of-focus cases, diffraction fringes will be widened and initially become more distinct. However, only the intensity of fringe A_1 is sufficient to be easily seen. To understand what takes place we consider once again the simplified situation of an electron beam interacting with an opaque edge (Fig. J–3). The undiffracted ray, *r*, is included in any image, whether it is under- or over-focused, so that the geometrical shadow of the opaque edge is always in position. The diffracted rays, which are the ones responsible for the first Fresnel fringes, are bent both away from the edge and into its shadow. Fringe A_1 is most obvious when the image is out-of-focus. Thus if the image is over-focused, fringe A_1, which is diffracted behind the opaque edge, will appear inside its geometrical shadow. If the objective current is decreased, the fringe will disappear only to reappear again when an under-focused image occurs. Under this condition, fringe A_1 appears outside the shadow of the edge.

By the use of carbon black particles or holes in a membrane, accurate focusing can be observed on the screen (Fig. J–3,4). In the under-focused position, (a), fringe A_1 is observed outside the geometrical shadow of the semi-opaque carbon film. At exact focus (b), as would be expected, the fringe is minimal and thus is not easily evident. In an over-focused condition (c), fringe A_1 is seen as a conspicuous bright band outlined by the black opaque edge of the hole which is the geometric shadow.

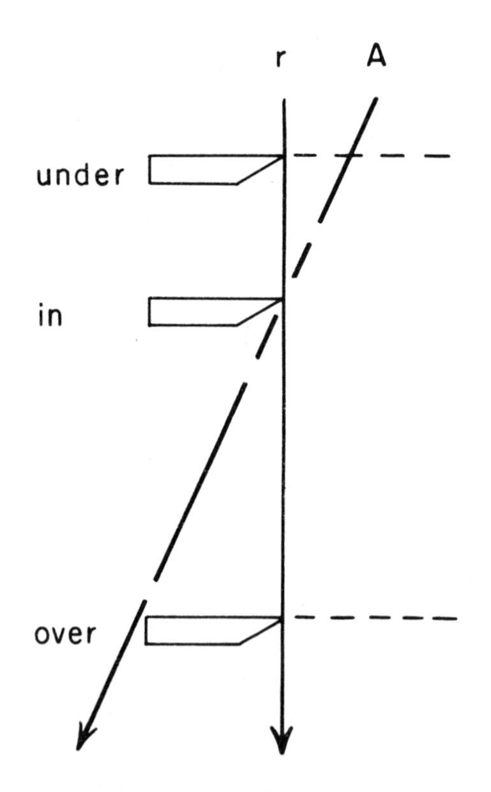

FIG. J–3. Schematic diagram of a Fresnel fringe (fringe A) originating at an opaque edge and its position with respect to the edge in the final image at different focal levels. *r* is an undiffracted ray.

When lens asymmetry is present, asymmetry of the fringe (see Figs. 67 and J–4b) can be seen only in positions other than the plane of the circle of minimum confusion which, however, is not an "in-focus" plane. The elimination of this aberration is discussed in the text (see p. 91).

At the beginning of this appendix, it was noted that under certain conditions fringe A_1 will have a markedly greater contrast than the other diffraction fringes. This increased contrast of fringe A_1 will now be explained.

If we have a point source at the first focal plane of a condenser lens, the condenser will produce a parallel beam. The objective will interrupt that beam and produce an image of the point in its second focal plane. If we place a *semitransparent* object in a plane near the focus of the objective and if the object has a different refractive index* than the unobstructed object-free area, the edges of the object will be imaged together with a set of fringes (distributed both outside *and inside* the edges and parallel to them) of rapidly decreasing contrast and decreasing separation as the distance from an edge increases. When the lens is not perfect (i.e., spherical aberration is present or, in the case of a perfect lens, if the object is slightly above or below focus) there will be an asymmetry in the intensities of the fringes such that with a slightly over-focused objective lens, the fringe inside the edge (fringe A_1) will be considerably brighter than the outside fringe (fringe A_2). This will be due to the fact that the undiffracted rays will pass exclusively through the very center of the aperture of the objective lens but the diffracted rays will use the full aperture of lens. In the presence of the *spherical aberration* of an over-focused magnetic lens, the optical path through the center will be greater than the paths through the outer zones. Because of the difference in refractive index of the object and its surrounds, the diffracted rays will have experienced an additional increase in path difference of 1/4 of a wavelength over the undiffracted rays (+1/4 for rays diffracted towards the region of higher refractive index and −1/4 for rays diffracted towards the region of lower index).

When the path difference through the outer and central zones of the objective lens also approximates 1/4 of a wavelength, the total change is zero for rays diffracted towards the region of higher refractive index and *constructive interference* results, producing a *bright phase contrast image* of the edge (A_1). For rays diffracted towards the region of lower refractive index, the net change is 1/2 wavelength and *destructive interference*

---

* All real objects which are transparent to electrons (i.e., scatter some electrons elastically) have an effective "refractive index" greater than that of the vacuum.

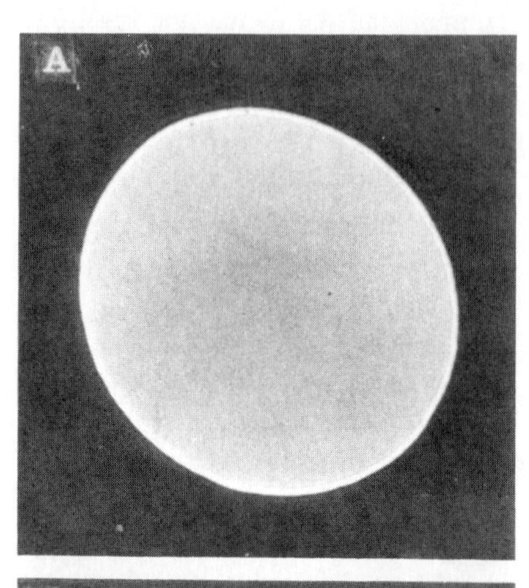

FIG. J–4. Diffraction patterns of a series of through-focus electron micrographs of a small hole in a thick carbon film under high magnification.

(A) An image in an *under-focused* condition shows the characteristic bright diffraction band outside of the edge of the geometric shadow.

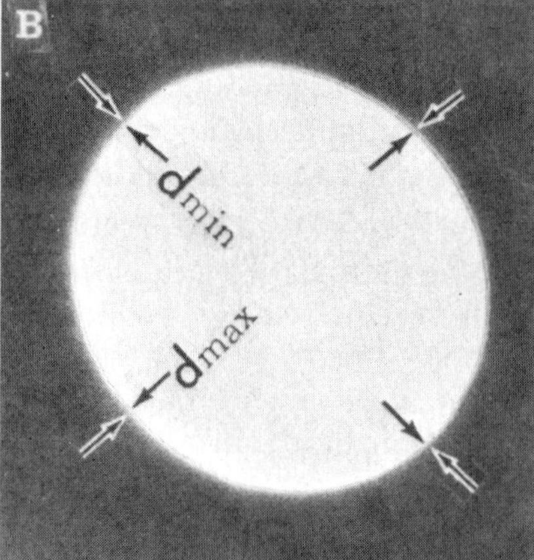

(B) An image that is very near to the *in-focus* condition as is indicated by the fine fringe. In addition the residual, (uncompensated) astigmatism of the objective lens is evident from the asymmetric character of the fringe. Quantitative determination of the astigmatic difference, A_s, is obtained using the following formula:

$$A_s = \frac{(d\ \max)^2 - (d\ \min)^2}{3.2\lambda} = D_{f_i}$$

where:

$d = \max 50$ Å $\qquad d = \min 25$ Å
$\lambda =$ wavelength of the electron beam

With an aperture angle of 5×10^{-3}, the astigmatic difference in focus is 1μ, which means that using equation:

$$D_{f_i} = \frac{2r}{\alpha_0} \text{(see p. 223)}$$

$$r = \frac{D_{f_i} \cdot \alpha_0}{2} = \frac{(10^4) \cdot (5 \times 10^{-3}) = 25}{2} = 25$$

that the objective lens has a potential resolving power of 25 Å, for $\lambda = 0.055$ Å (at 50 KV).

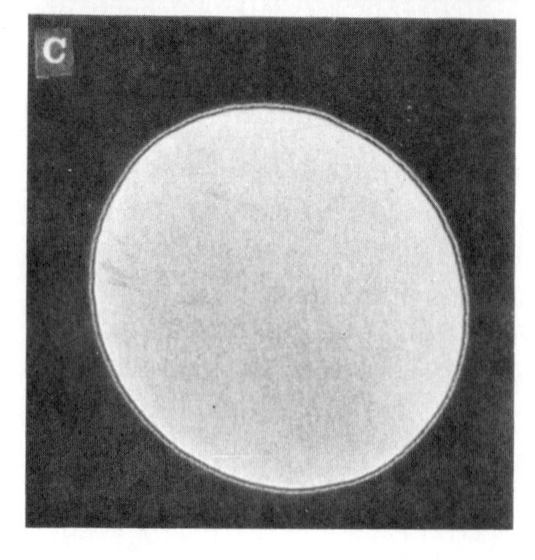

(C) An image in an *over-focused* condition is seen as a light band, outlined by a dark ring (the edge of the geometric shadow).

occurs producing a *dark phase contrast image* of the edge (e.g., the dark line between the edge and A_2).

In any real situation, the source is not a point source (i.e., the condenser aperture is usually a substantial fraction, e.g., 1/8 of the objective aperture) and the undiffracted rays will be imaged on and pass through a finite central area of the rear focal plane of the objective. Therefore, an appreciable proportion of the diffracted rays will also pass through this same central zone and there will be a net 1/4 wavelength path difference, and interference between these particular sets of diffracted and undiffracted bundles will produce less brightness.

The regions of the object which elastically* scatter electrons through relatively large angles (to the outer zones of the objective aperture) will be those which have fine details (e.g., edges). Most elastically scattered electrons associated with the coarse details will be diffracted through very small angles and will pass through the same central zone of the objective aperture as do the undiffracted (unscattered) electrons. Therefore, in an image plane, only the areas of the object which diffract rays at large angles (e.g., edges) will show an added contrast (e.g., bright inner fringes) over the surrounding areas. This explains the origin of the "halos" and the restriction of contrast to small detail in phase contrast microscopy, the Becke lines in ordinary light microscopy, and the especially contrasty "Fresnel diffraction fringe" (fringe A_1) in electron microscopy.

It now becomes obvious that if the illuminating aperture is increased so the undiffracted beam fills the objective aperture, then, since all the diffracted and a large portion of the undiffracted rays would pass through the same lens zones, there will be essentially no phase contrast effects and the fringe A_1 almost disappears. Conversely, fringe A_1 will show maximum contrast only when the illuminating aperture is substantially smaller than the objective aperture. Nonetheless, when the illuminating aperture equals the objective aperture, the resolution of objects which scatter appreciable numbers of electrons completely outside of the physical aperture of the objective can approach double that at low illuminating apertures (see p. 50 and footnote, p. 11).

* For definition, see p. 59.

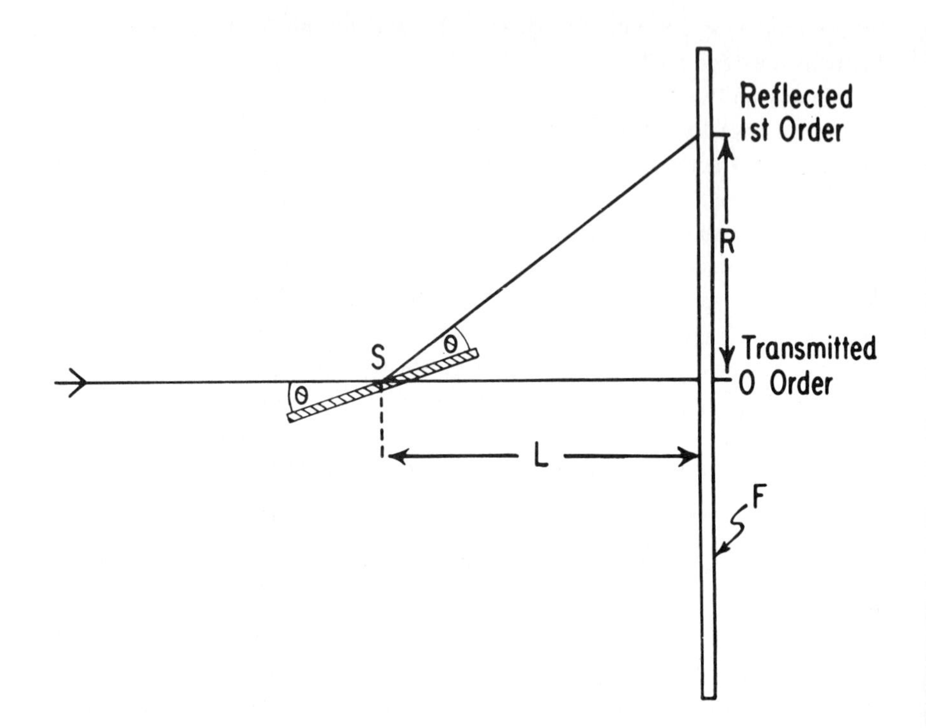

FIG. K-1. Diagrammatic outline for the definition of the terms in Bragg's Law (see discussion in text).

Appendix K

ELECTRON DIFFRACTION

In 1895, Roentgen discovered x-rays. In 1912, von Laue and others noted that an x-ray beam is acted upon by the atomic planes of a crystal in a manner similar to that by which a light beam is acted upon by a ruled grating. Thus the atoms in the planes of the specimen diffract x-rays just as the grating element diffracts light. Both the incident and diffracted radiation travel with a wave-like motion. The diffracted radiation can be detected only in those directions in which the portions sent out from the various planes throughout the crystal reinforce each other. The situation is also analogous in regard to the diffraction of electrons.

Thus diffraction effects, which arise when a monochromatic beam of electrons of wavelength λ reflects from surfaces of a crystal or passes through an ordered structure, can be ascribed to the regular path differences between the wavelets elastically scattered* by the regularly distributed atoms of the structure.

The geometry of the diffraction of x-rays and also of electrons is such that the reenforcement of the emitted waves occurs only when the wavelength of incident radiation has a relationship defined by Bragg's equation as:

$$n\lambda = 2d \sin \theta \qquad (1)$$

where:

n = the diffraction order (n = O, 1, 2 . .).
λ = wavelength of incident radiation.
d = the distance between atomic planes (in Å).
θ = the angle of incidence and also of reflection.

Thus, for example, with 100 KV, $\lambda = 0.037$ Å and d-spacing for a typical metallic crystal such as nickel, is 2.03 Å, $\sin \theta = 0.00911$ and the Bragg angle $\theta = 0.521$ degrees.

Suppose a crystalline specimen, S, is placed in the path of a narrow incident electron beam and diffraction occurs (Fig. K–1). The diffracted

* Many electrons which pass through a specimen suffer inelastic scattering and some energy is transferred to the specimen. These inelastically scattered electrons give rise to a general background to the diffraction patterns, which is most intense nearest the undeviated beam and rapidly diminishes in intensity at the larger angles.

rays *for the first orders* emerge so as to be located in the surface of a right circular cone with 2 θ equal to the semiangle. The cone intercepts a film plate, F, forming a circle with a radius, R, with the specimen to film plate distance being L. If a high voltage beam is used, its wavelength, λ, is much less than the common interplanar distances of crystalline materials. (Low voltage beams are of the same order of magnitude of x-ray beams.) Thus the diffraction angle θ for low orders (i.e., $n = 1 \rightarrow \approx 3$) is always very small. As a consequence of the smallness of these angles, the following derivation may be made from the Bragg equation:

$$n\lambda = 2d \sin\theta \qquad (1)$$

or

$$\frac{n\lambda}{2d} = \sin\theta \qquad (2)$$

and since for small angles $\sin\theta \simeq \theta$ (if expressed in radians), eq. (2) may be written as:

$$\frac{n\lambda}{2d} = \theta \qquad (3)$$

But, for small angles $\tan 2\theta \simeq 2\theta$ (in radians), then

$$2\theta = \frac{R}{L} \qquad (4)$$

or

$$\theta = \frac{R}{2L} \qquad (5)$$

Since θ has already been defined in eq. (3), eq. (5) can be rewritten as:

$$\frac{n\lambda}{2d} = \frac{R}{2L} \qquad (6)$$

and

$$d = \frac{nL\lambda}{R} \qquad (7)$$

Thus if $L\lambda$, the *camera constant* at a particular voltage, can be determined, then d would be a function of R and we could then determine the interplanar spacings of all diffraction planes in the crystalline specimen. The determination of L is usually made by means of a calibration technique using a known standard crystalline substance as a reference.*

* The calibration may be readily accomplished by taking a diffraction pattern of a specimen whose d values are known, measuring the radial distances of the spots produced by the diffracted beam of, for example, orders 1, 2 and 3, for a number of known d values and the preparing of a graph of distance R, versus interplanar spacing, d. The materials usually used for calibration standards are thinned gold foil or magnesium oxide crystals whose d values have been determined by x-ray diffraction.

The electron microscope contains the essential parts for electron diffraction. These are (1) an essentially monochromatic radiation source, the electron gun, (2) a collimation system to permit nearly parallel rays to strike the specimen, one or more electron lenses with associated apertures, (3) a specimen holder, and (4) a means of viewing or recording the diffraction pattern, the screen and camera.

In order for sharp diffraction patterns to form, it is essential that the electron beam striking the specimen be well-collimated, i.e., that all electrons striking a given spot on the specimen arrive from the approximate same direction. This requirement may be fulfilled, for example, by using a very small condenser aperture and by focusing the condenser to cross-over. In this way, only a small portion of the specimen is irradiated by an electron beam of low numerical aperture. In an intermediate image plane, an aperture is used to stop all electrons but those which pass through the specific region of the specimen of interest from reaching the screen (or film). The specimen is brought to focus and then the projector and intermediate lenses are adjusted to produce an image of the rear focal plane of the objective on the screen. This is the plane of the diffraction pattern. The image of the undiffracted or zero order beam (the image of the condenser aperture) appears on axis. Diffracted images of the condenser aperture appear as spots at distances proportional to R from the central spot (Figs. K–2 to K–5).

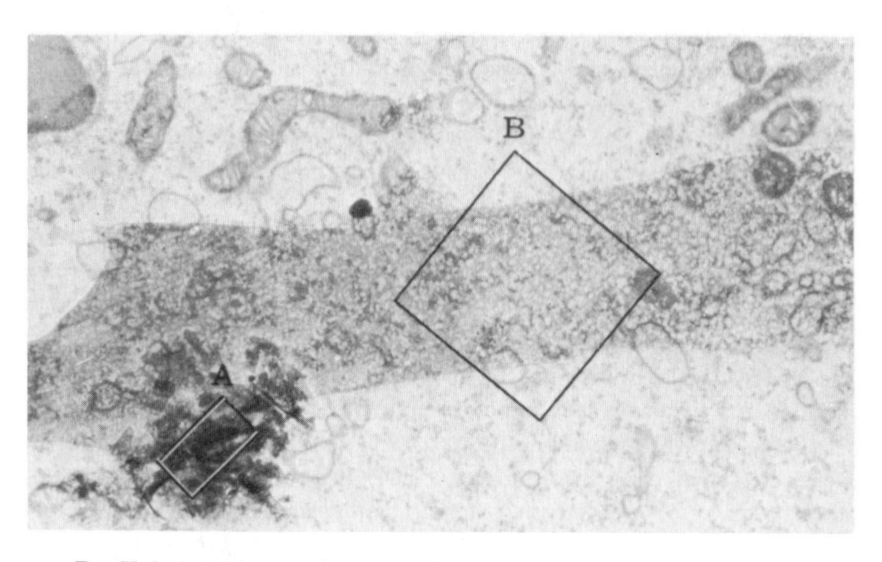

FIG. K-2. An electron micrograph of a portion of a thin section from a salamander oocyte. Superimposed on the section is a very thin, single crystal leaflet of mica with a small clump of mica crystals to the lower left. (The objective aperture was in its normal position.)

FIG. K-3. A diffraction micrograph of the clump of mica crystals seen in Fig. K-2. The micrograph was obtained by screening out, with an adjustable diaphragm, all electrons in an intermediate image from areas other than that indicated by the small rectangle (A). Note the rings of diffracted spots which are typical of polycrystalline materials. (The objective aperture was removed for this micrograph.)

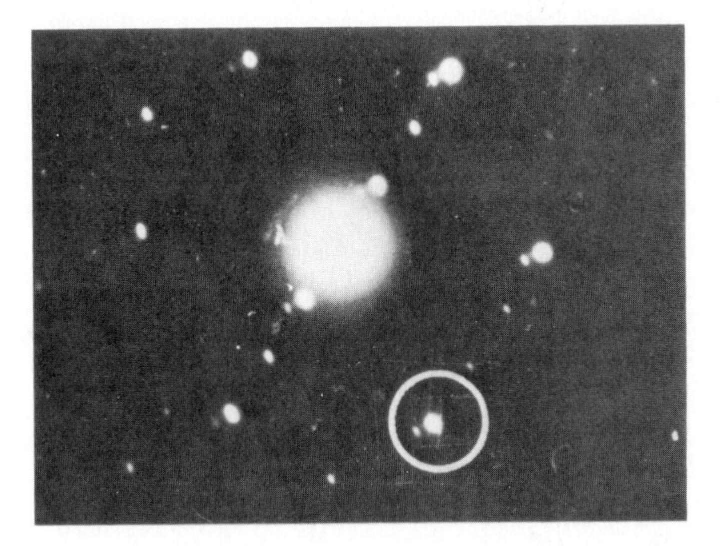

FIG. K–4. A diffraction micrograph of the portion of the "single" crystal indicated by the square (B) in Figure K–2. (The objective aperture was removed for this micrograph.)

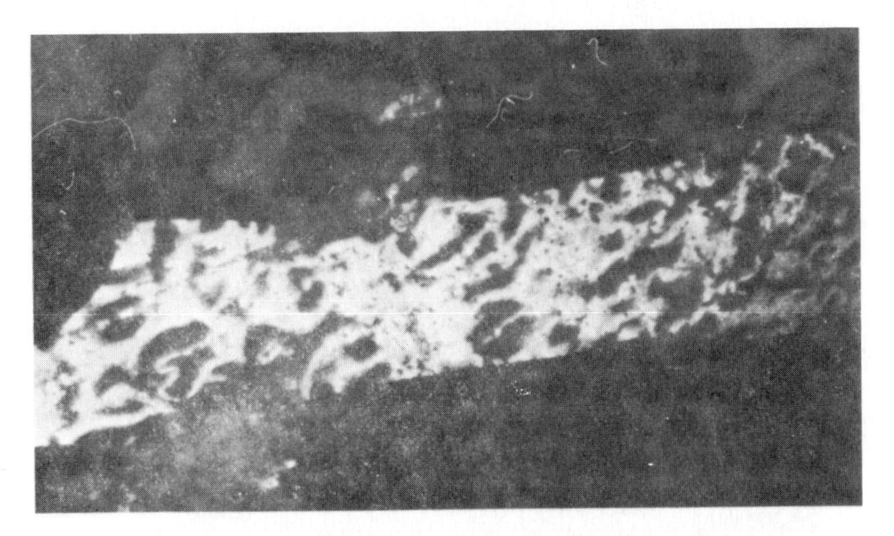

FIG. K–5. An electron micrograph of a dark field image formed by centering the objective aperture on the diffracted spot as indicated by the circle in Figure K–4 and then retracting the adjustable intermediate image diaphragm. Note that the clump of crystals and the tissue structures such as mitochondria (which diffracted few electrons to that position in the rear focal plane of the objective) appear in low contrast. Extensive regions of the somewhat wrinkled single crystal appear in bright contrast. These are all portions of the crystal which make exactly the same angle with the incident beam of electrons.

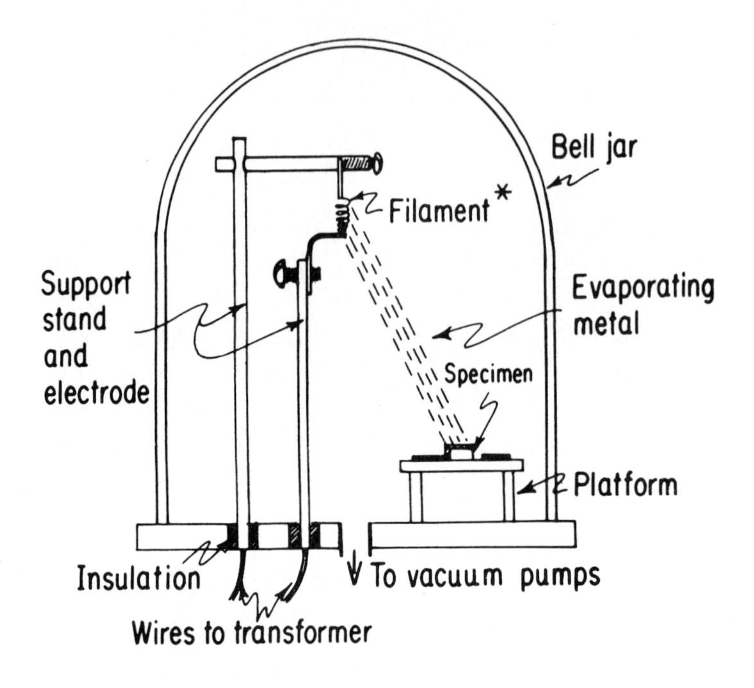

FIG. L–1. Representation of the mechanism of shadow casting.
*Tungsten filament coated with melted metal (e.g., platinum).

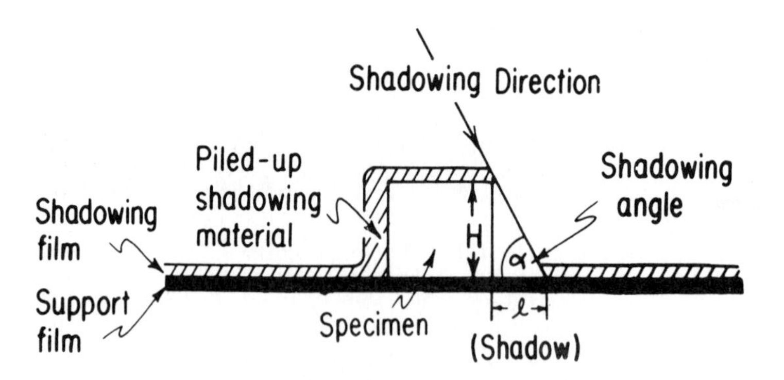

FIG. L–2. Enlarged representation of the specimen area.

Appendix L

SHADOW CASTING*

The method consists essentially of evaporating a metal from a source at an oblique angle to the specimen (Fig. L–1). If the specimen is not flat, metal piles up on the surfaces which face the source, but surfaces facing away from the source are shielded and receive no deposit. Thus the effect of oblique deposition is to cause relatively higher portions of the specimen to cast partial or total "shadows" in the direction away from the source. If a shadow cast specimen is observed in the electron microscope, the areas over which the metal coating is thickest will scatter the most electrons while the "shadows" will be highly transparent and bright. In photographic negatives the shadows appear black while the metalized surfaces are light, thus conforming to the more usual usage of the term shadow.

The height of a particle (Fig. L–2) can be calculated from the equation:

$$H = \tan\theta \cdot l \tag{1}$$

where:

H = height of the particle.
$\tan\theta$ = angle of shadowing.
l = length of the shadow.

The value for $\tan\theta$ can be determined by using measurements made on the angle subtended by the specimen from the source in the shadowing apparatus and substituting in the equation:

$$\tan\theta = \frac{c}{d} \tag{2}$$

where:

c = perpendicular height of the filament from the plane of the specimen.
d = distance from the point where the perpendicular from the filament intersects the plane of the specimen to the specimen.

The angle may be measured from the disposition of the source and specimen with a ruler or a protractor. The angle of shadowing used will vary according to the type of structure under examination. Thus for large objects an angle of deposition of 45° is desirable. (Under this condition $\tan\theta = 1$ and the length of the shadow will equal the height of the object.) For roughly spherical particles an angle of 20-30° usually will be satisfactory.

*The shadow casting technique was developed by R. C. Williams and R. W. G. Wycoff, *J. Appl. Phys.*, **15**, 712, 1944.

1. Shadowing Apparatus

A shadow casting unit can be purchased commercially or improvised from components. These include:

a. Vacuum chamber. A large glass bell jar and a base plate with vacuum and electrical feed troughs.
b. Vacuum pumps. A mechanical rotary and oil diffusion pump in series capable of producing and maintaining a vacuum of about 1×10^{-5} mm Hg.
c. Low voltage high amperage electrical system. This can be provided by a 6 to 12 V variable transformer capable of handling currents up to 30-50 amp.
d. Filament. Tungsten wire baskets or molybdenum boats.

2. Shadowing Metals

The chief requirements of the shadowing metal are chemical and thermal inertness, high density to provide maximum electron scattering for minimal deposit thickness, the capacity to be deposited in a uniform almost structureless form and to resist recrystalization and creep of deposit (which would change the apparent specimen structure).

Among the metals which have been found to meet these requirements and have been satisfactorily used are gold, palladium, chromium, platinum and uranium. The last four metals are the most popular. Thus chromium provides films, of a thickness down to about 50 Å, that adhere tightly to the substrates but may appear granular in the electron microscope. Platinum (used as platinum-palladium or platinum-carbon mixtures) provides useful deposits as thin as 3–5 Å. Uranium (as uranium-oxide) also can provide films of the same order of thickness and of small grain size.

While for most shadowing great precision in the amount of metal deposited is not required and can be visually estimated, when quantitation is desirable the following formula may be used to obtain the weight of metal needed to be evaporated.

$$M = \frac{4\pi r^2 td}{\sin\alpha} \qquad (3)$$

where:

M = weight of metal (g).
r = distance from source to specimen (cm).
t = thickness of deposit (Å).
d = density of material (g./cm^2).
α = angle of shadowing.

3. Applications of Shadow Casting

The usefulness of shadowing, aside from its straightforward application to a specimen mounted on a support film, has already been noted in the discussion of the mica substrate technique (p. 185), the surface spreading technique (p. 200), and in the study of small particles by dispersion or replication methods (p. 200).

Appendix M

INTENSITY RELATIONS IN ELECTRON BEAMS AND IMAGES

In utilizing the SEM it is desirable to obtain an electron image or spot of very high intensity. There are, however, limitations to the maximum intensity obtainable that are fixed by the specific emission of the source, the convergence of the electron paths at the image or spot, and the initial and final kinetic energies of the electrons.

The intensity at an image plane is related to the intensity at the object plane by the equation:*

$$n_o \, y_o \sin \alpha_o = n_i \, y_i \sin \alpha_i \tag{1}$$

where:

n_0 = refractive index in the object plane.
n_i = refractive index in the image plane.
y_o = object height.
y_i = image height.
$sin\alpha_0$ = sin of the aperture angle in object space.
$sin\alpha_i$ = sin of aperture angle in image space.

Where an object is imaged by an axially symmetric optical system, equation (1) establishes a general relation between the magnification (Y_o / Y_i) and the ratio of the sines of the convergence angles α_i and α_o (where α_i is the condenser aperture angle in image space).

Since intensity is inversely proportional to the square of the magnification, we have (where there is no loss of electron current in the system):

* This relationship is derivable from lens theory concepts (see Hall[23]), where it is known as Abbé's sine condition, or from the principles of thermodynamics (see Zworykin, et al., p. 264), where it is referred to as the Theorem of Clausius or Helmholtz-Lagrange.

$$\frac{I_i}{I_o} = \left(\frac{y_o}{y_i}\right)^2 = \frac{n_i^2 \sin^2 \alpha_i}{n_o^2 \sin^2 \alpha_o} \quad (2)$$

where:

I_i = current density (current/unit area) at the image plane.
I_o = current density at the object plane (electron source).

In equation (2), the relationship (n_i^2/n_o^2) represents the ratio of the energy in image space, eV_i, to the energy in object space (at the real or virtual emission surface), eV_o, which is approximately kT.

Substituting (eV_i/kT) for (n_i^2/n_o^2) in equation (2), we get:

$$\frac{I_i}{I_o} = \frac{eV_i \sin^2 \alpha_i}{kT \sin^2 \alpha_o} \quad (3)$$

where:

e = electronic charge
V_i = accelerating voltage
k = Boltzmann's constant
T = absolute temperature of the cathode

Equation (3) can be rewritten as:

$$I_i = I_o \frac{eV_i \sin^2 \alpha_i}{kT \sin^2 \alpha_o} \quad (4)$$

Optimally, if all electrons are collected from the filament, α_o, the aperture angle of the pencils leaving the cathode is $\pi/2$ and $\sin \alpha_o = 1$. Thus, equation (4) can be simplified as:

$$I_i = I_9 \frac{eV_i}{kT} \sin^2 \alpha_i \quad (5)$$

which is an approximate* derivation of the *Langmuir equation.*** This

---

* An exact derivation of this ratio, taking into account the Maxwellian distribution of the electrons leaving the cathode, is:

$$I_i = I_o \left[\frac{eV_i}{kT} + 1\right] \sin^2 \alpha_i$$

provided that $eV_i >> kT$

**Substituting the known values for 2 and k, equation (5) becomes:

$$I_i = I_o \frac{11{,}600V}{T} \sin \alpha_i^2$$

It should be noted that in order to increase I_i, the α_i must be increased, and lens aberration factors set practical limits on the extent to which this can be done.

equation sets an upper limit for the intensity or current density in the image plane, i.e., in the focused electron spot.

Brightness is defined as the intensity per unit solid angle (or current per unit area per unit solid angle); for small angles, brightness of the image, β_i. can be expressed as:

$$\beta_i = \frac{I_i}{\pi \sin^2 \alpha_i} \tag{6}$$

Since β_i has an upper limit, independent of probe diameter, reduction of source diameter (and therefore probe diameter at a fixed α_i) results in a reduction of signal current in proportion to the change in probe area. If the current becomes too low, its statistical fluctuations set the limit to the signal/noise ratio, and therefore limit resolution as well. To improve resolution by reducing probe diameter, it is essential also to improve source current density I_0. This is accomplished by using a LaB_6 or a field emission gun.

Appendix N

VACUUM PUMPS

A perfect vacuum, by definition, is a space entirely devoid of matter. The term, vacuum, is normally used to denote gas pressures below the "normal" atmospheric, sea level pressure of 760 torr (1 torr = 1 millimeter of mercury). The quality of the vacuum attained is indicated by the total pressure of gases in the vessel which is pumped. The quality of the vacuum can be roughly classified in one of four categories as shown below:

Pumping range	*Pressure range (torr)*
Rough pumping	10^3 - 10^{-3}
High vacuum	10^{-3} - 10^{-7}
Very high vacuum	10^{-7} - 10^{-9}
Ultra high vacuum	better than 10^{-9}

The concept of pumping out air from a closed vessel to create a vacuum was thought of centuries ago. While initially vacuum pumps were solely laboratory instruments, they have found very wide commercial application over the years.

1. Mechanical Pump

Figure N-1 represents a mechanical pump (e.g., a bicycle pump) in its simplest form. It consists of a metal cylinder, a tightly fitted piston, and two valves arranged to let air into and out of the cylinder, respectively. When the piston is raised on the upstroke, it reduces air pressure in the cylinder allowing air to enter through the inlet valve from the container that is being evacuated. When the piston is pushed down on the downstroke, the air in the cylinder is compressed, closing the inlet valve and opening the outlet valve through which the air is then forced out. Thus, the air in a container attached to the inlet valve will be removed. As a result, with each double stroke, a certain fraction of the air in the container is being removed, this fraction being the ratio of the volume of the cylinder to the combined volume of the cylinder and container. The pump will cease to be effective when the air pressure in the container being emptied is too low to open the inlet valve (on the upstroke).

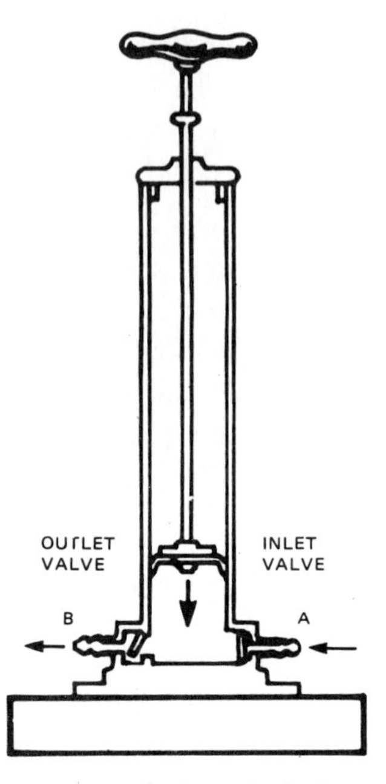

FIG. N-1. A simple mechanical pump.

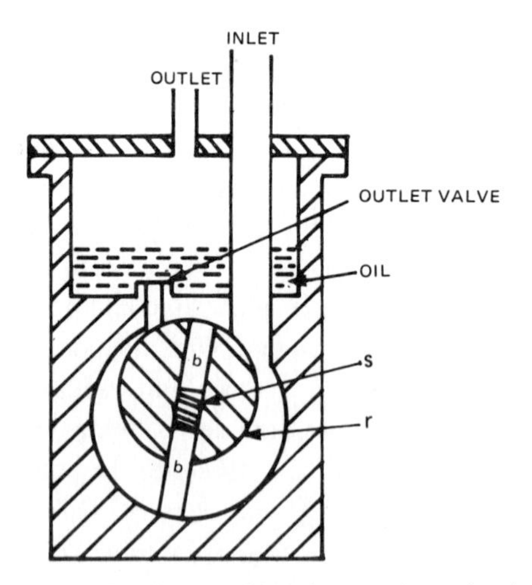

FIG. N-2. A rotary pump. This is a two-bladed pump (*b*), spring loaded (*s*) with an eccentric rotor (*r*).

It is apparent that an incomplete vacuum will result with such a simple pump because, with each stroke, only a certain fraction of the air is removed and the remainder expands to fill the container. In addition, leakage of the outlet valve contributes to restricting the efficiency of the pump. To produce a preliminary vacuum with a pressure at a level of ca. 10^{-2} torr, a *rotary mechanical pump* is often used.

In a typical rotary pump (Fig. N-2) air molecules are continually moving from the container to be emptied into the space between a revolving eccentric cylinder and the walls of a stationary cylinder. Such air molecules are trapped and swept into the outer air through a valve. All small clearance spaces are filled with an oil having a low vapor pressure. Figure N-2 shows the two bladed rotary pump where gas is trapped between the blades and swept out through the exhaust valve. The oil functions as a lubricant and sealant between the moving parts of the pump. The ultimate pressures attainable are limited by leakage between the high and low pressure sides of the pump (due mainly to carryover of gases and vapors dissolved in the oil that flash off when exposed to the low inlet pressures) and by decomposition of the oil exposed to high temperature spots generated by friction.

When it is essential to avoid contamination from rotary pump oil vapor, *sorption pumps* may be used. These pumps are based on the high gas uptake of charcoal or molecular sieve material at liquid nitrogen temperatures. Thus, they may prove particularly helpful in systems such as electron microscopes utilized under zero-leak, contamination-free conditions where the amount of gas to be evacuated is limited.

2. Diffusion Pump (see discussion, p. 78).

3. Sputter-ion Pumps

These pumps provide clean and dry pumping in the ultrahigh vacuum range without the use of pumping fluids. Pumping action is achieved through a combination of ionization, sputtering, and ion burial. The two major components of this pump system are a power supply and the pump. The latter consists of an "envelope" containing an open structured anode sandwiched between two cathode plates, as well as a magnet that lies just outside the envelope (Fig. N-3). The "envelope" and the magnet are both enclosed in a common iron yoke.

Pumping is initiated by a suitable voltage (2 to 10 KV) between the anode grid and cathode plates. The electrons, which tend to flow toward the

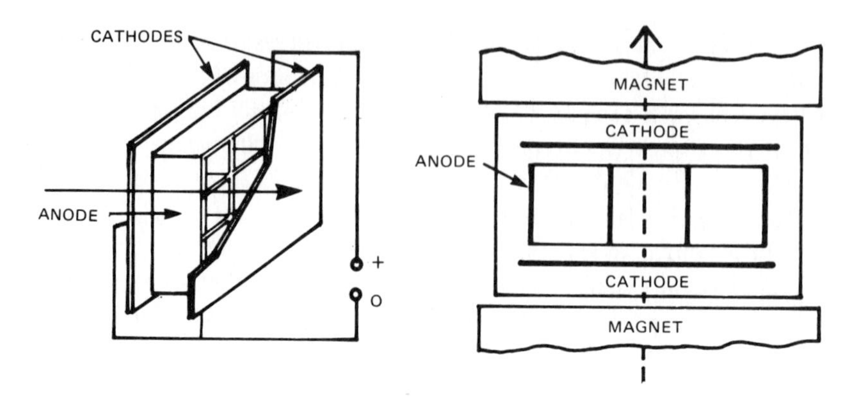

FIG. N-3. A sputter-ion pump.

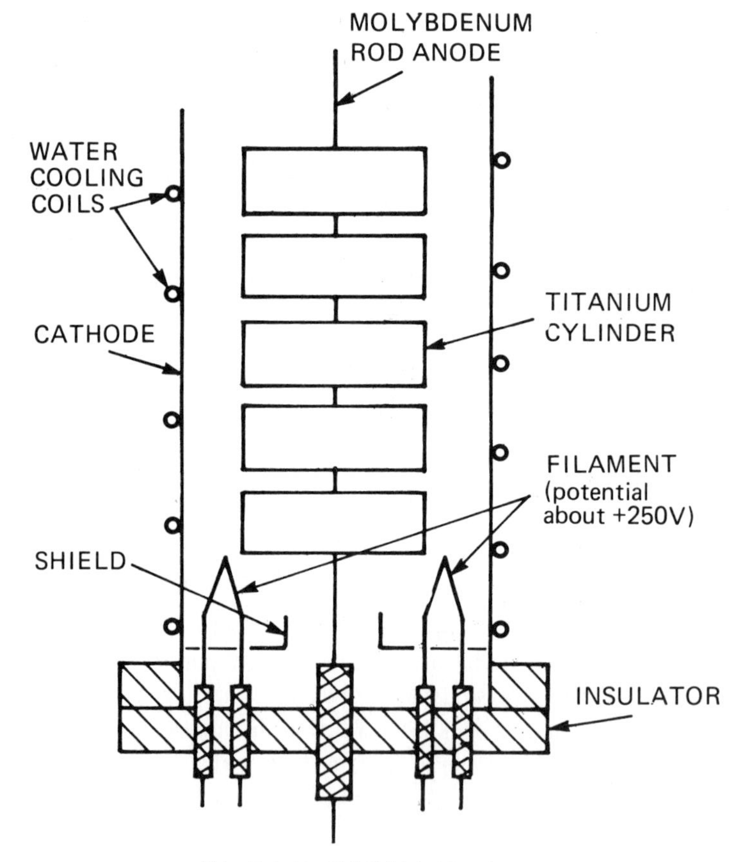

FIG. N-4 A radial field (orbitron) pump.
(The cathode and shield are at ground potential.)

anode, are forced into long helical paths involving multiple reflections at the cathodes under the influence of the strong magnetic field that is present. The greatly increased electron path length results in a high probability of collision between free electrons and gas molecules. Such collisions produce positively charged gas ions (ionization) and more free electrons. The gas ions then bombard the titanium cathode plates causing titanium atoms to be knocked out of the plates (sputtering). These titanium atoms are then deposited on the anode and later form chemically stable, nonvolatile compounds with active gas atoms (e.g., N, O, H, etc.). When a metal readily binds molecules of such gases it is called a "getter." Chemically inert gases such as argon, on the other hand, are ionized, accelerated toward the cathodes, and driven into and buried under the titanium of the cathodes.

The sputter-ion pump is effective in the ultrahigh vacuum range because each collision produces an increasing number of electrons with long, effective path lengths. Thus, pumping action is maintained down to very low pressure (below 10^{-9} torr). Also, since the titanium consumption results from sputtering, no heat is required and the ultimate vacuum is not limited by the release of gases trapped in the titanium itself. Finally, self-regulation of titanium consumption is achieved because the sputtering rate is virtually a linear function of the number of electron-atom collisions.

4. Radial Field (Orbitron) Pumps

These vacuum pumps also belong to the oil-free group. They differ from sputter-ion pumps by: (1) using an electrostatic (rather than electromagnetic) field to achieve high pumping, and (2) using sublimation (rather than sputtering) to free getter material. Pumping action is achieved through a combination of ionization, sublimation, and ion burial.

The pump consists of a central anode and a molybdenum rod, that supports a titanium cylinder of larger diameter. This is mounted concentrically inside an outer cylindrical cathode that acts as the vacuum envelope. This container is air or water cooled. The central electrode is held at a high positive potential while the outer cylinder is grounded (Fig. N-4).

Electrons from two filaments, biased at about +250V above the cathode, are randomly emitted into the electrostatic field that exists in the annular region between the central electrode and outer cylinder. All electrons with initial direction other than radial will travel in orbits around the central electrode. These electrons travel along the axis and some are reflected back

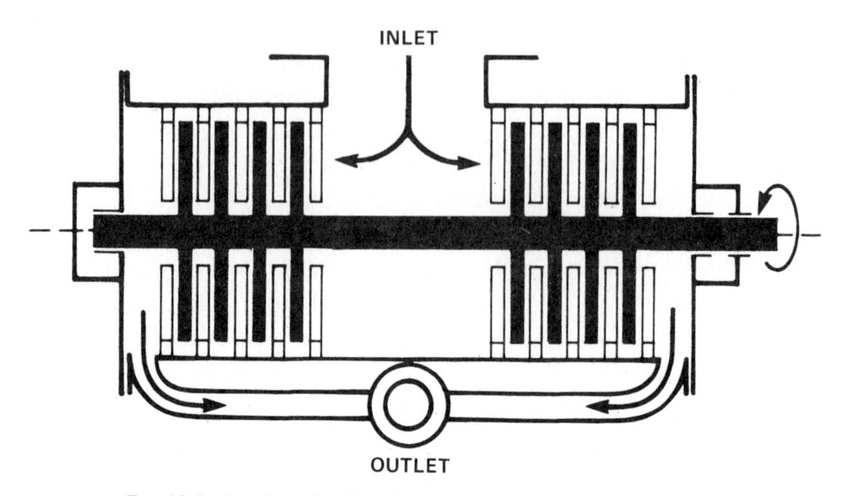

FIG. N-5. A turbomolecular pump (seen in schematic crosssection).

into the field at both ends. They have fairly large mean free paths and ionize gas molecules in the region between the electrodes. Since the titanium cylinders are larger in diameter than their supporting rod, they finally intercept most of the electrons and are heated to sublimation temperature by this electron bombardment. Titanium vapors are sublimed onto the walls of the outer cylinder and all active gases are continuously pumped by reaction with this film. Inert gases are ionized by the electrons in the annular region between the electrodes and then driven to the outer cylinder to be buried by the fresh titanium deposition. Ions are trapped and covered by the continuously subliming titanium. Re-emission of inert gases is thus eliminated.

The basic differences between the radial field and sputter-ion pump follow directly from the incorporation of the hot filament and therefore a supply of electrons that is independent of pressure. This means that: (a) the rate of titanium sublimation is independent of pressure or of the amount of ionization that occurs and hence the power dissipated in the pump is independent of pressure; (b) pump life as defined by the usage of the titanium charge (which is replaceable) is determined by filament emission; and (c) the pumping speed characteristics for active gases which are pumped without ionization will be similar to that found with a titanium sublimation pump. Electron path lengths are shorter than those in a sputter-ion pump and electrons can reach the anode without making an ionizing collision. If the roughing pressure is less than 1×10^{-3} torr, then rapid starting is achieved, but at pressures above 1×10^{-3} torr, a glow discharge prevents adequate heating of the titanium to cause sublimation.

The radial field pump has, by design, many of the advantages of the sputter-ion pump, such as freedom from vibration and organic contamination and the ability to act as a pressure gauge, pressure switch, and leak detector. Compared to the sputter-ion pump, its application may be limited as a result of reduction in pumping speed at pressures above 10^{-6} torr and high power dissipation at low pressures.

5. Titanium Sublimation Pump

These pumps are equipped with titanium sublimation sources, that consist of a dense array of cooled panels arranged in a cylindrical structure upon which titanium vapor is deposited. This pump is designed to ensure both high speed getter-pumping of active gases and long-term operation. To facilitate operation of such pumps, a liquid nitrogen cryopump is used for high speed pumping of water vapor.

6. Turbomolecular Pumps

Turbomolecular pumps are hydrocarbon-free, purely mechanical pumps used for attaining from high to ultra high vacuum.

These pumps obtain their pumping action by virtue of the relative velocities between gas molecules and moving and stationery slotted discs resembling stages of a turbine. The gas is mechanically driven from the inlet to the exhaust port of the pump with resultant compression ratios of 1 million to 1 for air (and even higher for oil vapor). The rotor speed of the pump is about 15,000 rpm. In a schematic diagram of the pump (Fig. N-5), it is seen that oblique slotted discs are arranged alternately in the housing and on the rotor; thus, the pumping action is from the center toward both ends (fore-vacuum) of the pump. Each of the moving and stationary discs forms a pressure stage. The oblique slots are designed so that acceleration is forcibly imparted to the gas molecules in an axial direction, with an optimum relationship between pumping speed and pressure ratio.

Turbomolecular pumps have the advantage of not using pumping fluid, not permitting backstreaming during operation, and not being damaged if exposed to high pressure during operation. Any leaks around the rotating seals are into the high pressure ends at the pump. They remove hydrocarbons (in unlimited quantities) and all gases without selectivity (and without the use of cold traps). The system can be rough-pumped directly through the turbomolecular pump without elaborate valving. Other useful features include reliability, safety, low down-time, and ease of leak detection.

7. Cryogenic Pumps

In cryogenic pumping (usually prepumped to about 10^{-2} torr), a surface is reduced to such low temperatures as to cause condensation of gas molecules that impinge upon it. Thus, for example, if the surface is maintained at the temperature of liquid helium (−269°C), all other gases have insignificantly low vapor pressures at this temperature and molecules of these gases impinging on the surface would remain there. Liquid-helium cooled metallic surfaces are being actively investigated with the possibility of providing pumping speeds of up to 10^6 liters/sec.

LITERATURE CITED

1. DE BROGLIE, L. *Phil. Mag.*, **47**, 446, 1924.
2. BUSCH, H. *Ann. Physik.*, **81** (Ser. 4), 974, 1926.
3. KNOLL, M. AND RUSKA, E. *Z. Physik.*, **78**, 318, 1932.
4. RÜDENBERG, R. *J. Appl. Phys.*, **14**, 434, 1943.
5. FREUNDLICH, M. *Science*, **142**, 185, 1963.
6. BRÜCHE, E. AND JOHANNSON, H. *Ann. Physik.*, **15** (Ser. 5), 145, 1932.
7. RUSKA, E. *Z. Physik.*, **87**, 580, 1934.
8. DRIEST, E. AND MULLER, H. O. *Z. Wiss. Mikroskop.*, **52**, 53, 1935.
9. VON BORRIES, B. AND RUSKA, E. *Wiss. Veroff. Siemens Werke* **17**, 99, 1938.
10. VON BORRIES, B. AND RUSKA, E. *Z. Wiss. Mikroskop.* **56**, 314, 1939.
11. MAHL, H. *Z. tech. Physik.*, **20**, 316, 1939.
12. BURTON, E. F., HILLIER, J. AND PREBUS, A. *Phys. Rev.*, **56**, 1171, 1939.
13. HILLIER, J. AND VANCE, A. W. *Proc. Inst. Radio Engrs.*, **29**, 167, 1941.
14. VON ARDENNE, M. *Kolloid. Z.*, **108**, 195, 1944.
15. HILLIER, J. *J. Appl. Phys.*, **17**, 307, 1946.
16. HILLIER, J. AND RAMBERG, E. G. *J. Appl. Phys.*, **18**, 48, 1947.
17. HEIDENREICH, R. D. *J. Elect. Micr.*, **16**, 23, 1967.
18. KOMODA, T. AND OTSUKI, M. *Jap. J. Appl. Phys.*, **3**, 666, 1964.
19. DYKE, W. P. *Sci. Amer.*, **210**, 108, 1964.
20. FERNÁNDEZ-MORÁN, H. In *Proc. VI Inter. Cong. Electr. Micr. Kyoto*, 2nd ed., **2**, p. 13, Maruzen, Tokyo, 1966.
21. WACHTEL, A. W., GETTNER, M. E. AND ORNSTEIN, L. In *Physical Techniques in Biological Research* (G. Oster and A. W. Pollister, Eds.), Vol. 3A, p. 173, Academic Press, New York, 1966.
22. HALL, C. E. *J. Biophys. Biochem. Cytol.*, **1**, 1, 1955.
23. HALL, C. E. *Introduction to Electron Microscopy*, 2nd ed., p. 118, McGraw-Hill, New York, 1966.
24. VALENTINE, R. C. *Lab. Invest.*, **14**, 596, 1965.
25. WILSKA, A. P. *J. Roy. Micr. Soc.* **82**, 287, 1964.
26. COLEMAN, J. W. AND REISNER, J. H. *Sci. Instr. News*, **10**, 19, 1965.
27. LEDBETTER, M. C. AND DELL, R. D. *J. Appl. Phys.*, **36**, 2604, 1965.
28. WATSON, J. H. L. *J. Appl. Phys.*, **18**, 153, 1947.
29. HEIDE, H. G. *Z. Angew. Phys.*, **17**, 70, 1964.
30. HEIDE, H. G. *Lab. Invest.*, **14**, 1140, 1965.
31. KNOLL, M. *Z. Techn. Phys.*, **16**, 467, 1935.
32. FERNÁNDEZ-MORÁN, H. AND FINEAN. J. B. *J. Biophys. Biochem. Cytol.*, **3**, 725, 1957.
33. FINEAN, J. B. *J. Cell Biol.*, **8**, 13, 1960.
34. CLAUDE, A. *Harvey Lec.*, **43**, 121, 1947.
35. PEASE, D. C. AND BAKER, R. F. *Proc. Soc. Exptl. Biol. Med.*, **67**, 470, 1948.
36. NEWMAN, S. B., BORYSKO, E. AND SWERDLOW, M. *J. Res. Natl. Bur. Standards*, **43**, 183, 1949.
37. GETTNER, M. E. AND HILLIER, J. *J. Appl. Physics*, **21**, 68, 1950.
38. LATTA, H. AND HARTMAN, J. F. *Proc. Soc. Exptl. Biol. Med.*, **74**, 436, 1950.
39. SJÖSTRAND, F. S. In *Freezing and Drying* (R. J. C. Harris, Ed.), p. 177, Inst. Biol., London, 1951.
40. PALADE, G. E. *J. Exptl. Med.*, **95**, 285, 1952.
41. PORTER, K. R. AND BLUM, J. *Anat. Rec.*, **117**, 685, 1953.
42. FERNÁNDEZ-MORÁN, H. *Exptl. Cell Research*, **5**, 255, 1953.
43. WATSON, M. L. *J. Biophys. Biochem. Cytol.*, **1**, 183, 1955.
44. SHELDON, H., ZETTERQUIST, H. AND BRANDES, D. *Exp. Cell. Res.*, **9**, 592, 1955.
45. ESSNER, E., NOVIKOFF, A. B. AND MASEK, B. *J. Biophys. Biochem. Cytol.*, **4**, 711, 1958.
46. NOVIKOFF, A. B. AND GOLDFISCHER, S. *Proc. Nat. Acad. Sci.*, **47**, 802, 1961.
47. HALL, C. E. *J. Biophys. Biochem. Cytol.*, **1**, 1, 1955.
48. BRENNER, S. AND HORNE, R. W. *Biochim. Biophys. Acta*, **34**, 103, 1959.
49. ZEITLER, E. AND BAHR, G. F. *Sci. Instr. News*, **7**, 2, 1962.
50. MORGAN, C., MOORE, D. H. AND ROSE, H. M. *J. Biophys. Biochem. Cytol.*, **2**, (Suppl.), 21, 1956.

51. BORYSKO, E. *J. Biophys. Biochem. Cytol.*, **2,** (Suppl.), 3, 1956.
52. MAALØE, O. AND BIRCH-ANDERSEN, A. *Symp. Soc. Gen. Microbiol.*, **6,** 261, 1956.
53. GLAUERT, A. M., ROGERS, G. E. AND GLAUERT, R. H. *Nature*, **178,** 803, 1956.
54. LUFT, J. H. *J. Biophys. Biochem. Cytol.*, **2,** 799, 1956.
55. GIBBONS, I. R. AND BRADFIELD, J. R. G. *Proc. First European Reg. Conf. Elect. Micr.*, 121, 1956.
56. PALADE, G. E. AND SIEKEVITZ, P. *J. Biophys. Biochem. Cytol.*, **2,** 671, 1956.
57. NOVIKOFF, A. B. *J. Biophys. Biochem. Cytol.*, **2,** (Suppl.), 65, 1956.
58. KELLENBERGER, E., SCHWAB, W. AND RYTER, A. *Experientia*, **12,** 421, 1956.
59. STEERE, R. L. *J. Biophys. Biochem. Cytol.*, **3,** 45, 1957.
60. MOOR, H. AND MÜHLETHALER, K. *J. Cell Biol.*, **17,** 609, 1963.
61. WATSON, M. L. *J. Biophys. Biochem. Cytol.*, **4,** 727, 1958.
62. MILLONIG, G. *J. Appl. Phys.*, **32,** 1637, 1961.
63. KARNOVSKY, M. *J. Biophys. Biochem. Cytol.*, **11,** 729, 1961.
64. REYNOLDS, E. A. *J. Cell Biol.*, **17,** 208, 1963.
65. SWIFT, H. AND RASCH, E. *Sci. Instr. News*, **3,** 1, 1958.
66. WISSIG, S. L. *Anat. Rec.*, **130,** 467, 1958.
67. LEHRER, G. M. AND ORNSTEIN, L. *J. Biophys. Biochem. Cytol.*, **6,** 399, 1959.
68. WACHTEL, A., LEHRER, G. M., MAUTNER, W., DAVIS, B. J. AND ORNSTEIN, L. *J. Histochem. Cytochem.*, **7,** 291, 1959.
69. HOLT, S. J. AND HICKS, R. M. In *The Interpretation of Ultrastructure*, (R. J. C. Harris, Ed.), p. 193, Academic Press, New York, 1962.
70. SINGER, S. J. *Nature*, **183,** 1523, 1959.
71. FERNÁNDEZ-MORÁN, H. *J. Appl. Phys.*, **30,** 2038, 1959.
72. REBHUN, L. I. *J. Biophys. Biochem. Cytol.*, **9,** 785, 1961.
73. VAN HARREVELD, A., CROWELL, J. AND MALHOTRA, S. K. *J. Cell Biol.*, **25,** 117, 1965.
74. KLEINSCHMIDT, A. K. AND ZAHN, R. K. *Z. Naturforsch.*, **146,** 770, 1959.
75. ROSENBERG, M., BARTL, P. AND LESKO, J. *J. Ultrastruct. Res.*, **4,** 298, 1960.
76. LEDUC, E. H. AND BERNHARD, W. In *The Interpretation of Ultrastructure* (R. J. C. Harris, Ed.,) p. 21. Academic Press, New York, 1962.
77. LUFT, J. H. *J. Biophys. Biochem. Cytol.*, **9,** 409, 1961.
78. WATSON, M. L. AND ALDRIDGE, W. G. *J. Biophys. Biochem. Cytol.*, **11,** 257, 1961.
79. CARO, L. G. *J. Biophys. Biochem. Cytol.*, **10,** 37, 1961.
80. SALPETER, M. M. AND BACHMANN, L. *J. Cell Biol.*, **22,** 469, 1964.
81. WOLFE, D. E., AXELROD, J., POTTER, L. J. AND RICHARDSON, K. C. In *Electron Microscopy* (S.S. Breese, Jr., Ed.), Vol. 2, p. L–12, Academic Press, New York, 1962.
82. PALAY, S. L., MCGEE-RUSSELL, S. M., GORDON, S. AND GRILLO, M. *J. Cell Biol.*, **12,** 385, 1962.
83. SABATINI, D. D., BENSCH, K. AND BARRNETT, R. J. *J. Cell Biol.*, **17,** 19, 1963.
84. KARNOVSKY, M. *J. Cell Biol.*, **23,** 217, 1964.
85. KARNOVSKY, M. *J. Cell Biol.*, **27,** 49A, 1965 (abs.).
86. PEASE, D. C. *Anat. Rec.*, **151,** 492, 1965.
87. KARNOVSKY, M. J. AND REVEL, J. P. *J. Cell Biol.* **31,** 56A , 1966 (abs.).
88. NAKANE, P. K. AND PIERCE G. B. Jr. *J. Hist. Cyto. Chem.*, **14,** 929, 1967.
89. ITO, S. In *The Interpretation of Ultrastructure* (R. J. C. Harris, Ed.), p. 129, Academic Press, New York, 1962.
90. FAWCETT, D. W. In *Modern Developments in Electron Microscopy* (B. M. Siegel, Ed.), p. 257, Academic Press, New York, 1964.
91. BENNETT, H. S. AND LUFT, J. H. *J. Biophys. Biochem. Cytol.* **6,** 113, 1959.
92. MILLONIG, G. *J. Appl. Phys.*, **32,** 1637, 1961.
93. CAULFIELD, J. B. *J. Biophys. Biochem. Cytol.*, **3,** 827, 1957.
94. DALTON, A. J. *Anat. Rec.*, **121,** 281, 1955.
95. EHRLICH, H. G. *Exptl. Cell Res.*, **15,** 463, 1958.
96. AFZELIUS, B. A. In *The Interpretation of Ultrastructure* (R. J. C. Harris, Ed.), p. 1, Academic Press, New York, 1962.
97. LOW, F. N. AND FREEMAN. J. A. *J. Biophys. Biochem. Cytol.*, **2,** 629, 1956.
98. GLAUERT, A. M. In *Techniques for Electron Microscopy* (D. Kay, Ed.), 2nd ed., p. 167, Davis, Philadelphia, 1965.
99. LUFT, J. H. *Anat. Rec.*, **133,** 305, 1959.

100. TORMEY, J. McD. *J. Cell Biol.*, **23**, 658, 1964.
101. REWCASTLE, N. B. *Nature*, **205**, 207, 1965.
102. LEDBETTER, M. C. AND GUNNING, B. E. S. *J. Roy. Micr. Soc.*, **83**, 331, 1964.
103. MOLLENHAUER, H. H. *J. Biophys. Biochem. Cytol.*, **6**, 427, 1959.
104. ROBERTSON, J. D. *Sci. Amer.*, **206**, 64, 1962.
105. BRADBURY, S. AND MEEK, G. A. *Quart. J. Micr. Sci.*, **101**, 241, 1960.
106. WOOD, R. L. AND LUFT, J. H. *J. Ultrastruct. Res.*, **12**, 22, 1965.
107. ERICSSON, J. L. E., SALADINO, A. J. AND TRUMP, B. F. *Z. Zellforsch.*, **66**, 161, 1965.
108. PEASE, D. C. *J. Ultrastruct. Res.*, **14**, 356, 1966.
109. PEASE, D. C. *J. Ultrastruct. Res.*, **14**, 374, 1966.
110. KURTZ, S. M. *J. Ultrastruct. Res.*, **5**, 468, 1961.
111. BORYSKO, E. AND SAPRANAUSKAS, P. *Bull. Johns Hopkins Hosp.*, **95**, 68, 1954.
112. MOORE, D. H. AND GRIMLEY, P. M. *J. Biophys. Biochem. Cytol.*, **3**, 255, 1957.
113. WARD, R. T. *J. Histochem. Cytochem.*, **6**, 393, 1958.
114. WATSON, M. L. *J. Biophys. Biochem. Cytol.*, **3**, 1017, 1957.
115. LEDUC, E. H., MARINOZZI, V. AND BERNHARD, W. *J. Roy. Micr. Soc.*, **81**, 119, 1963.
116. LEDUC, E. H. AND HOLT, S. S. *J. Cell Biol.*, **26**, 137, 1965.
117. WINBORN, W. B. *Anat. Rec.*, **148**, 422, 1964.
118. LOCKWOOD, W. R. *Anat. Rec.*, **150**, 129, 1964.
119. PEASE, D. C. *Histological Techniques for Electron Microscopy*, 2nd ed., Academic Press, New York, 1964.
120. FREEMAN, J. A. AND SPURLOCK, B. O. *J. Cell Biol.*, **13**, 437, 1962.
121. SPURLOCK, B. O., KATTINE, V. C. AND FREEMAN, J. A. *J. Cell Biol.*, **17**, 203, 1963.
122. ERLANDSON, R. A. *J. Cell Biol.*, **22**, 704, 1964.
123. KUSHIDA, H. *J. Electron Micr.*, **11**, 135, 1962.
124. HELLSTROM, B. *Sci. Tools*, **7**, 10, 1960.
125. HUXLEY, H. E. *J. Roy. Micr. Soc.*, **78**, 30, 1958.
126. FAHRENBACH, W. H. *J. Cell Biol.*, **18**, 475, 1963.
127. SATIR, P. G. AND PEACHEY, L. D. *J. Biophys. Biochem. Cytol.*, **4**, 345, 1958.
128. BURGE, R. E. AND SILVESTER, N. R. *J. Biophys. Biochem. Cytol.*, **8**, 1, 1960.
129. WATSON, M. L. *J. Biophys. Biochem. Cytol.*, **4**, 475, 1958.
130. WATSON, M. L. *J. Biophys. Biochem. Cytol.*, **4**, 727, 1958.
131. MILLONIG, G. *J. Appl. Phys.*, **32**, 1637, 1961.
132. KARNOVSKY, M. J. *J. Biophys. Biochem. Cytol.*, **11**, 729, 1961.
133. REYNOLDS, E. A. *J. Cell Biol.*, **17**, 208, 1963.
134. BJÖRKMAN, N. AND HELLSTROM, B. *Stain Tech.*, **40**, 169, 1965.
135. STEMPAK, J. G. AND WARD, R. T. *J. Cell Biol.*, **22**, 697, 1964.
136. FRASCA, J. M. AND PARKS, V. R. *J. Cell Biol.*, **25**, 157, 1965.
137. VALENTINE, R. C. AND HORNE, R. W. In *The Interpretation of Ultrastructure* (R. J. C. Harris, Ed.), p. 263, Academic Press, New York, 1962.
138. BERNSTEIN, M. H. *J. Biophys. Biochem. Cytol.*, **2**, 633, 1956.
139. HUXLEY, H. E. AND ZUBAY, G. *J. Biophys. Biochem. Cytol.*, **11**, 273, 1961.
140. ZOBEL, C. R. AND BEER, M. *J. Biophys. Biochem. Cytol.*, **10**, 335, 1961.
141. ALBERSHEIM, P. AND KILLIAS, U. *J. Cell Biol.*, **17**, 93, 1963.
142. WATSON, M. L. AND ALDRIDGE, W. G. *J. Biophys. Biochem. Cytol.*, **11**, 257, 1961.
143. COLEMAN, J. R. AND MOSES, M. J. *J. Cell. Biol.*, **23**, 63, 1964.
144. SWIFT, H. In *The Interpretation of Ultrastructure* (R. J. C. Harris, Ed.), p. 213, Academic Press, New York, 1962.
145. GOLDFISCHER, S., ESSNER, E. AND NOVIKOFF, A. *J. Histochem. Cytochem.*, **12**, 72, 1964.
146. SMITH, C. W. AND FARQUHAR, M. G. *Sci. Instr. News*, **10**, 9, 1965.
147. HANKER, J. S., SEAMAN, A. R., WEISS, L. P., VENO, H., BERGMAN, R. A. AND SELIGMAN, A. M. *Science*, **146**, 1039, 1964.
148. FARRANT, J. L. *Biochim. Biophys. Acta*, **13**, 569, 1954.
149. SINGER, S. J. AND SCHICK, A. F. *J. Biophys. Biochem. Cytol.*, **9**, 519, 1961.
150. TAWDE, S. S. AND SU RAM, J. *Arch. Biochem. Biophys.*, **97**, 430, 1962.
151. RIFKIND, R. A., HSU, K. C. AND MORGAN, C. *J. Histochem. Cytochem.*, **12**, 131, 1964.
152. STERNBERGER, L. A., DONATI, E. J., CUCULIS, J. J. AND PETRALI, J. P. *Exp. Mol. Path.*, **4**, 112, 1965.

153. STERNBERGER, L. A., DONATI, E. J., PETRALI, J. P., HANKER, J. S. AND SELIGMAN, A. M. *J. Histochem. Cytochem.*, **14,** 711, 1966.
154. NAKANE, P. AND PIERCE, G. B., JR. *J. Cell Biol.*, **33,** 307, 1967.
154a. STERNBERGER, L. A. AND CUCULIS, J. J. *J. Histochem. Cytochem.*, **17,** 190, 1969 (abs).
154b. KAWARAI, Y. AND NAKANE, P. *J. Histochem. Cytochem.*, **17,** 191, 1969 (abs.).
155. REVEL, J. P. AND HEY, E. D. *Exp. Cell Res.*, **25,** 474, 1961.
156. SALPETER, M. M. *J. Cell Biol.*, **32,** 379, 1967.
157. HALL, C. E. In *Modern Developments in Electron Microscopy* (B. M. Siegel, Ed.), p. 395, Academic Press, New York, 1964.
158. FERNÁNDEZ-MORÁN, H. *Ann. N.Y. Acad. Sci.*, **85,** 689, 1960.
159. RUSKA, E. *J. Roy. Micr. Soc.*, **84,** 77, 1965.
160. EISENHANDLER, C. B. AND SIEGEL, B. M. *Appl. Phys. Letters*, **9,** 217, 1966.
161. FARRANT, J. L. *J. Appl. Phys.*, **35,** 3074, 1964.
162. WOOD, R. L. AND LUFT, J. H. *J. Ultrastruct. Res.*, **12,** 22, 1965.
163. ERICSSON, J. L. E., SALADINO A. J. AND TRUMP, B. F. *Z. Zellforsch.*, **66,** 161, 1965.
164. HACKENBROCK C. R. *J. Cell Biol.* **30,** 269 1966.
165. FEDER N. AND SIDMAN, R. L. *J. Biophys. Biochem. Cytol.*, **4,** 593, 1958.
166. DAVIS, J., ORNSTEIN, L., TALEPOROS, P. AND KOULISH, S. *J. Histochem. Cytochem.*, **7,** 291, 1959.
167. REBHUN, L. I. *Fed. Proc.*, **24,** S217, 1965.
168. FRENÁNDEZ-MORÁN, H. In *The Interpretation of Ultrastructure* (R. J. C. Harris, Ed.), p. 411, Academic Press, New York.
169. MOOR, H., MUHLETHALER, K., WALDNER, H. AND FREY-WYSSLING, A. *J. Cell Biol.*, **10,** 1, 1961.
170. BULLIVANT, S. AND AMES, A., 3rd. *J. Cell Biol.*, **29,** 435, 1966.
171. BRADLEY, D. E. In *Techniques for Electron Microscopy* (D. H. Kay, Ed.), 2nd ed., p. 75, Davis, Philadelphia, 1965.
172. BAHR, G. F. AND ZEITLER, E. *Lab. Invest.*, **14,** 955, 1965.
173. ZEITLER, E. AND BAHR, G. F. *Lab. Invest.*, **14,** 208, 1965.
174. THOMSON, J. J. *Phil. Mag.* **4,** 293, 1897.
175. VON ARDENNE, M. *Z. Physik,* **109,** 553, 1938.
176. ZWORYKIN, V. K., HILLIER, J. AND SNYDER, R. L. *ASTM BULL.*, **117,** 15, 1942.
177. OATLEY, C. W., NIXON, W. C. AND PEASE, R. F. W. *Adv. Electron Phys.*, **21,** 181, 1965.
178. MCMULLEN, D., *Ph.D. Dissertation,* Cambridge Univ., England, 1953.
179. SMITH, K. C. A. AND OATLEY, C. W. *Brit. J. Appl. Phys.*, **6,** 391, 1955.
180. EVERHART, T. E. AND THORNLEY, R. M. F. *J. Sci. Instr.*, **37,** 246, 1960.
181. PEASE, R. F. W. *Ph.D. Dissertation,* Cambridge Univ., England, 1963.
182. LANGMUIR, D. B. *Proc. Inst. Radio Engrs.*, **25,** 977, 1937.
183. LAFFERTY, J. M. *J. Appl. Phys.*, **22,** 299, 1951.
184. BROERS, A. N. *Physics,* E2, 273, 1969.
185. MURATA, K., MATSUKAWA, T. AND SHIMIZU, R. *Jap. J. Appl. Phys.*, **10,** 684, 1971.
186. CREWE, A. V. *Sci. Am.*, **224,** 26, 1971.
187. MOSELEY, H. G. J. *Phil. Mag.*, **26,** 1024, 1913.
188. CASTAING, R. AND GUINIER, A., In *Proc. Cong. Electr. Micr. Delft,* p. 60, 1950.
189. COSSLETT, V. E. AND DUNCUMB, P. *Nature,* **177,** 1172, 1956.
190. BOYDE, A. AND STEWART, A. D. G. *J. Ultrastructure Res.*, **7,** 159, 1962.
191. STEWART, A. D. G. AND BOYDE, A. *Nature,* **196,** 81, 1962.
192. JAQUES, W. E., COALSON, J. AND ZERVINS, A. *Exper. Molec. Path.*, **4,** 576, 1965.
193. PEASE, R. F. W. AND HAYES, T. L. *Nature,* **210,** 1049, 1966.
194. PEASE, R. F. W., HAYES, T. L. AND CAMP, A. S. *Science,* **154,** 1185, 1966.
195. BARBER, V. C. AND BOYDE, A. *Z. Zellforsch.*, **84,** 269, 1968.
196. BOYDE, A. AND WOOD, C. *J. Micr.* **90,** 221, 1969.
197. ECHLIN, P., PADEN, R., DRONZEK, B. AND WAYTE, R. *Proc. Third Ann. SEM Symp.* p. 49, IITRI, Chicago, 1970.
198. SCHEID, W. AND TRAUT, H. *Mutation Res.*, **11,** 253, 1971.
199. HAGGIS, G. H. In *Proc. V European Cong. Elect. Micr.*, p. 250, Inst. Phys., London, 1972.
200. LOBUGLIO, A. F., RINEHART, J. J. AND BALCERZAK, S. P. *Proc. Fifth Ann. SEM Symp.*, p. 313, IITRI, Chicago, 1972.

201. FLOOD, P. R. *Proc. Eighth Ann. SEM Symp.*, p. 287, IITRI, Chicago, 1975.
202. MILLER, M. M. AND REVEL, J. P. *Anat, Rec.*, **183,** 339, 1975.
203. BOYDE, A. *Proc. Fifth Ann. SEM Symp.*, p. 257, IITRI, Chicago, 1972.
204. BOYDE, A. AND STEWART, A. D. G. *J. Ultrastructure Res.*, **7,** 159, 162.
205. GLAUERT, A. M. (Ed.) *Practical Methods in Electron Microscopy,* North-Holland, New York, 1974.
206. HAYAT, M. A. *Fixation for Electron Microscopy.* Van Nostrand Reinhold, New York, 1978.
207. PARDUCZ, B. *Int. Rev. Cytol.*, **21,** 91, 1967.
208. KARNOVSKY, M. J. *J. Cell Biol.*, **27,** 137A, 1965.
209. MAKITA, T. AND SANDBORN, E. B. *Exp. Cell Res.*, **60,** 477, 1970.
210. MOHAPATRA, S. C. AND JOHNSON, W. H. *Tob. Sci.*, **18,** 80, 1974.
211. MYKELBUST, R., DALEN, H. AND SAWTERSDAL, T. S. *J. Micro.*, **105,** 57, 1975.
212. PACHTER, B. R., PENHA, D., DAVIDOWITZ, J. AND BREININ, G. M. *Proc. Sixth Ann. SEM Symp.*, p. 387, IIRTII, Chicago, 1973.
213. HAGGIS, G. H., BOND, E. F. AND PHIPPS, B. *Proc. Ninth Ann. SEM Symp.*, p. 281, IITRI, Chicago, 1976.
214. HUMPHREYS, W. J., SPURLOCK, B. D., AND JOHNSON, J. S. *Proc. Seventh Ann. SEM Symp.*, p. 275, IITRI, Chicago, 1974.
215. FRISCH, B. AND LEWIS, S. M. In Principles and Techniques of Scanning Electron Microscopy: Biological Applications. (M. A. Hayat, Ed.) Vol. 6, p. 100, Van Nostrand Reinhold, New York, 1978.
216. NEMANIC, M. K. *Proc. Eight Ann. SEM Symp.*, p. 342, IITRI, Chicago 1975.
217. MOLDAY, R. S. *Proc. Tenth Ann. SEM Symp.*, p. 59, IITRI, Chicago, 1977.
218. LOBUGLIO, A. F., RINEHART, J. J. AND BALCERZAK, S. P. *Proc. Fifth Ann. SEM Symp.*, p. 313, IITRI, Chicago.
219. LINTHICUM, D. S. AND SELL, S. J. *Ultrastructure Res.*, **51,** 55, 1975.
220. MOLODAY, R. S., DRYER, W. J., REMBAUM A. AND YEN, S. P. S. *Nature*, **249,** 81, 1974.
221. PETERS, K. R., GSCHWENDER, H. H., HALLER, W. AND RUTTER, G. *Proc. Ninth Ann. SEM Symp.*, p. 75, IITRI, Chicago, 1976.
222. HATTORI, A., MATSUKURA, Y., ITO, S., FUGITA, T., AND TIKUNAGA, J., *Arch. Histol. JAP.*, **39,** 105, 1976.
223. HAMMERLING, V., POLLIACK, A., LAMPEN, N., SABETY, M. AND DE HARVEN, E. *J. Exp. Med.*, **141,** 518, 1975.
224. KUMON, H., UNO, F., AND TAWARA, J. *Proc. Ninth Ann. SEM Symp.*, p. 85, IITRI, Chicago, New York, 1976.
225. WELLER, N. K. *J. Cell Biol.*, **63,** 699, 1974.
226. HORISBERGER, M., ROSSET, J., AND BAUER, H. *Experientia*, **31,** 1147, 1975.
227. DUPOUY, G., PERRIER, F. AND DURRIEU, L. *Compt. Rend.*, **251,** 2836, 1960.
228. HARTMEN, H. AND HAYES, T. L. *J. Hered.*, **62,** 41, 1971.
229. WICKHMAN, M. G. AND WORTHEN, D. M. *Stain Tech.*, **48,** 63, 1973.
230. FONTE, V. G. AND PORTER, K. R. *Proc. Seventh Ann. SEM Symp.*, p. 827, IITRI, Chicago, 1974.
231. BRUMMER, M. E. G., LOWRIE, P. M. AND TYLER, W. S. *Proc. Eighth. Ann. SEM Symp.*, p. 333, IITRI, Chicago, 1975.
232. TANAKA, K. AND SMITH, C. A. *Am. J. Anat.*, **153,** 251, 1978.
233. CUTZ, E. AND COWEN, P. E. *Am. J. Anat.*, **151,** 87, 1978.

REFERENCES*

SECTION ONE: TRANSMISSION ELECTRON MICROSCOPY

PART ONE: INSTRUMENTATION

CHAPTER 1. INTRODUCTION

COSSLETT, V. E. The future of the electron microscope. *J. Roy. Micr. Soc.*, **87**, 53–76, 1967.
COWLEY, J. M. Electron microscopy. *Anal. Chem.*, **50**, 76R–80R, 1978.
JACOBSEN, E. H., KING, J. G., THOMASON, M. G. R., WEAVER, J. C. Electron and molecular microscopy: possibilities rather than limitations. *Science*, **173**, 751–2, 1971.

CHAPTER 2. HISTORICAL REVIEWS[5]

GABOR, D. Die Entwicklungsgeschichte des Elektronemikroskops. *Elektrotech. Z.*, *A***78**, 522–30, 1957.
MARTON, L. *Early History of the Electron Microscope*. San Francisco Press, San Francisco, 1968.
MULVEY, T. The history of the electron microscope. In *Historical Aspects of Microscopy* (S. Bradbury and G. L' E. Turner, Eds.), pp. 210–27. Heffer, Cambridge, England, 1967.
RUSKA, E. 25 Jahre Elektronenmikroskopie. *Elektrotech. Z.*, **A78**, 531–43, 1957.
THOMSON, G. The early history of electron diffraction. *Contempor. Phys.*, **9**, 1–15, 1968.

CHAPTER 3. B. RESOLUTION[17,18]

ARDENNE, M. v. The determination of the resolving power of electron microscopes. *Z. Physik*, **42**, 72–4, 1941.
BREEDLOVE, J. R. JR., AND TRAMMELL, G. T. Molecular microscopy: fundamental limitations. *Science*, **170**, 1310–13, 1970.
COSSLETT, V. E. The variation of resolution with voltage in the magnetic electron microscope. *Proc. Phys. Soc.*, **B58**, 443–55, 1946.
LIEBMANN, G. The limiting resolving power of the electron microscope. *Phil. Mag.*, **37**, 677–85, 1946.
MARTON, L. Resolving power in electronic microscopy. *Physica*, **3**, 959–67, 1936.
REBSCH, R. The theoretical resolving power of the electron microscope. *Ann. Physik*, **31**, 551–60, 1938.
RUSKA, E. Past and present attempts to attain the resolution limit of the transmission electron microscope. *Adv. Opt. Electron Micr.*, **1**, 115–79, 1966.
ZEITLER, E. Resolution in electron microscopy. *Adv. Electronics Electron Phys.*, **25**, 277–332, 1968.

CHAPTER 4. A. 1. ELECTRON GUN

BLOOMER, R. N. The lives of electron microscope filaments. *Brit. J. Appl. Phys.*, **8**, 83–5, 1957.
COLEMAN, J. W. Design parameters for the telefocus grid cap. *Sci. Instr. News*, **5**, 1–4, 1960.
JOYCE, R. E. A simple technique for repair of TEM Filaments. *Microstructures*, June/July, 22–24, 1972.
HAIME, M. E., AND EINSTEIN, P. A. Characteristics of the hot cathode electron microscope gun. *Brit. J. Appl. Phys.*, **3**, 40–6. 1952.

CHAPTER 4. B. IMAGING SYSTEM

DOSSE, J. Optical constants of strong electron lenses. *Z. Physik*, **117**, 722–53, 1941.
GRIVET, P. Electron lenses. *Adv. Electronics*, **2**, 47–100, 1950.
HARDY, D. F. Superconducting electron lenses. *Adv. Opt. Elect. Micr.*, **5**, 201–37, 1973.

*The references listed in this section are to selected classical and recent papers relevant to the subjects discussed in the text. They are arranged under the same section headings that are listed in the Table of Contents. The superscripts placed adjacent to the section headings refer to the original papers cited in the text dealing with the subject in question, and they are identified under "Literature Cited" (pp. 343–347). References to papers published in conference proceedings are more fully identified on p. 362 where all of them are listed.

HILLIER, J. A removable lens for extending the magnification range of an electron microscope. *J. Appl. Phys.*, **21**, 785–90, 1950.
RUSKA, E. A magnetic objective for the electron microscope. *Z. Physik*, **89**, 90–128, 1934.
RUSKA, E. Design and calculation of pole piece lenses for high resolution electron microscopy. *Arch. Elektrotech.*, **38**, 102–30, 1944.

CHAPTER 4. B. 1. b. LENS ABERRATIONS

GLASER, W. Chromatic aberration of electron lenses. *Z. Physik*, **116**, 56–67, 1940.
HALL, C. E. Method of measuring spherical aberration of an electron microscope objective. *J. Appl. Phys.*, **20**, 631–32, 1949.
HILLIER, J. A study of distortion in electron microscope projection lenses. *J. Appl. Phys.*, **17**, 411–19, 1946.
LENZ, F., AND WILSKA, A. P. Electron optical systems with annular apertures and with corrected spherical aberration. *Optik*, **24**, 383–96, 1966/67.
LIEBMANN, G. The magnetic electron microscope objective lens of lowest chromatic aberration. *Proc. Phys. Soc.*, **B65**, 188–92, 1952.
MISELL, D. L., AND CRICK, R. A. An estimate of the effect of chromatic aberration in electron microscopy. *J. Phys.*, **4**, 1668–74, 1971.
NAGATA, T., AND HAMA, K. Chromatic aberration and the electron microscope image of biological sectioned specimen. *J. Electron Micr.*, **20**, 172–76, 1971.
RAMBERG, E. G. Variation of axial aberrations of electron lenses with lens strength. *J. Appl. Phys.*, **13**, 582–94, 1942.
ROGOWSKI, W. Aberrations of electron lenses. *Arch. Elektrotech.*, **31**, 555–93, 1937.
SCHERZER, O. On some aberrations of electron lenses. *Z. Physik*, **101**, 593–603, 1936.
SEPTIER, A. The struggle to overcome spherical aberration in electron optics. *Adv. Opt. Electron Micr.*, **1**, 204–74, 1966.

CHAPTER 4. B. 1. d. CONTRAST AND IMAGE FORMATION

BARNETT, M. E. Image formation in optical and electron transmission microscopy. *J. Micr.*, **102**, 1–28, 1974.
BECKER, H., AND WALLRAFF, A. Spherical aberration of magnetic lenses. *Arch. Elektrotech.*, **32**, 664–75, 1938.
BOERSCH, H. Image formation in the electron microscope. *Nat. Bur. Standards Circ.*, **527**, 127–44, 1954.
BUAGE, R. F. Mechanisms of contrast and image formation of biological specimens in the transmission electron microscope. *J. Micr.*, **98**, 251–85, 1973.
DOSTEN, A. C. VAN, The role of accelerating voltage in image formation. *Lab. Invest.*, **14**, 819–24, 1965.
HANSZEN, K. J. Neue Erkenntnisse über Auflosung und Kontrast im Elektronemikroskopischen Bild. *Naturwissenschaften*, **54**, 125–33, 1967.
HAWKES, P. W. Computer processing of electron micrographs. *Int. Rev. Cytol.*, **42**, 103–26, 1975.
LOCKE, M., KRISHNAN, N., MCMAHON, J. T. A routine method for obtaining high cell contrast without staining sections. *J. Cell Biol.*, **50**, 540–44, 1971.
MISELL, D. L. Image formation in the electron microscope with particular reference to the defects in electron optic images. *Adv. Electronics Electron Phys.*, **32**, 63–91, 1973.
MARTON, L. Image formation mechanism. In *The Encyclopedia of Microscopy* (G. L. CLARK, Ed.), pp.159–63, Reinhold, New York, 1961.
MISELL, D. L. The phase problem in electron microscopy. *Adv. Opt. Electron Micr.*, **7**, 185–279, 1979.
REISNER, J. H. Instrumental factors involved in the improvement of contrast in electron microscopy. *Sci. Instr. News*, **4**, 7–13, 1959.
ZEITLER, E., AND BAHR, G. F. The interpretation of contrast in an electron micrograph. *Sci. Instr., News*, **5**, 5–16, 1960.

CHAPTER 4. C. 2. PHOTOGRAPHIC RECORDING

FARNELL, G. C., AND FLINT, R. B. The response of photographic materials to electrons with particular reference to electron microscopy. *J. Micr.*, **97,** 271–91, 1972.

FARNELL, G. C., AND FLINT, R. B. Exposure level and image quality in electron micrographs. *J. Micr.*, **103,** 319–32, 1975.

FRIEDMAN, M. A. A reevaluation of the Markham rotation technique using model systems. *J. Ultrastructure Res.*, **32,** 226–36, 1970.

HAMILTON, J. F., AND MARCHANT, J. C. Image recording in electron microscopy. *J. Opt. Soc., Am.*, **57,** 232–39, 1967.

KURTZ, S. M. A consideration of accelerating voltage and photographic emulsion in biological electron microscopy. *Norelco Rep.*, **7,** 14–17, 1960.

VALENTINE, R. C. The response of photographic emulsions to electrons. *Adv. Opt. Electron Micr.*, **1,** 180–203, 1966.

WOOD, R. L. Considerations of photographic material for electron microscopy with special reference to the use of sheet film. *Norelco Rep.*, **7,** 23, 1960.

ZEILER, E. H. Characteristics of photographic emulsions for electron microscopy. *J. Opt. Soc., Am.*, **55,** 204–05, 1965.

CHAPTER 4. D. 2. OBJECTIVE APERTURES[27]

PLANTHOLT, B. A., AND WEHNAN, H. J. A method of cleaning platinum apertures for electron lenses by surface polishing and sonification. *Stain Tech.*, **45,** 299–300, 1970.

SCHABTACH, E. A method for the fabrication of thin-foil apertures for electron microscopy. *J. Micr.*, **101,** 121–26, 1974.

CHAPTER 4. E. 1. ALIGNMENT

LEISEGANG, S. Zur Zentrierung magnetischer Elektronenlensen. *Optik*, **11,** 397–406, 1954.

CHAPTER 4. E. 2. DETECTION OF LENS ASYMMETRY

BOERSCH, H. v. Fresnelsche Beugungserscheinung im Übermikroskop. *Naturwissenschaften*, **28,** 71–12, 1940.

HAINE, M. E., AND MULVEY, T. Applications and limitations of the edge-diffraction test for astigmatism in the electron microscope. *J. Sci Instr.*, **31,** 326–32, 1954.

HILLIER, J., AND RAMBERG, E. The magnetic electron microscope objective: contour phenomena and the attainment of high resolving power. *J. Appl. Physics*, **18,** 48–71, 1947.

LEISEGANG, S. Zum astigmatismus von Elektronenlinsen, *Optik*, **10,** 5–14, 1953.

LEISEGANG, S. Ein einfacher Stigmator für magnetische Electronenlinsen, *Optik*, **11,** 49–60, 1954.

LEPOOLE, J. B., AND RUTTE, W. A. Correction and detection of astigmatism. *Proc. Conf. Electron Micr.*, Delft, pp. 84–8, 1950.

REISNER, J. H. Practical aspects of lens correction for astigmatism. Parts I and II. *Sci. Instr. News*, **1,** 5–8, 8–12, 1956.

WEGMANN, L. Zur Bestimmung des Auflosungsvermogens durch Fresnelische Beugung im Elektronenmikroskop. *Helv. Phys. Acta*, **23,** 1–26, 1950.

CHAPTER 4. E. 3. e. CONTAMINATION[29]

BENNETT, P. M. Decrease in section thickness on exposure to the electron beam; the use of tilted sections in estimating the amount of shrinkage. *J. Cell. Sci.*, **15,** 693–701, 1974.

GLAESER, R. M. Limitations to significant information in biological electron microscopy as a result of radiation damage. *J. Ultrastructure Res.*, **36,** 466–82, 1971.

HARTMAN, R. E., AND HARTMAN, R. S. Elimination of potential sources of contaminating material. *Lab. Invest.*, **14,** 1147–54, 1965.

JAMES, R., AND BEER, M. The reduction of contamination in the electron microscope. *Norelco Rep.*, **11,** 80, 1964.

LEISGANG, S., AND SCHOTT, O. Der Einfluss der Bestrahlungsbedingungen auf die Object-verschmutzung. *Proc. First European Reg. Conf. Electron Micr.*, Stockholm, pp. 20–7, 1957.

PATTERSON, R. L., AND WAYMAN, C. M. Study of contamination rates in the electron microscope. *Rev. Sci. Instr.*, **34,** 1213–15, 1963.
STENN, K., AND BAHR, G. F. Specimen damage caused by the beam of the transmission electron microscope, a correlative reconsideration. *J. Ultrastructure Res.*, **31,** 526–50, 1970.

CHAPTER 4. G. 2. IMAGE INTENSIFIERS

ANONYMOUS. TV-Display systems and image intensifier for electron microscopes. *Norelco Rept.*, **20,** 18–22, 1973.
HAINE, M. E., AND EINSTEIN, P. A. Image intensifiers. *Proc. Second European Reg. Conf. Electron Micr.*, Delft, **1,** 97–100, 1961.
NATHAN, R. Image processing for electron microscopy. I. Enhancement procedures. *Adv. Opt. Electron Micr.*, **4,** 85–125, 1971.
REYNOLDS, G. T. Image intensification applied to microscope systems. *Adv. Opt. Electron Micr.*, **2,** 1–40, 1968.
SADASHIGE, K. Image intensifier TV camera system, its performance and application. *Appl. Opt.*, **6,** 2179–90, 1967.

CHAPTER 4. G. 4. ELECTRON DIFFRACTION ACCESSORIES

ALDERSON, R. H., AND HALLIDAY, J. S. Electron diffraction. In *Techniques for Electron Microscopy* (D. H. Kay, Ed.), 2nd ed., pp. 478–524. Davis, Philadelphia, 1965.
GLAESER, R. M., AND THOMAS, G. Application of electron diffraction to biological electron microscopy. *Biophys. J.*, **9,** 1073–93, 1969.

CHAPTER 4. H. 2. SCANNING ELECTRON MICROSCOPY (see Chapter 14).

CHAPTER 4. H. 3. HIGH VOLTAGE ELECTRON MICROSCOPY[17, 18, 157, 161]

BEESTROM, B. E. P. High voltage microscopy of biological specimens, some practical considerations. *J. Micr.*, **98,** 402–16, 1973.
COSSLETT, V. E. Current developments in high voltage electron microscopy, *J. Micr.*, **100,** 233–46, 1974.
FAVARD, P., AND N. CARASSO The preparation and observation of thick biological sections in the high voltage electron microscope. *J. Micr.*, **97,** 59–81, 1973.
PARSONS, D. F. Problems in high resolution electron microscopy of biological materials in their natural state. In *Some Biological Technique in Electron Microscopy* (D. F. Parsons, Ed.), pp. 1–68. Academic Press, New York, 1970.
PANDE, A. High voltage electron microscopy. *Lab. Prac.*, **19,** 1125–32, 1970.
THOMAS, G., AND LACAZE, J. C. Transmission electron microscopy at 2.5 MeV. *J. Micr.*, **97,** 301–08, 1973.

PART TWO: METHODOLOGY

CHAPTER 5. INTRODUCTION

BILS, R. F. Tissue preparation for electron microscopy. *Am. J. Med. Technol.*, **33,** 35–45, 1967.
CHAPMAN, J. A. New techniques in conventional electron microscopy in biology. *Lab. Pract.*, **19,** 477–81, 1971.
DAVID, H. Theorie und Prakitsche Anwendbarkeit der Fixations-und Einbettungsmethoden für die Elktronenmikroskopie. *Z. Med. Labor. Tech.*, **5,** 169–216, 1964.
FAWCETT, D. W. In histology and cytology. In *Modern Developments in Electron Microscopy* (B. M. Siegel, Ed.), pp. 257–333. Academic Press, New York, 1964.
HAVDENSCHILD, C., AND TSCHIRKY, H. Automated specimen processing for electron microscopy: a new apparatus. *Experientia*, **28,** 1389–91, 1972.
PARSONS, D. F., MATRICARDI, V. R., MORETZ, R. C., AND TURNER, J. N. Electron microscopy and diffraction of wet unstained and unfixed biological objects. *Adv. Biol. Med. Phys.*, **15,** 162–270, 1974.

CHAPTER 6. HISTORICAL REVIEW

FARQUHAR, M. G. Preparation of ultrathin tissue sections for electron microscopy. *Lab. Invest.*, **5,** 317–37, 1956.

SJÖSTRAND, F. W. Electron microscopy of tissues and cells. In *Physical Techniques in Biological Research* (G. Oster and A. W. Pollister, Eds.), vol. 3, pp. 241–98. Academic Press, New York, 1956.

CHAPTER 7. C. PERFUSION[82]

ABRUNHOSA, R. Microperfusion fixation of embryos for ultrastructural studies. *J. Ultrastructure Res.*, **41,** 176–88, 1972.

FAHIMI, H. D. Perfusion and immersion fixation of rat liver with gluteraldehyde. *Lab Invest.*, **16** 736–50, 1967.

LEGATO, M. J., SPIRO, D., AND LANGER, G. A. Ultrastructural alterations produced in mammalian myocardium by variation in perfusate ionic composition. *J. Cell Biol.*, **37,** 1–12, 1968.

JOHNSTON, P. V., AND ROOTS, B. I. Fixation of the central nervous system by perfusion with aldehydes and its effect on the extracellular space as seen by electron microscopy. *J. Cell Sci.*, **2,** 377–86, 1967.

MAUNSBACH, A. B. The influence of different fixatives and fixation methods on the ultrastructure of rat kidney proximal tubule cells. I. Comparison of different perfusion methods and of gluteraldehyde, formaldehyde and osmium tetroxide fixatives. *J. Ultrastructure Res.*, **15,** 242–82, 1966.

CHAPTER 7. D. HANDLING OF SMALL SPECIMENS

ANDERSON, T. F. Electron microscopy of microorganisms. In *Physical Techniques in Biological Research* (G. Oster and A. W. Pollister, Eds.), 2nd ed., vol. 3A, pp. 319–87. Academic Press, New York, 1966.

CHANG, J. H. T. Fixation and embeddment, *in situ,* of tissue culture cells for electron microscopy. *Tissue and Cell,* **4,** 561–74, 1972.

FLICKINGER, C. J. Methods for handling small numbers of cells for electron microscopy. In *Methods in Cell Physiology* (D. M. Prescott, Ed.), vol. 2, pp. 311–21. Academic Press, New York, 1966.

FRANKS, L. M., AND WILSON, P. D. Origin and ultrastructure of cells *in vitro*. *Int. Rev. Cytol.*, **48,** 55 139, 1977.

LAVAIL, M. M., III. A method of embedding selected areas of tissue cultures for electron microscopy. *Texas Rep. Biol. Med.*, **26,** 215–22, 1968.

NELSON, B. K., AND FLAXMAN, B. A. *In situ* embedding and vertical sectioning for electron microscopy of tissue cultures grown on plastic petri dishes. *Stain Tech.*, **47,** 261–65, 1972.

WEBER, G. Electron microscopy of free floating cells: thin-layering technique and selection of individual cells for ultramicrotomy. *Stain Tech.*, **52,** 25–9, 1977.

CHAPTER 7. E. HANDLING OF HUMAN MATERIAL

FARQUHAR, M. G. Electron microscopy of renal biopsies. *Bull. N. Y. Acad. Med.*, **36,** 419–23, 1960.

LYNN, A., MARTIN, J. H., AND RACE, G. J. Recent improvements of histologic technics for the combined light and electron microscopic examination of surgical specimens. *Am. J. Clin. Path.*, **45,** 704–13, 1966.

MOSES, H. L. Comparative fine structure of the trigeminal ganglia, including human autopsy studies. *J. Neurosurg.*, **26,** 112–16, 1967.

PRICE, H. M., HOWES, E. L. JR, SHELDON, D. B., HUTSON, O. D., FITZGERALD, R. T., AND BLUMBERG, J. M. An improved biopsy technique for light and electron microscopic studies of human skeletal muscles. *Lab Invest.*, **14,** 194–99, 1965.

CHAPTER 8. FIXATION[40]

GENERAL METHODS

BROWN, J. N. The avian erythrocyte. A study of fixation for electron microscopy. *J. Micr.*, **104,** 293–305, 1975.

MILLONIG, G., AND MARINOZZI, V. Fixation and embedding in electron microscopy. *Adv. Opt. Electron Micr.*, **2,** 252–341, 1968.

SABBATH, M., AND ANDERSSON, B. *In situ* fixation and embedding for electron microscopy. *Meth. Cell. Biol.*, **15,** 435–44, 1977.

SIMKINS, T. Preservation of biological tissues for TEM by chemical fixation. *Microstructures*, **3,** 23–8, 1972.

FIXATION MECHANISMS AND EFFECTS

BONE, Q., AND DENTON, E. S. The osmotic effects of electron microscope fixatures. *J. Cell Biol.*, **49,** 571–81, 1971.

EISENBERG, B. G., AND MOBLEY, B. A. Size changes in single muscle fibers during fixation and embedding. *Tissue and Cell*, **7,** 383–87, 1975.

LARSSON, W. H. Effects of different fixatives on the ultrastructure of developing proximal tubules in the rat kidney. *J. Ultrastructure Res.*, **51,** 140–51, 1975.

LITMAN, R. B. AND BARRNETT, R. J. The mechanism of the fixation of tissue components by osmium textroxide via hydrogen bonding. *J. Ultrastructure Res.*, **38,** 23–8, 1972.

PERFUSION FIXATION[82]

DODSON, R. F. S-Collidine as a buffering system for glutaraldehyde perfusion of the central nervous system. *J. Neuropathol. Exper. Neurol.*, **30,** 714–22, 1971.

JOHNSON, W. H., LATTA, H., AND OSVALDO, L. Variations in glomerular ultrastructure in rat kidneys fixed by perfusion. *J. Ultrastructure Res.*, **45,** 149–67, 1973.

ROSSI, G. L. Simple apparatus for perfusion fixation for electron microscopy. *Experientia*, **31,** 998–1000, 1975.

WILLIAMS, T. H., AND JEW, J. Y. An improved method for perfusion fixation of neural tissues for electron microscopy. *Tissue and Cell*, **7,** 407–18, 1975.

CHAPTER 8. B. 1. a. FORMALIN FIXATION

BURGOS, M. H., VITALE-CALPE, R., AND TELLEX DE INON, M. T. Studies on paraformaldehyde fixation for electron microscopy. I. Effect of concentration on ultrastructure. *J. Micr.*, **6,** 457–67, 1967.

HOLT, S. J., AND HICKS, R. M. Studies on formalin fixation for electron microscopy and cytochemical staining. *J. Biophys. Biochem. Cytol.*, **11,** 31–45, 1961.

KARNOVSKY, M. J. A formaldehyde-gluteraldehyde fixative of high osmolarity for use in electron microscopy. *J. Cell Biol.*, **27,** 137–38A, 1965 (Abs.).

CHAPTER 8. B. 1. b. GLUTARALDEHYDE[83,99]

ERICSSON, J. L. E., AND BIBERFELD, P. Studies on aldehyde fixation. Fixation rates and their relation to fine structure and some histochemical reactions in liver. *Lab Invest.*, **17,** 281–98, 1967.

MCMILLAN, P. N., AND LUFTIG, R. B. Preservation of membrane ultrastructure with aldehyde or imidate fixatives. *J. Ultrastructure Res.*, **52,** 243–60, 1975.

PERACCHIA, C., AND MITTLER, B. S. New glutaraldehyde fixation procedures. *J. Ultrastructure Res.*, **39,** 57–64, 1972.

RASMUSSEN, K. E. Fixation in aldehydes. A study on the influence of the fixative, buffer and osmolarity upon the fixation of the rat retina. *J. Ultrastructure Res.*, **46,** 87–102, 1974.

SABATINI, D. D., MILLER, F., AND BARNETT, R. J. Aldehyde fixation for morphological and enzyme histochemical studies with the electron microscope. *J. Histochem. Cytochem.*, **12,** 57–71, 1964.

WEAKLEY, B. S. A comparison of three different electron microscopical grade glutaraldehyde used to fix ovarian tissue. *J. Micr.*, **101,** 127–41, 1974.

CHAPTER 8. B. 2. PERMANGANATE[54]

ROSENBLUTH, J. Contrast between osmium-fixed and permanganate-fixed toad spinal ganglia. *J. Cell Biol.*, **16,** 143–57, 1963.

SUTTON, J. S. Potassium permanganate staining of ultrathin sections for electron microscopy. *J. Ultrastructure Res.*, **21,** 5–6, 1967.

WETZEL, B. K. Sodium permanganate fixation for electron microscopy. *J. Biophys. Biochem. Cytol.*, **9,** 711–16, 1961.

CHAPTER 10. EMBEDDING MEDIA[36]

COULTER, H. D. Rapid and improved methods for embedding biological tissues in Epon 812 and Araldite 502. *J. Ultrastructure Res.*, **20,** 346–55, 1967.

DELLMAN, D.-H., AND PEARSON, C. B. Better epoxy resin embedding for electron microscopy at low relative humidity. *Stain Tech.*, **52,** 5–8, 1977.

ESTES, L. W., AND APICELLA, J. V. A rapid embedding technique for electron microscopy. *Lab. Invest.*, **20,** 159–63, 1969.

GEISELMAN, C. W., AND BURKE, C. N. Exact anhydride: epoxy percentages for araldite and araldite-epon embedding. *J. Ultrastructure Res.*, **43,** 220–27, 1973.

LEDUC, E. H., AND BERNHARD, W. Recent modifications of the glycol methacrylate procedure. *J. Ultrastructure Res.*, **10,** 196–99, 1967.

LOCKWOOD, W. R. A reliable and easily sectioned epoxy embedding medium. *Anat. Rec.*, **50,** 129–40, 1964.

MCGEE-RUSSELL, S. M., AND DE BRYN, W. C. Experiments on embedding media for electron microscopy. *Quart. J. Micr. Sci.*, **105,** 231–44, 1964.

RAMPLEY, D. N. Embedding media for electron microscopy. *Lab. Pract.*, **16,** 591–93, 1967.

WINBORN, W. B. Dow epoxy resin with triallyl cyanurate and similarly modified Araldite and Maraglas mixtures, as embedding media for electron microscopy. *Stain Tech.*, **40,** 227–31, 1965.

CHAPTER 11. MICROTOMY[12,38,41,42]

BERNARD, W., AND VIRON, A. Improved techniques for the preparation of ultrathin frozen sections. *J. Cell Biol.*, **49,** 731–46, 1971.

BLACK, J. T. Ultramicrotomy of embedding plastics *Am. Chem. Soc. J.*, **30,** 314–28, 1970.

GORYCKI, M. A. Methods for precisely trimming block faces for ultramicrotomy. *Stain Tech.*, **53,** 63–66, 1978.

HELANDER, H. F. Some observations on knife properties and sectioning mechanics during ultramicrotomy of plastic embedding media. *J. Micr.*, **101,** 81–95, 1974.

WARD, R. T. Some observations on glass-knife making. *Stain Tech.*, **52,** 305–9, 1977.

ICLESIAS, J. R., BERNIER, R., AND SIMARD, R. Ultracryotomy: a routine procedure. *J. Ultrastructure Res.*, **36,** 271–89, 1971.

WALLSTROM, A. C., AND ISEN, D. A. Ultrasonic cleaning of diamond knives. *J. Ultrastructure Res.*, **41,** 561–2, 1972.

WILLIAMS, M. A., AND MEEK, G. A. Studies on thickness variation in ultrathin sections for electron microscopy. *J. Roy. Micr. Soc.*, **85,** 337–52, 1965.

CHAPTER 12. B. POSITIVE STAINING[55,61]

BAUMSTER, L. H. Lanthanum as an intracelluar stain. *J. Micr.*, **95,** 413–19, 1972.

BEER, M. Selective staining for electron microscopy. *Lab Invest.*, **14,** 1020–25, 1965.

BJÖRKMAN, N., AND HELLSTRÖM, B. Lead-ammonium acetate; a staining medium for electron microscopy free of contamination by carbonate. *Stain Tech.*, **40,** 169–71, 1965.

LAWN, A. M. The use of potassium permanganate as an electron-dense stain for sections of tissue embedded in epoxy resin. *J. Biophys. Biochem. Cytol.*, **7,** 197–99, 1960.

LOCKE, M. Hot alcoholic phosphotungstic acid and uranyl acetate as routine stains for thick and thin sections. *J. Cell Biol.*, **50,** 550–57, 1971.

PARSONS, D. F. A simple method for obtaining increased contrast in Araldite sections by using post-fixation staining of tissues with potassium permanganate. *J. Cell Biol.*, **11,** 492–97, 1961.

PETERS, A., LOWARY-HINDS, P., AND VAUGHN, J. E. Extent of stain penetration in sections prepared for electron microscopy. *J. Ultrastructure Res.*, **36**, 37–45, 1971.

ZOBEL, C. R., AND ROE, J. Y. K. Positive staining of protein molecules for electron microscopy. *Biochem. Biophys. Acta,* **133,** 157–61, 1967.

CHAPTER 12. C. NEGATIVE STAINING[47,48,137]

GLAUERT, A. M. Factors influencing the appearance of biologic specimens in negatively stained preparations. *Lab Invest.*, **14,** 1069–79, 1965.

HORNE, R. W. Recent advances in the application of negative staining techniques to the study of virus particles examined in the electron microscope. *Adv. Opt. Electron Micr.*, **6,** 227–301, 1975.

MUNN, E. A. The application of the negative staining technique to the study of membranes. *Meth. Enzymol.*, **32,** 20–35, 1974.

CHAPTER 13. A. LOCALIZATION OF NUCLEOPROTEINS[65,78]

EVENSON, D. P. Electron microscopy of viral nucleic acids. *Meth. Virol.*, **6,** 220–64, 1977.

GAUTIER, A. Ultrastructural localization of DNA in ultrathin tissue sections. *Int. Rev. Cytol.*, **44,** 113–91, 1976.

MOYNE, G. Feulgen-derived techniques for electron microscopical cytochemistry of DNA. *J. Ultrastructure Res.*, **45,** 102–23, 1973.

SWIFT, H. Nucleoprotein localization in electron micrographs: metal binding and radioautography. *Symp Int. Soc. Cell Biol.*, **1,** 213–32, 1962.

WATSON, M. L., AND ALDRIDGE, W. G. Selective electron staining of nucleic acids. *J. Histochem. Cytochem.*, **12,** 96–103, 1964.

CHAPTER 13. B. ULTRASTRUCTURAL ENZYME CYTOCHEMISTRY[44,45,46]

BARRNETT, R. S. Localization of enzymatic activity at the fine structural level. *J. Roy. Micr. Soc.*, **83,** 143–51, 1964.

DAEMS, W. T., AND PERSUN, J. P. Enzyme histochemistry in electron microscopy. In *Enzymes in Clinical Chemistry* (R. Ruyssen and L. Vandendreissche, Eds.), pp. 75–104, Elsevier, New York, N. Y. 1965.

HOLT, S. J., AND HICKS, R. M. Combination of cytochemical staining methods for enzyme localization with electron microscopy. *Symp. Int. Soc. Cell Biol.*, **1,** 193–211, 1962.

LEWIS, P. R. Electron microscopical localization of enzymes. *Adv. Opt. Electron Micr.*, **6,** 171–226, 1975.

MOLBERT, E. R. G. Ultrastructural cytochemistry. *Meth. Achievements Exper. Path.*, **2,** 1–29, 1967.

SCARPELLI, D. G., AND KANCZAK, N. M. Ultrastructural cytochemistry: principles, limitations, and applications. *Int. Rev. Exp. Path.*, **4,** 55–126, 1965.

SHNITKA, T. K., AND SELIGMAN, A. M. Ultrastructure cytochemistry of enzymes and some applications. *Meth. Cancer Res.*, **10,** 37–84, 1974.

CHAPTER 13. C. TRACERS IN ELECTRON MICROSCOPY[70,85,87]

AINSWORTH, S. K., AND KARNOVSKY, M. J. An ultrastructural staining method for enhancing the size and electron opacity of ferritin in thin sections, *J. Histochem. Cytochem.*, **20,** 225–9, 1972.

BECKER, N. H., NOVIKOFF, A. B., AND ZIMMERMAN, H. M. Fine structure observations of the uptake of intravenously injected peroxidase by the rat choroid plexus. *J. Histochem. Cytochem.*, **15,** 160–65, 1967.

BRUNS, R. R., AND PALADE, G. E. Studies on blood capillaries. II. Transport of ferritin molecules across the wall of muscle capillaries. *J. Cell Biol.*, **37,** 227–99, 1968.

COTRAN, R. S., AND KARNOVSKY, M. J. Ultrastructural studies on the permeability of the mesothelium to horseradish peroxidase. *J. Cell. Biol.*, **37,** 123–37, 1968.

KISHIDA, Y., OLSEN, B. R., BERG, R. A., AND PROCOP, D. J. Two improved methods for preparing ferritin-protein conjugates for electron microscopy. *J. Cell Biol.*, **64,** 331–39, 1975.

SHARABADI, M. S., AND YAMAMOTO, T. A method for staining intracellular antigens in thin sections with ferritin-labeled antibody. *J. Cell. Biol.*, **50,** 246–50, 1971.

CHAPTER 13. D. ULTRASTRUCTURAL IMMUNO-ELECTRON MICROSCOPY[70,88,150,154]

BUSTIN, M., GOLDBLATT, D., AND SPERLING, R. Chromatin structure visualization by immuno-electron microscopy. *Tissue and Cell,* **7,** 297–304, 1975.

HATTORI, A., MATSUKURA, Y., ITO, S., FUGITA, T., AND TOKUNAGA, J. Ferritin as a surface marker for immunoscanning electron microscopy. Observation of individual ferritin particles on erythrocytes. *Arch. Histol. Jap.*, **39,** 105–15, 1976.

KRAEHENBOHL, J. P., GRANDI, P. B. DE, AND CAMPICHE, M. A. Ultrastructural localization of intracellular antigen using enzyme-labeled antibody fragments. *J. Cell Biol.*, **50,** 432–48, 1971.

KHULMANN, W. D., AND MILLER, H. R. P. A comparative study of the technique for ultrastructural localization of antienzyme antibodies. *J. Ultrastructure Res.*, **35,** 370–85, 1971.

MILNE, R. G., AND LUISONI, E. Rapid immune electron microscopy of virus preparations. *Meth. Virol.,* **6,** 265–81, 1977.

MORGAN, C. The use of ferritin-conjugated antibodies in electron microscopy. *Int. Rev. Cytol.,* **32,** 291–326, 1972.

STERNBERGER, L. A. Some new developments in immunocytochemistry. *Mikroskopie,* **25,** 346–61, 1969.

CHAPTER 13. E. ELECTRON AUTORADIOGRAPHY[79,80]

BUDD, G. E. Recent developments in light and electron microscope radioautography. *Int. Rev. Cytol.,* **31,** 21–56, 1971.

GRANBOULAN, P. Autoradiographic methods for electron microscopy. *Meth. Virol.,* **3,** 618–37, 1971.

HÜLSER, D. F., AND RAJEWSKY, M. F. Autoradiography with the electron microscope: properties of photographic emulsions. *Meth. Cell Physiol.,* **3,** 293–306, 1971.

JACOB, J. The practice and application of electron microscope autoradiography, *Int. Rev. Cytol.,* **30,** 90–181, 1971.

ROGERS, A. W. Recent developments in the use of autoradiographic techniques with electron microscopy. *Phil. Trans. Roy. Soc. Lond. B* **261,** 159–71, 1971.

WILLIAMS, M. A. The assessment of electron microscopic autoradiographs. *Adv. Opt. Electron Micr.,* **3,** 219–72, 1969.

CHAPTER 13. F. HIGH-RESOLUTION ELECTRON MICROSCOPY (see Chapter 4. H. 3.).

CHAPTER 13. G. ELECTRON MICROSCOPIC ANALYSIS OF CELL[56,57]

BAUDHUIN, P., EVRARD, P., AND BERTHET, J. Electron microscopic examination of subcellular fractions. II. Quantitative analysis of the mitochondrial population isolated from rat liver. *J. Cell Biol.,* **35,** 631–48, 1967.

DALEN, H. A filtration technique for preparing cells in suspension for electron microscopy. *J. Micro.,* **91,** 213–15, 1970.

LEIGHTON, F., POOLE, B., BEAUFAY, H., BAUDIN, P., COFFEY, J. W., FOWLER, S., AND DEDUVE, C. The large-scale separation of peroxisomes, mitochondria, and lysosomes from the livers of rats injected with Triton WR-1339. Improved isolation procedures, automated analysis, biochemical and morphological properties of fractions. *J. Cell Biol.,* **37,** 482–513, 1968.

SADOWSKI, P. D., AND STEINER, J. W. Electron microscopic and biochemical characteristics of nuclei and nucleoli isolated from rat liver. *J. Cell Biol.,* **37,** 147–61, 1968.

STOCKING, C. R., SHUMWAY, L. K., WEIER, T. E., AND GREENWOOD, D. Ultrastructure of chloroplasts isolated by nonaqueous extraction. *J. Cell Biol.,* **36,** 270–75, 1968.

CHAPTER 13. H. CRYOFIXATION FOR ELECTRON MICROSCOPY[59,60,71,72]

BULLIVANT, S. Freeze substitution and supporting techniques. *Lab. Invest.,* **14,** 1178–95, 1965.

HEREWARD, F. V., AND NORTHCOTE, D. H. A simple freeze-substitution method for the study of ultrastructure of plant tissue. *Exper. Cell Res.*, **70,** 73–80, 1972.

KOEHLER, J. K. The technique and application of freeze-etching in ultrastructural research. *Adv. Biol. Med. Phys.*, **12,** 1–84, 1968.

MOOR, H. Recent progress in freeze-etching technique. *Phil Trans. Roy. Soc. Lond. B.*, **261,** 121–131, 1971.

MUHLETHALER, K. Studies on freeze-etching of cell membranes. *Int. Rev. Cytol.*, **31,** 1–19, 1971.

STEERE, R. L. Freeze-etching simplified. *Cytobiology,* **5,** 306–23, 1969.

CHAPTER 13 I. SURFACE EXAMINATION OF SMALL SPECIMENS[74,171]

BAILEY, G. W., AND ELLIS, J. R. Particle replication, techniques and applications. *Micr. Crystal Front,* **14,** 306–18, 1965.

PREUSS, L. E. Shadow-casting and contrast. *Sci. Instr. News,* **4,** 13–22, 1959.

CHAPTER 13. J. QUANTITATIVE ELECTRON MICROSCOPY[172, 173]

FRITSCH, S., AND GRIMM, H. Methodische Untersuchungen zur feinstrukturellen. Morphometric and Modell leukotischer Zellen. *Virchows Arch.*, **B3,** 127–36, 1969.

WEIBEL, E. R. Sterological principles for morphometry in electron microscopic cytology. *Int. Rev. Cytol.*, **26,** 235–302, 1969.

SECTION TWO: SCANNING ELECTRON MICROSCOPY*

PART ONE: INSTRUMENTATION

CHAPTER 14. INTRODUCTION

CARR, K. E. Application of scanning electron microscopy in biology. *Int. Rev. Cytol.*, **27,** 283–348, 1970.

CREWE, A. V. The current state of high resolution scanning electron microscopy. *Quart. Rev. Biophys.*, **3,** 137–75, 1970.

ECHLIN, P. The scanning electron microscope and its application to research. *Micr. Acta,* **73,** 189–204, 1973.

HAYES, T. L. Trends and prospects in scanning electron microscopy. *J. Micr.*, **100,** 133–42, 1974.

HOLLENBERG, M. J., AND ERICKSON, A. M. The scanning electron microscope: potential usefulness to biologists. A review. *J. Histochem. Cytochem.*, **21,** 109-30, 1973.

LIN, P. S. D., AND LAMRIK, M. K. High resolution scanning electron microscopy at the sub-cellular level. *J. Micr.*, **103,** 249–57, 1975.

MCDONALD, L. W., PEASE, R. F. W., AND HAYES, T. L. Scanning electron microscopy of sectioned tissues. *Lab. Invest.*, **16,** 532–38, 1967.

OATLEY, C. W., NIXON, W. C., AND PEASE, R. F. W. Scanning electron microscopy. *Adv. Electronics Electron Phys.*, **21,** 181–247, 1965.

CHAPTER 16. BASIC THEORY

EVERHART, T. E., AND HAYES, T. L. The scanning electron microscope. *Sci. Am.*, **226,** 55–69, 1972.

PEASE, R. F. Fundamentals of scanning electron microscopy. *Proc. Fourth Ann. SEM Symp.*, 11–16, Chicago, 1971.

*Only in the past decade has there been a rapid application of the SEM in biological research. Thus, compared with TEM, a very extensive literature has not yet developed concerning both the instrument, its operation, and its application. Nevertheless, significant information is available in specialized SEM texts (see Bibliography D. 2. B), Atlases (E) and in the proceedings of the annual SEM symposia published by IIT Research Institute (10 West 35 Street, Chicago, Ill., 60616). Representative papers dealing with subjects discussed in the text are also listed in this section under the same headings as in the Table of Contents. The subscripts adjacent to the headings are identified under "Literature Cited" (pp. 343-347).

CHAPTER 17. A. 1 ELECTRON GUN[183,184]

BROERS, A. N. Electron sources for scanning electron microscopy. *Proc. Eighth Ann. SEM Symp.*, 661–70, Chicago, 1975.

CREWE, A. V., EGGENBERGER, D. N., WALL, J., AND WELTER, L. M. Electron gun using a field emission source. *Rev. Sci. Instr.*, **39,** 576–83, 1968.

CHAPTER 17. B. 2. a. BEAM-SPECIMEN INTERACTION

EVERHART, T. E. Simple theory concerning the reflection of electrons from solids. *J. Appl. Phys.*, **31,** 1483–90, 1960.

MURATA, K. Monte Carlo calculations on electron scattering and secondary electron production in the SEM. *Proc. Sixth Ann. SEM Symp.*, pp. 267–69, Chicago 1973.

REIMER, L. Electron-specimen interactions and applications in SEM and STEM. *Proc. Ninth Ann. SEM Symp.*, **I,** 1–8, Chicago, 1976.

CHAPTER 17 B. 2. b. ELECTRON COLLECTION[178]

EVERHART, T. E., AND THORNLEY, R. M. E. Wide band detector for microampere low-energy electron currents. *J. Sci. Instr.*, **37,** 246–48, 1960.

LIN, P. S. D., AND BECKER, R. P. Detection of backscattered electrons with high resolution. *Proc. Eighth Ann. SEM Symp.*, 61–70, Chicago, 1975.

WELLS, O. C. Measurements of low-loss electron emission from amorphous targets. *Proc. Eighth Ann. SEM Symp.*, 43–50, Chicago, 1975.

CHAPTER 17. B. 2. d. SIGNAL DISPLAY

BECK, V. Slow scan display system for a scanning electron microscope. *Rev. Sci. Instrum.* **44,** 1064–66, 1973.

CHAPTER 18. B. RESOLUTION

BROERS, A. N. Factors affecting resolution in the scanning electron microscope. *Proc. Third Ann. SEM Symp.*, 1–8, Chicago, 1970.

CATTO, C. J. D., AND SMITH, K. C. A. Resolution limits in the surface scanning electron microscope. *J. Micr.*, **98,** 417–35, 1973.

SIMON, R. Resolving power and information capacity in scanning electron microscopy. *J. Appl. Phys.*, **41,** 4632–41, 1970.

CHAPTER 19. A. LOW VOLTAGE MICROSCOPY

BROWN, A. C., AND SWIFT, J. A. Low voltage scanning electron microscopy of keratin fiber surfaces. *Proc. Seventh Ann. SEM Symp.*, 67–74, Chicago, 1974.

WELTER, L. M., AND COATES, V. J. High resolution scanning electron microscopy at low accelerating voltages. *Proc. Seventh Ann. SEM Symp.* 59–66, Chicago, 1974.

CHAPTER 19. B. SCANNING TRANSMISSION ELECTRON MICROSCOPY

CREWE, A. V. A high resolution scanning electron microscope. *Sci. Am.*, **224,** 26–35, 1971.

CHAPTER 19. C. SCANNING ELECTRON MICROPROBE ANALYSIS[188,189]

BAJAJ, Y. P. S., RASMUNSSEN, H. P., AND ADAMS, M. W. Electron-microprobe analysis of isolated plant cells. *J. Exp. Bot.*, **22,** 749–52, 1971.

HEINRICH, K. F. J. Scanning electron probe microanalysis. *Adv. Opt. Electron Micr.*, **6,** 275–301, 1975.

HUTCHINSON, T. E. Determination of subcellular elemental concentration through ultra-high resolution electron microprobe analysis, *Int. Rev. Cytol.*, **58,** 115–158, 1978.

KIERSZENBAUM, A. L., LIBANATI, C. M., AND TANDLER, C. J. The distribution of inorganic cations in mouse testis—electron microscope and microprobe analysis. *J. Cell Biol.*, **48,** 314–23, 1971.

PART TWO: SEM METHODOLOGY

CHAPTER 22. INTRODUCTION

BOYDE, A. Biological specimen preparation for the scanning electron microscope: An overview. *Proc. Fifth Ann. SEM Symp.*, 257–64, Chicago, 1972.

MUNGER, B. L., AND MUMAW, V. Specimen preparation for SEM study of cells and cell organelles in uncoated preparation. *Proc. Ninth SEM Symp.*, 275–80, Chicago, 1976.

PEASE, R. F. W., AND HAYES, T. L. Scanning electron microscopy of biological materials. *Nature*, **210**, 1049, 1966.

CHAPTER 24. PREFIXATION HANDLING[202,203,204]

EISENSTAT, L. F., LEVIN, B., GOLOMB, H. M., AND RIDDELL, R. H. A technique for removing mucus and debris from mucosal surfaces. *Proc. Ninth Ann. SEM Symp.*, **II**, 263–8, Chicago, 1976.

EVAN, A., DAIL, W. G., DAMMROSE, D., AND PALMER, C. Scanning electron microscopy of tissues following removal of basement membrane and collagen. *Proc. Ninth Ann. SEM Symp.*, **II**, 203–8, Chicago, 1976.

FRISCH, B., LEWIS, S. M., STUART, P. R., AND OSBORN, J. S. A transmission electron microscope study of the effects of ion etching on cells. *J. Micr.*, **104**, 219–33, 1975.

LEDBETTER, M. C. Practical problems in observation of unfixed, uncoated plant surfaces by SEM. *Proc. Ninth Ann. SEM Symp.*, **II**, 453–60, Chicago, 1976.

LIM, D. J. Scanning electron microscopic observation on non-mechanically cryofractured biological tissues. *Proc. Fourth Ann. SEM Symp.*, 259–64, Chicago, 1971.

MARCHANT, H. J. Processing small delicate biological specimens for scanning electron microscopy. *J. Micr.*, **97**, 369–71, 1973.

MARSZALEK, D. S., AND SMALL, E. B. Preparation of soft tissue materials for scanning electron microscopy. *Proc. Sec. Ann. SEM Symp.*, 213–9, Chicago, 1969.

MCDONALD, L. W., PEASE, R. F. W., AND HAYES, T. L. Scanning electron microscopy of sectioned tissue. *Lab. Invest.*, **16**, 532–38, 1967.

MILLER, M. M., AND REVEL, J. P. Scanning electron microscopy of epithelia prepared by blunt dissection. *Anat. Rec.*, **183**, 339–58, 1975.

POH, R. L., ALTENHOFF, J., ABRAHAM, S., AND HAYES, T. L. Scanning electron microscopy of myocardial sections originally prepared for light microscopy. *Exp. Mol. Pathol.*, **14**, 404–7, 1971.

SPECTOR, M. Ion beam etching in a scanning electron microscope. Red blood cell etching. *Micron*, **5**, 263–73, 1975.

SWENEY, L. R., AND SHAPIRO, B. L. Rapid preparation of uncoated biological specimens for scanning electron microscopy. *Stain Tech.*, **52**, 221–27, 1977.

VITAL, J., AND PORTER, K. R. Scanning microscopy of dissociated tissue cells. *J. Cell. Biol.*, **67**, 345–60, 1975.

WOODS, P. S., AND LEDBETTER, M. C. Cell organelles at uncoated cryofractured surfaces as viewed with the scanning electron microscope. *J. Cell Sci.*, **21**, 47–58, 1976.

CHAPTER 25. FIXATION (See Chapter 8)[207,208]

CHAPTER 26. B. 2. CRITICAL POINT DRYING

BARTLETT, A. A., AND BURSTYN, H. P. A review of the physics of critical point drying. *Proc. Eighth Ann. SEM Symp.*, 305–16, Chicago, 1975.

COHEN, A. L. A critical look at critical point dying — theory, practice and artifacts. *Proc. Tenth Ann. SEM Symp.*, **I**, 525–36, Chicago, 1977.

LEWIS, E. R., AND NEMANIC, M. K. Critical point drying techniques. *Proc. Sixth Ann. SEM Symp.*, 767–74, Chicago, 1973.

RICE, R. M., HEGYELL, A. F., AND BREENDEN, L. G. Specimen holders for simultaneous critical point drying of multiple biological specimens. *Stain Tech.*, **51**, 51–4, 1976.

TANAKA, K., AND IINO, A. Critical point drying method using dry ice. *Stain Tech.*, **49**, 203–6, 1974.

CHAPTER 26. D. SPECIMEN COATING

CHATTERJI, S., MOORE, N., AND JEFFREY, J. W. Preparation of nonconducting samples for the scanning electron microscope. *J. Phys. E.*, **5**, 118–19, 1972.

DE NEE, P. B., AND WALKER, E. R. Specimen coating technique for the SEM-2 comparative study. *Proc. Eighth Ann. SEM Symp.*, 225–32, Chicago, 1975.

MUNGER, B. L. The problem of specimen conductivity in electron microscopy. *Proc. Tenth Ann. SEM Symp.*, pp. 481–83, Chicago, 1977.

RAMPLEY, D. N. The effect of the electron beam on various mounting and coating media in the SEM II. *J. Micr.*, **107**, 99–102, 1976.

CHAPTER 26. F. CELL SURFACE LABELING[216-226]

LOBUGLIO, A. F., RINEHART, J. J., AND BALCERZAK, S. P. A new immunologic marker for scanning electron microscopy. *Proc. Fifth Ann. SEM Symp.*, 313–20, Chicago, 1972.

MOLDAY, R. S. Cell surface labeling technique for SEM. *Proc. Tenth Ann. SEM Symp.*, **II**, 59–74, Chicago, 1977.

NEMANIC, M. K. On cell surface labeling for the SEM. *Proc. Eighth Ann. SEM Symp.*, 341–50, Chicago, 1975.

PETERS, H. R., GSCHWENDER, H. H., HALLER, W., AND RUTTER, G. Utilization of a high resolution spherical marker for labeling of virus antigens at the cell membrane in conventional scanning electron microscopy. *Proc. Ninth Ann. SEM Symp.*, **II**, 75–84, Chicago, 1976.

BIBLIOGRAPHY

This is a selective bibliography that aims to provide the reader with a basis for delving deeper into an area of personal interest. The bibliography consists of references to six types of publications: general reference sources, proceedings of international congresses, proceedings of regional congresses, books, atlases, and papers. Three groups of articles have been included in the latter category: major contributions, review articles, and recent publications.

The detailed classification of the bibliographic references is as follows:

A. General Reference Sources

B. Proceedings of International Congresses

C. Proceedings of Regional Conferences

D. Books

1. Electron optics
2. Electron microscopes: TEM and SEM
3. Electron microscopy: preparation methods
4. Electron microscopy: biomedical applications

E. Atlases

F. Papers

General
1. Microscopes and microscopy
2. Preparation methods
3. General cytology

The Cell
4. Cell membrane
5. Cell membrane specializations
6. Mitochondria
7. Golgi complex
8. Endoplasmic reticulum
9. Centrioles
10. Lysosomes and microbodies
11. Microtubules and microfilaments
12. Annulate lamellae
13. Inclusion bodies
14. Nucleus
15. Nuclear envelope
16. Nucleolus
17. Chromosomes, mitosis, and meiosis

Tissues
18. Epithelium and glands
19. Connective tissue
20. Cartilage and bone
21. Nervous tissue
22. Muscular tissue
23. Blood
24. Hemopoeisis

Organs and Systems
25. Digestive system and glands
26. Respiratory system
27. Circulatory system
28. Urinary system
29. Male genital system
30. Female genital system
31. Placenta, umbilical cord and mammary gland
32. Integumentary system
33. Nervous system
34. Sensory system
35. Lymphoid organs
36. Endocrine organs

Others
37. Bacteria, viruses, and macromolecules
38. Botany
39. Tumors
40. Medicine

A. General Reference Sources

1. Cumulative Bibliographies

Bulletin Signalétique, Section 6ME, Microscopie Electronique, Vol. 25–, C. N. R. S., Paris, 1964–.

MOORE, D. H. *The International Bibliography of Electron Microscopy*. Vol. 1, 1950–155; Vol. 2, 1956–1961. N.Y. Soc. Electron Microscopists, New York, 1959 and 1962.

2. Serial Publications

Advances in Biological and Medical Physics.
Advances in Electronics and Electron Physics.
Advances in Optical and Electron Microscopy.
International Review of Cytology.
Symposia of the International Society for Cell Biology.

3. Journals

American Journal of Anatomy.
Anatomical Record.
Cytobiologie.
Experimental Cell Research.
Histochemie.
Journal of Anatomy.
Journal of Applied Physics.
Journal of Cell Biology.
Journal of Cell Science.
Journal of Electron Microscopy.
Journal of Histochemistry and Cytochemistry.
Journal de Microscopie.
Journal of Microscopy.
Journal of Molecular Biology.
Journal of Morphology.
Journal of the Optical Society of America.
Journal of Ultrastructure Research.
Histochemie.
Laboratory Investigation.
Mikroskopie.
Micron.
Protoplasma.
Stain Technology.
Tissue and Cell.
Zeitschrift für Zellforschung und Mikroskopische Anatomie.

B. Proceedings of International Congresses

1948 Conference on Electron Microscopy, Delft. Nijhoff, Hague (1950).*

1950 First International Congress on Electron Microscopy, Paris. Revue d'optique Théorique et Instrumentale, Paris (1953).

1954 Third International Congress on Electron Microscopy, London. Roy. Micr. Soc., London (1956).

*This was actually the first congress; the second was held in Paris in 1950. The confusion arose because the Proceedings of the Delft Congress were published without a numerical label. (See the discussion in the Preface to the Proceedings of the 3rd Congress.)

1958 Fourth International Conference on Electron Microscopy, Berlin. Springer, Berlin (1960).
1962 Fifth International Congress on Electron Microscopy, Philadelphia. Academic Press, New York (1962).
1966 Sixth International Congress on Electron Microscopy, Kyoto. Maruzen, Tokyo (1966).
1970 Seventh International Congress on Electron Microscopy, Grenoble. Soc. Française Micr., Paris (1970).
1974 Eighth International Congress on Electron Microscopy, Canberra. Australian Acad. Sc., Canberra (1974).
1978 Ninth International Congress on Electron Microscopy, Toronto. Micr. Soc. Canada, Toronto (1978).

C. Proceedings of Regional Congresses

1954 European Congress on Electron Microscopy, Ghent. Vandermeerssche, Brussels (1954).
1956 First European Regional Congress on Electron Microscopy, Stockholm. Academic Press, New York (1957).
1956 First Regional Conference on Electron Microscopy in Asia and Oceania, Tokyo. Electrotechnical Lab., Tanashi-mochi, Kaminuko-dai, Tokyo (1957).
1960 Second European Regional Conference on Electron Microscopy, Delft. Nederlandse Vereniging voor Electronen-Microscopie, Delft (1961).
1964 Third European Regional Conference on Electron Microscopy, Prague. Roy. Micr. Soc., London (1965).
1965 Second Regional Conference on Electron Microscopy in Far East and Oceania, Calcutta. Electron Micr. Soc. of India, Calcutta (1965).
1968 Fourth European Regional Conference on Electron Microscopy, Rome. Institut Superiore Di Sanita, Rome (1968).
1972 Fifth European Regional Conference on Electron Microscopy, Manchester. Institute of Physics, London (1972).
1972 First Latin American Congress of Electron Microscopy, Maracaibo.
1976 Sixth European Regional Conference on Electron Microscopy, Israel.
1980 Seventh European Regional Conference on Electron Microscopy, Holland.

D. BOOKS

1. Electron Optics

BRÜCHE, E., AND SCHERZER, O. *Geometrische Elektronenoptik*. Springer, Berlin, 1934.
BUSCH, H., AND BRÜCHE, E. *Beitrage zur Elektronenoptik*. Barth, Leipzig, 1937.
COSSLETT, V. E. *Introduction to Electron Optics*. Oxford, Clarendon Press, Oxford, 1946.
DAHL, P. *Introduction to Electron and Ion Optics*. Academic Press, New York, 1973.
EL-KAREH, A. B., AND EL-KAREH, J. C. J. *Electron Beams, Lenses, and Optics*, Vol. 1. Academic Press, New York, 1970.
GLASER, W. *Grundlagen der Elektronenoptik*. Springer, Vienna, 1952.
GRIVET, P. *Electron Optics*. 2nd ed., Pergamon Press, New York, 1972.
HAWKES, P. W. *Electron Optics and Electron Microscope*. Barnes & Nobel, New York, 1972.
JACOB, L. *Introduction to Electron Optics*. Methuen, London, 1951.
KLEMPERER, O. *Electron Optics*. 2nd ed., Cambridge Univeristy Press, Cambridge, 1971.
MYERS, L. M. *Electron Optics*. Cmapman & Hall, London, 1939.
PASZKOWSKI, B. *Electron Optics*. Elsevier, New York, 1969.
RUSTERHOLZ, A. *Electronenoptic*. Barkhäuser, Basel, 1950.
WAINRIB, E. A., AND MILIUTIN, W. I. *Elektronenoptik*. EBV Verlag Technik, Berlin, 1954.
ZWORYKIN, V. K., MORTON, G. A., RAUBERG, E. G., HILLIER, J., AND VANCE, A. W. *Electron Optics and the Electron Microscope*. John Wiley, New York, 1945.

2. Electron Microscopes: TEM and SEM

Transmission Electron Microscope

ARDENNE, M. v. *Electronen Übermikroskopie*. Springer, Berlin, 1940.
BORRIES, B. v. *Die Übermikroskopie*. Aulendorf Wurtt., Cantor, 1949.
BURTON, E. F., AND KOHL, W. H. *The Electron Microscope*. Reinhold, New York, 2nd ed. 1946.
COSSLETT, V. E. *Practical Electron Microscopy*. Academic Press, New York, 1952.
COSSLETT, V. E. *Modern Microscopy or Seeing the Very Small*. Cornell University Press, Ithaca, 1966.
DE BROGLIE, L. *Optique Electronique et Corpuscularies*. Herman, Paris, 1950.
FISCHER, R. B. *Applied Electron Microscopy*. Indiana University Press, Bloomington, 1954.
FRYER, J. R. *The Chemical Applications of Transmission Electron Microscopy*. Academic Press, New York, 1978.
GABOR, D. *The Electron Microscope*. Chemical, New York, 1948.
GONZALEZ-SANTANDER, R. *Manual de Microscopia Electrónica*. I. *Elementos de Microscopia Electrónica*. Monographia de Ciencia Moderna, No. 75, Madrid, 1966.
GREGOIRE, C. *Microscope Electronique et Recherche Biologique*. Masson, Paris, 1950.
HAINE, M. E., AND COSSLETT, V. E. *The Electron Microscope*. Spon, London, 1961.
HALL, C. B. *Introduction to Electron Microscopy*. 2nd ed. McGraw-Hill, New York, 1966.
HEIDENREICH, R. D. *Fundamentals of Transmission Electron Microscopy*. Wiley, New York 1964.
MAHL, H., AND GOLZ. E. *Electronenmikroskopie*. VEB Biographisches Inst., Leipzig, 1951.
MEEK, G. A. *Practical Electron Microscopy for Biologists*. 2nd ed. Wiley, New York, 1976.
OCKENDEN, F. F. J. *Introduction to the Electron Microscope*. Williams and Northgate, London, 1946.
PALACIOS DE BORAO, G. *Supermicroscopia Electronica*. Monographia de Ciencia Moderna, No. 21, Madrid, 1950.
PICHT, J. *Das Electronenmikroskop*. Fachbuchverlag, Leipzig, 1955.
PORTOCALA, R., AND IONESCU, N. I. *Microscopia Electronica in Biologie si Inframicrobiologie*. Acad. Republ. Populare Romine, Bucharest, 1962.
REIS, T. *Le Microscope Electronique et ses Applications*. Presses Univ. de France, Paris, 1951.
ROCHOW, T. G. *An Introduction to Microscopy by Means of Light, Electrons, X-rays or Ultrasound*. Plenum, New York, 1978.
RUHLE, R. *Das Electronenmikroskop*. Schwab, Stuttgart, 1949.
SELME, P. *Le Microscope Electronique*. Presses Univ. de France, Paris, 1963.
SAXTON, O. D. *Computer Techniques for Image Processing in Electron Microscopy*. Academic Press, New York, 1978.
SWANN, P. R., HUMPHREYS, C. J., AND GORINGE, M. J. (eds.) *High Voltage Electron Microscopy*. Academic Press, London, 1974.
SWIFT, J. A. *The Electron Microscope*. Page, London, 1970.
VENABLES, J. A. *Developments in Electron Microscopy and Analysis*. Academic Press, New York, 1976.
TERRIERE, G. *Microscopie Electronique*. Tech. Ingr. Mesures et Anal., Paris, 1962.
WYCKOFF, R. W. G. *The World of the Electron Microcope*. Yale University Press, New Haven, 1958.

Scanning Electron Microscope

ERASMUS, D. A. (ed.) *Electron Probe Microanalysis in Biology*. Chapman and Hall, London, 1978.
GILMORE, C. P. *The Scanning Electron Microscope: World of the Infinitely Small*. N.Y. Graphic Soc., Greenwich, Conn., 1972.
GOLDSTEIN, J. I., AND YAKOWITZ, H. (eds.) *Practical Scanning Electron Microscopy: Electron and Ion Microprobe Analysis*. Plenum Press, New York, 1975.
HALL, T., ECHLIN, P., AND KAUFMANN, R. (eds.) *Microprobe Analysis as Applied to Cells and Tissues*. Academic Press, New York, 1974.
HAYAT, M. A. *Introduction to Biological Scanning Electron Microscopy*. University Park Press, Baltimore, 1978.

HAYAT, M. A. (ed.) *Principles and Techniques of Scanning Electron Microscopy. Biological Applications,* Vols. 1–6. Van Nostrand Rheinhold, New York 1970–1978.
HEARLE, J. W. S., SPARROW, J. T., AND CROSS, P. M. *The Use of the Scanning Electron Microscope.* Pergamon Press, Elmsford, N.Y., 1972.
HEYWOOD, V. H. (ed.) *Scanning Electron Microscopy: Systematic and Evolutionary Applications.* Academic Press, New York, 1971.
HOLT, D. B., MUIR, M. D., GRANT, P. R., AND BOSWARVA, I. M. *Quantitative Scanning Electron Microscopy.* Academic Press, New York, 1974.
NIXON, W. C. (ed.) *Scanning Electron Microscopy. Systems and Applications.* Dawson, London, 1974.
OATLEY, C. W. *The Scanning Electron Microscope.* Part I. *The Instrument,* Cambridge University Press, Cambridge, 1972.
OHNSORGE, J., AND HOLM, R. *Rastenelektronenmikroskopie.* Thieme, Stuttgart, 1972.
POSTEK, M., HOWARD, K., JOHNSON, A., AND MCMICHAEL, J. *Scanning Electron Microscopy, A Students Handbook.* Ladd Research Industries, Burlington, Vt., 1979.
REIMER, L., AND PFEFERKORN, O. *Rastenelektronenmikroskopie.* Springer, Berlin, 1973.
SIEGEL, B., AND BEAMAN, D. R. *Physical Aspects of Electron Microscopy and Microbeam Analysis.* Wiley, New York.
WELLS, O. C. *Scanning Electron Microscopy.* McGraw-Hill, New York, 1976.

3. Electron Microscopy: Preparation Methods

BENEDETTI, E. L., AND FARVARD (eds.) *Freeze Etching: Techniques and Applications.* Soc. Franc. Micro. Elect., Paris, 1973.
DAWES, C. J. *Biological Techniques in Electron Microscopy.* Barnes & Noble, New York, 1971.
GEYER, G. *Ultrahistochemie.* 2nd ed. Fisher, Jena, 1973.
GLAUERT, A. M. (ed.) *Practical Methods in Electron Microscopy.* Vols. 1–7, Elsevier, N. Y., 1972–1979.
GONZÁLEZ-SANTANDER, R. *Technicas de Microscopia Electronica en Biologia.* Aquilar, Madrid, 1969.
HAYAT, M. A. (ed.) *Principles and Techniques for Electron Microscopy: Biological Applications.* Vols. 1–8, Van Nostrand Reinhold, New York, 1970–1978.
HAYAT, M. A. (ed.) *Basic Electron Microscope Techniques,* Van Nostrand Reinhold, New York, 1972.
HAYAT, M. A. (ed.) *Electron Microscopy of Enzymes: Principles and Methods.* Vols. 1–5, Van Nostrand Reinhold, New York, 1973–1979.
HAYAT, M. A. (ed.) *Positive Staining for Electron Microscopy.* Van Nostrand Reinhold, New York, 1975.
JUNIPER, B. E., GILCHRIST, A. J., AND WILLIAMS, P. R. *Techniques for Plant Electron Microsopy.* Lippincott, Philadelphia, 1970.
KAY, D. *Techniques for Electron Microscopy.* 2nd ed., Davis, Philadelphia, 1965.
KOEHLER, J. K. (ed.) *Advanced Techniques in Biological Electron Microscopy.* I and II. Springer, New York, 1973 and 1978.
MERCER, E. H., AND BIREBECK, M. S. C. *Electron Microscopy: A Handbook for Biologists.* 2nd ed. Davis, Philadelphia, 1966.
MULLER, H. *Präparation von Technisch-physikalischen Objekten für die Elektronenmikroskopische Untersuchung.* Geest Portig, Leipzig, 1962.
PARSONS, D. F. (ed.) *Some Biological Techniques in Electron Microscopy.* Academic Press, New York, 1970.
PEASE, D. C. *Histological Techniques for Electron Microscopy.* 2nd ed. Academic Press, New York, 1964.
RASH, J. E. (ed.) *Freeze Fracture: Methods, Artifacts and Interpretations.* Raven Press, New York, 1979.
REIMER, L. *Electronenmikroscopische Untersuchungs-und Präparationsmethoden.* 2nd ed. Springer, Berlin, 1966.
SCHÄFER, H. *Immunoelektronenmikroskopie.* Fischer, Stuttgart, 1971.
SCHIMMEL, G., AND VOGELL, W. (eds.) *Methodensammlung der Elektronenmikroskopie.* Wissenshaftliche Verlaggesellschaft, Stuttgart, 1970.

SJÖSTRAND, S. F. *Electron Microscopy of Cells and Tissues,* Vol. 1, *Instrumentation and Techniques*. Academic Press, New York, 1967.

STOLINSKIK, C. AND BREATHNACH, A. S. *Freeze-Fracture Replication of Biological Tissues*. Academic Press, London, 1975.

WEAKLEY, B. S. *A Beginner's Handbook in Biological Electron Microscopy*. Williams and Wilkins, Baltimore, 1973.

WITTING, G. *Manual and Study Guide for the Introductory Electron Microscopy Laboratory*. Stipes, Champaign, Ill., 1972.

ZAPF, K., AND LUDVIK, J. *Einführung in die Electronenmikroskopische Prapariertechnik in der Mikrobiologie*. Fischer, Jena, 1961.

4. Electron Microscopy: Biomedical Applications

ALEXANDROWICZ, J., BLICHARSKI, J., AND FELTYNOWSKI, A. *Electron Microscopy of Blood*. (In Polish) Panstwowe Wydawnictwo Naukowe, Warsaw, 1955.

ANDRÉ, J. A. (ed.) *Problèmes de Structures d'Ultrastructures et de Fonctions Cellulaires*. Masson, Paris, 1955.

ANDRÉ, J. A. (ed.) *Problèmes d'Ultrastructure et de Fonctions Nucleaires*. Masson, Paris, 1960.

AUSTIN, C. R. *Ultrastructure of Fertilization*. Holt, Rinehart and Winston, New York, 1968.

BERNHARD, W., AND LEPLUS, R. *Fine Structure of the Normal and Malignant Human Lymph Node*. Pergamon Press, Elmsford, N.Y., 1965.

BESSIS, M. *Living Blood Cells and their Ultrastructure*. Springer, New York, 1973.

BOURNE, G. H., AND DANIELLI, J. F. (eds.) *Studies in Ultrastructure*. Academic Press, New York, 1978.

BLOODWORTH, J. M. B., AZAR, H. A., AND YODAIKEN, R. E. (eds.) *Symposium on Electron Microscopy in Diagnostic Pathology*. Saunders, Philadelphia, 1975.

BRIEGER, E. M. *Structure and Ultrastructure of Microorganisms*. Academic Press, New York, 1963.

BRINKLEY, B. R., AND PORTER, K. R. *International Cell Biology* 1976–1977. Rockefeller University Press, New York, 1977.

CHALLICE, C. E., AND Viragh, S. (eds.) *Ultrastructure of the Mammalian Heart*. Academic Press, New York, 1973.

CONSTANTINIDES, P. *Functional Electron Histology. A Correlation of Ultrastructure and Function in all Mammalian Tissues*. Elsevier, New York, 1974.

COSSEL, L. *Die Menschliche Leber im Elektronenmikroskop*. Fischer, Jena, 1964.

COTÉ, W. A., JR. (ed.) *Cellular Ultrastructure of Woody Plants*. Syracuse University Press, Syracuse, 1965.

DALTON, A. J., AND HAGUENAU, F. (eds.) *Ultrastructure of the Kidney*. Academic Press, New York, 1967.

DALTON, A. J., AND HAGUENAU, F. (eds.) *Ultrastructure in Biological Systems*. Academic Press, New York, 1968.

DAVID, H. *Submicroscopic Ortho- and Patho-Morphology of the Liver*. Pergamon Press, Elmsford, New York, 1964.

DODGE, J. D. *The Fine Structure of Algal Cells*. Academic Press, New York, 1973.

DUSTIN, P. *Microtubules*. Springer, New York, 1978.

FELIX, H. D., HAEMMERLI, G., AND STRAULI, P. *Dynamic Morphology of Leukemia Cells: A Comparative Study by Scanning Electron Microscopy and Microcinematography*. Springer, New York, 1978.

FERNÁNDEZ-MORÁN, H., AND BROWN, R. (eds.) *The Submicroscopic Organization and Function of Nerve Cells*. Academic Press, New York, 1958.

FRIEDMAN, I. (ed.) *The Ultrastructure of Sensory Organs,* Elsevier, New York, 1973.

FREY-WYSSLING, A., AND MUHLETHALER, K. *Ultrastructural Plant Cytology*. Elsevier, New York, 1965.

FULLER, R., AND LOVELOCK, D. W. *Microbial Ultrastructure*. Academic Press, New York, 1976.

GAILLARD, P. J., TALMAGE, R. V., AND BUDY, A. M. (eds.) *The Parathyroid Glands: Ultrastructure, Secretion, and Function,* University of Chicago Press, Chicago, 1965.
GHADIALLY, F. H., AND ROY, S. *Ultrastructure of Synovial Joints in Health and Disease.* Butterworth, London, 1969.
GHADIALLY, F. H. *Ultrastructural Pathology of the Cell.* Butterworth, London, 1975.
GIESKING, R. *Mesenchymale Gewebe und ihre Reaktionsformen im Elektronenoptischen Bild.* Fischer, Stuttgart, 1966.
GRAN, F. C. (ed.) *Structure and Function of the Endoplasmic Reticulum in Animal Cells.* Academic Press, New York 1968.
GRIMSTONE, A. V. *The Electron Microscope in Biology.* 2nd ed., University Park Press, Baltimore, 1977.
HAGGIS, G. H. *The Electron Microscope in Molecular Biology.* Wiley, New York, 1966.
HARRIS, P. J. *Biological Ultrastructure: the Origin of Cell Organelles.* Oregon State University Press, Corvallis, 1969.
HESS, M. (ed.) *Electron Microscopic Concepts of Secretion: Ultrastructure of Endocrine and Reproductive Organs.* Wiley, New York, 1974.
HUHN, D., AND STICH, W. *Fine Structure of Blood and Bone Marrow. An Introduction to Electron Microscopic Hematology.* Hafner, New York, 1969.
IURATO, S. *Submicroscopic Structure of the Inner Ear.* Pergamon Press, Elmsford, N.Y., 1967.
JAKUS, M. H. *Ocular Fine Structure,* Vol. 1. Little Brown, Boston, 1964.
JENSEN, W. A., AND PARK, R. B. *Cell Ultrastructure.* Wadsworth, Belmont, California, 1967.
JOHANNESSEN, J. V. (ed.) *Electron Microscopy in Human Medicine.* Vol. I–X, McGraw-Hill, New York, 1978.
JORGENSEN, F. *The Ultrastructure of the Normal Human Glomerulus.* Munksgaard, Copenhagen, 1966.
KING, D. W. *Ultrastructural Aspects of Disease.* Harper & Row, New York, 1967.
KISCH, B. *Electron Microscopy of the Cardiovascular System.* Thomas, Springfield, Ill., 1960.
KURTZ, S. M. (ed.) *Electron Microscopic Anatomy.* Academic Press, New York, 1964.
LEVADITI, C. *Images Electroniques en Microbiologie.* Librairie Maloine, Paris, 1949.
LONGACRE, J. J. (ed.) *The Ultrastructure of Collagen,* Thomas, Springfield, Ill., 1976.
LUPULESCU, A., AND PETROVICI, A. *Ultrastructure of the Thyroid Gland.* Williams & Wilkins, Baltimore, 1968.
MANDAL, A. K. *Electron Microscopy of the Kidney in Renal Disease and Hypertension.* Plenum, New York, 1979.
MCGEE-RUSSEL, S. M., AND ROSS, K. F. A. *Cell Structure and Its Interpretation.* St. Martin's Press, New York, 1968.
ORCI, L. *Freeze-etch Histology.* Springer, Berlin, 1975.
PALAY, S. L., AND CHAN-PALAY, V. *Cerebellar Cortex: Cytology and Organization.* Springer, New York, 1974.
PETERS, A., PALAY, S. L., AND WEBSTER, H. F. *Fine Structure of the Nervous System.* Harper & Row, New York, 1969.
PEKLOV, A. P., *Electronmicroscopical Investigation of Bacteria and Phages: Submicroscopic Anatomy.* (In Russian) Gos. Ise. Med. Lit., Moscow, 1962.
PITELKA, D. R. *Electron-microscopic Structure of Protozoa.* Pergamon Press, Elmsford, N.Y., 1063.
POLICARD, A., AND BAUD, C. A. *Les Structures Inframicroscopiques Normales et Pathologiques des Cellules et des Tissues; Signification Physiologique et Pathogenique.* Masson, Paris, 1958.
SCHULZ, H. *Die Submicroscopische Anatomie und Pathologie der Lunge.* Springer, Berlin, 1959.
SCHULTZ, H. *Thrombocyten und Thrombose im Elektronenmikroskopischen Bild.* Springer, Berlin, 1968.
SJÖSTRAND, F. S. *Electron Microscopy of Cells and Tissues.* Academic Press, New York, 1967.
SPITZNAS, M. (ed.) *Current Research in Ophthalmic Electron Microscopy.* Springer, New York, 1977.
TANIKAWA, K. *Ultrastructural Aspects of the Liver and Its Disorders.* Springer, New York, 1969.

THREADGOLD, L. T. *The Ultrastructure of the Animal Cell*. Pergamon Press, Elmsford, N.Y., 1967.
TIKHONENKO, A. S., *Ultrastructure of Bacterial Viruses*. Plenum, New York, 1970.
TONER, P. G., AND CARR, K. E. *Cell Structure. An Introduction to Biological Electron Microscopy*. 2nd ed., Elsevier, New York, 1971.
TÖRÖ, I., AND RAPPAY, G. (eds.) *Ultrastructural Features of Cells and Tissues in Culture*, Akademiai Kiado, Budapest, 1972.
TRUMP, B. F., AND JONES, R. T. *Diagnostic Electron Microscopy*. Wiley, N.Y., 1978.
YAMADA, E., MIZUHIRA, V., KUROSOMI, K., AND NAGANO, T. *Recent Progress in Electron Microscopy of Cells and Tissues*. University Park Press, Baltimore, 1976.
YATES, R. D., AND GORDON, S. *Male Reproductive System: Fine Structure Analysis by Scanning and Transmission Electron Microscopy*. Masson, New York, 1977.
ZELICKSON, A. S. *Ultrastructure of Normal and Abnormal Skin*. Lea & Febiger, Philadelphia, 1967.

E. Atlases

ALDAMIZ, H. O. *Ultrastructura Celular*. Editorial Paz Montaivo, Madrid, 1964.
BABEL, J., BISCHOFF, A., AND SPOENDLIN, H. *Ultrastructure of the Peripheral Nervous System and Sense Organs. An Atlas of Normal and Pathologic Anatomy*. Thieme, Stuttgart, 1970.
BESSIS, M. *The Ultrastructure of Cells*. Sandoz Monographs, Basel, 1960.
BESSIS, M. *Atlas of Red Blood Cells and their Shapes*. Springer, New York, 1974.
BJÖRKMAN, N. *An Atlas of Placental Fine Structure*. Bailliere, Tindall and Cassell, London, 1970.
BREATHNACH, A. S. *Atlas of the Ultrastructure of the Human Skin*. Churchill, London, 1971.
CAWLEY, J. C., AND HAYHOE, F. G. J. *Ultrastructure of Haemic Cells: A Cytological Atlas of Normal and Leukaemic Blood and Bone Marrow*. Saunders, Philadelphia, 1973.
DALTON, J. D., AND HAGUENAU, F. (eds.) *Ultrastructure of Animal Viruses and Bacteriophases: An Atlas*. Academic Press, New York, 1973.
DODGE, J. D. *An Atlas of Biological Ultrastructure*. Williams & Williams, Baltimore, 1968.
EBE, T., AND KOBAYASHI, S. *Fine Structure of Human Cells and Tissues*. Wiley, New York 1972.
FAWCETT, D. W. *An Atlas of Fine Structure: The Cell*. Saunders, Philadelphia, 1966.
FELDMAN, D. G. *Electron Micrographs in Teaching of Biology*. Teachers College, Columbia University, New York, 1962.
FERENCZY, A., AND RICHART, R. M. *Female Reproductive System. Dynamics of Scan and Transmission Electron Microscopy*. Wiley, New York, 1974.
FREEMAN, J. A. *Cellular Fine Structure: An Introductory Student Text and Atlas*. Blakiston Division, McGraw-Hill, New York, 1964.
FUJITA, T., TOKUNAGA, J., AND INOUE, H. *Atlas of Scanning Electron Microscopy*. Elsevier, New York, 1971.
HERZBERG, M., AND RIVEL, M. *Atlas de Biologie Moléculaire Microscopie Electronique des Molécules Informatives*. Herman, Paris, 1972.
HIGASHI, N. (ed.) *World Through the Electron Microscope*. Vol. I, 1961, Vol. II, 1964; Vol. III, 1967. Japan Electron Optics Lab., Tokyo.
HIROHATA, K., AND MORIMOTO, K. *Ultrastructure of Bone and Joint Disease*. Grune and Stratton, New York, 1972.
HURRY, S. W. *The Microstructure of Cells*. Houghton Mifflin, Boston, 1964.
JENSEN, W. A., AND PERK, R. B. *Cell Ultrastructure*. Wadsworth, Belmont, Calif., 1967.
KESSEL, R. G., AND SHIH, C. Y. *Scanning Electron Microscopy in Biology. A Students Atlas on Biological Organization*. Springer, New York, 1974.
KESSEL, R. G., AND KARDON, R. H. *Tissues and Organs: A Text–Atlas of Scanning Electron Microscopy*. Freeman, San Francisco, 1979.
LAGUENS, R. P., AND GOMEZ-DUMN, C. L. A. *Atlas of Human Electron Microscopy*. Mosby, St. Louis, 1969.
LEDBETTER, M. C., AND PORTER, K. R. *Introduction to The Fine Structure of Plant Cells*. Springer, New York, 1970.

LOW, F. N., AND FREEMAN, J. *Electron Microscopic Atlas of Normal and Leukemic Blood.* McGraw-Hill, New York, 1958.
LUDWIG, H., AND METZGER, H. *The Human Female Reproductive Tract: A Scanning Electron Microscopic Atlas.* Springer, New York, 1976.
MAIR, W. G. P., AND TOME, F. M. S. *Atlas of the Ultrastructure of Diseased Muscle.* Williams and Wilkens, Baltimore, 1972.
MATSUMIYA, S., AND TAKUMA, S. *Atlas of Electron Micrographs of the Human Dental Tissues.* Tokyo Dental College Press, Tokyo, 1954.
MORI, Y., AND LENNERT, K. *Electron Microscopic Atlas of Lymph Node Cytology and Pathology.* Springer, New York, 1970.
MOTTA, P., ANDREWS, P. M., AND PORTER, K. R. *Micro-anatomy of Cell and Tissue Surface: An Atlas of Scanning Electron Microscopy.* Lea and Ferbiger, Philadelphia, 1977.
MOTTA, P., MUTO, M., AND FUJITA, T. *The Liver: An Atlas of Scanning Electron Microscopy.* Igaku-Shoin, New York, 1978.
PFEIFFER, C. J., ROWDEN, G., AND WEIBEL, J. *Gastrointestinal Ultrastructure: An Atlas of Scanning Transmission Electron Micrographs.* Academic Press, New York, 1974.
POLLIACK, A. *Normal Transformed and Leukemic Leukocytes: A Scanning Electron Microscopy Atlas.* Springer, New York, 1977.
POON, T. P. *Electron Microscopic Atlas of Brain Tumors.* Grune and Stratton, New York, 1971.
PORTER, K. R., AND BONNEVILLE, M. A. *An Introduction to the Fine Structure of Cells and Tissues.* 3rd ed., Lea & Febiger, Philadelphia, 1968.
RHODIN, J. A. G. *An Atlas of Ultrastructure.* Saunders, Philadelphia, 1963.
RHODIN, J. A. G. *Histology: A Text and Atlas.* Oxford University Press, New York, 1974.
ROBARDS, A. W. *Ultrastructure der Planzlichen Zelle.* Thieme, Stuttgart, 1974.
SANDBORN, F. B. *Cells and Tissues by Light and Electron Microscopy.* Vols. I and II. Academic Press, New York, 1970.
SANDBORN, E. B. *Light and Electron Microscopy of Cells and Tissues: An Atlas for Students in Biology and Medicine.* Academic Press, New York, 1972.
SCANGA, F. *Atlas of Electron Microscopy: Biological Applications.* Elsevier, New York, 1964.
TANAKA, Y., AND GOODMAN, J. R. *Electron Microscopy of Human Blood Cells.* Harper and Row, New York, 1972.
TERADA, M., AND FUKAI, K. (eds.) *Atlas of Electron Monographs.* Honda, Kyoto, 1952.
TONER, P. G., CARR, K. E., AND WYNBURN, G. M. *The Digestive System. An Ultrastructural Atlas and Review.* Appleton-Century-Crofts, New York, 1971.
VEHARA, Y., CAMPBELL, G., AND BURNSTOCK, G. *Muscle and its Innervation. An Atlas of Fine Structure.* Arnold, London, 1976.
WILLIAMS, R. C., AND FISCHER, H. W. *An Electron Micrographic Atlas of Viruses.* Thomas, Springfield, Ill., 1974.
YAMADA, E., AND HARNA, K. (eds.) *Electron Micrographs. Biology–1.* Hitachi, Tokyo, 1964.
YAMADA, E., AND SHIKANO. *Electron Microscopic Atlas in Ophthamology,* Igaku-Shoin, Tokyo, 1972.
ZAMBONI, L. *Fine Morphology of Mammalian Fertilization,* Harper & Row, New York, 1971.

F. Papers

1. Electron Microscopes and Electron Microscopy

BOURNE, P. The application of electron microscopy techniques. *Can. J. Med. Tech.,* **30,** 217–34, 1968.
BYSTRICKY, V. The electron microscope. An introduction to applied electron microscopy. *Arch. Immunol. Therap. Exper.,* **10,** 187–341, 1962.
JANSEN, P. C. An introduction to the electron microscope. *Science and Industry,* **7,** 31–51, 1960.
MOLLENSTEDT, G. Actual problems in electron microscopy. *J. Micr.,* **4,** 413–28, 1965.
ROBERTSON, J. D. The electron microscope. In *Tools of Biological Research.* Vol. 1, (H. J. B. Atkins, Ed.), pp. 72–121. Thomas, Springfield, Ill., 1959.
SIEGEL, B. The physics of the electron microscope. In *Modern Developments in Electron Microscopy,* (B. M. Siegel, Ed.), pp. 1–79. Academic Press, New York, 1964.

SWERDLOW, M., BOTTY, M. C., ROCHOW, T. G., FELTON, C. D., GREIDER, M. H., FISCHER, R. M., LALLY, J. S., BAIN, E. C., AND BEER, M. Electron microscopy. *Anal. Chem.*, **32**, 93R–103R, 1960; **34**, 127–43R, 1962; **36**, 173–99R, 1964; **38**, 209–50R, 1966; **40**, 554R–60R, 1968; **42**, 362R–66R, 1970; **44**, 97R–100R, 1972; **46**, 428–30R, 1974; **48**, 93R–5R, 1976, **50**, 76R–80R, 1978.

2. Electron Microscope Preparation Methods

BAIN, J. M., AND GOVE, D. W. Rapid preparation of plant tissue for electron microscopy. *J. Micr.*, **93**, 159–62, 1971.

BENCOSME, S. A., AND TSUTSOMI, V. A fast method for processing biologic material for electron microscopy. *Lab. Invest.*, **23**, 447–50, 1970.

CHAPMAN, J. A. New technique in conventional electron microscopy in biology. *Lab. Pract.*, **19**, 477–81, 1970.

DEHARVEN, F. Methods in electron microscopic cytology. *Meth. Cancer, Res.*, **1**, 3–44, 1967.

GAHAN, P. B., GREENOAK, G. C., AND JAMES, D. Preparation of ultra-thin frozen sections of plant tissues for electron microscopy. *Histochemie*, **24**, 230–35, 1970.

IDLE, D. B. Preparation of plant material for scanning electron microscopy. *J. Micr.*, **93**, 77–9, 1971.

LOCKE, M., KRISHNAN, N., AND MCNAHON, J. T. A routine method for obtaining high contrast without staining sections. *J. Cell Biol.*, **50**, 540–44, 1971.

SCHABTACH, E., AND PARKENING, T. A. A method for sequential high resolution light and electron microscopy of selected areas of the same material. *J. Cell Biol.*, **61**, 261–64, 1974.

TURNER, R. H., AND SMITH, C. B. A simple technique for examining fresh, frozen biological specimens in the scanning electron microscope. *J. Micr.*, **102**, 209–14, 1974.

WATTERS, W. B., AND BUCK, R. C. An improved simple method of specimen preparation for replicas or scanning electron microscopy. *J. Micr.*, **94**, 185–87, 1971.

WHITTAKER, D. K., AND WILLIAM-JONES, D. G. Biopsy techniques for electron microscopy. *J. Pathol.*, **103**, 61–3, 1971.

3. General Cytology

FERNANDEZ-MORAN, H. Cell fine structure and function—past and present. *Exp. Cell. Res.*, **62**, 90–101, 1970.

FRANKS, L. M., AND WILSON, P. D. Origin and ultrastructure of cells *in vitro*. *Int. Rev. Cytol.*, **48**, 55–139, 1977.

LESSON, T. Cell structure and function. *Can. Med. Ass. J.*, **93**, 921–32, 1965.

OBERLING, C. The structure of cytoplasm. *Int. Rev. Cytol.*, **8**, 1–32, 1959.

PALADE, G. Structure and function at the cellular level. *J. Am. Med. Ass.*, **198**, 815–25. 1966.

4. Cell Membrane

ELBERS, P. F. The cell membrane: image and interpretation. *Recent Progr. Surface Sci.*, **2**, 443–503, 1964.

GLAUBERT, A. M. Electron microscopy of lipids and membranes. *J. Roy. Micr. Soc.*, **88**, 49–70, 1968.

FINAGAN, J. B. The development of ideas on membrane structure. *Sub-cell. Biochem.*, **1**, 363–73, 1972.

HENDLER, R. W. Biological membrane ultrastructure. *Physiol. Rev.*, **51**, 66–97, 1971.

KORN, E. D. Structure of biological membranes. *Science*, **153**, 1491–98, 1966.

PARSONS, D. F. Ultrastructural and molecular aspects of cell membrane. *Proc. 7th Canadian Cancer Conf.*, pp. 193–246. Pergamon Press, Elmsford, N. Y., 1967.

STOECKENIUS, W., AND ENGELMAN, D. M. Current models for the structure of biological membranes. *J. Cell Biol.*, **42**, 613–46, 1969.

THE, G. de. Ultrastructural cytochemistry of the cellular membranes. In *The Membranes* (A. J. Dalton, and F. Haguenau, Eds.), pp. 122–50. Academic Press, New York, 1968.

WEISS, L. The cell periphery. *Int. Rev. Cytol.*, **26**, 63–105, 1969.

YAMAMOTO, T. On the thickness of the unit membrane. *J. Cell Biol.*, **17**, 413–22, 1963.

5. Cell Membrane Specializations

Cell Junctional Specializations

ALBERTINI, D. F., FAWCETT, D. W., AND OLDS, P. S. Morphological variations in gap junctions of ovarian granulosa cells. *Tissue and Cell*, **7**, 389–40, 1975.

DECKER, R. J., AND FRIEND, D. S. Assembly of gap junctions during amphibian neurulation. *J. Cell Biol.*, **62**, 32–47, 1974.

FRIEND, D. S., AND GILUDA, N. B. Variations in tight and gap junctions in mammalian tissues. *J. Cell Biol.*, **53**, 758–776, 1972.

STAEHELIN, L. A. Further observations on the fine structure of freeze-cleaned tight junctions. *J. Cell Sci.*, **13**, 763–786, 1973.

WADE, J. B., AND KARNOVSKY, M. J. The structure of the zonula occludens. *J. Cell Biol.*, **60**, 168–180, 1974.

Transient Cell Membrane Specializations

FAWCETT, D. W. Surface specializations of absorbing cell. *J. Histochem. Cytochem.*, **13**, 75–91, 1965.

HOLTER, H. Pinocytosis. *Int. Rev. Cytol.*, **8**, 481–505, 1959.

PALADE, G. E., AND BRUNS, R. R. Structural modulations of plasmalemmal vesicles. *J. Cell Biol.*, **37**, 633–49, 1968.

Motile Cell Processes: Cilia and Flagella

ALLEN, R. D. A. Reinvestigation of cross-sections of cilia. *J. Cell Biol.*, **37**, 825–31, 1968.

GIBBONS, I. R. The structure and composition of cilia. *Symp. Int. Soc. Cell Biol.*, **6**, 99–114, 1967.

GRIMSTONE, A. V. Observations on the substructure of flagellar fibres. *J. Cell Sci.*, **1**, 351–62, 1966.

PARDUCZ, B. Ciliary movement and coordination in ciliates. *Int. Rev. Cytol.*, **21**, 91–128, 1967.

PITELKA, D. R. Ciliate ultrastructure, some problems in cell biology. *J. Protozool.*, **17**, 1–10, 1970.

WEBBER, W. A., AND LEE, J. Fine structure of mammalian renal cilia. *Anat. Rec.*, **182**, 339–44, 1975.

Microvilli

ANDERSON, J. H., AND TAYLOR, A. B. Scanning and transmission electron microscopic studies of jejunal microvilli of the rat, hamster and dog. *J. Morph.*, **141**, 281–91, 1973.

CRANE, R. K. Structural and functional organization of an epithelial cell brush border. *Symp. Int. Soc. Cell. Biol.*, **5**, 71–102, 1966.

MUKHERJEE, T. M., AND WILLIAMS, A. W. A comparative study of the ultrastructure of microvilli in the epithelium of small and large intestine of mice. *J. Cell Biol.*, **34**, 447–61, 1967.

6. Mitochondria

BARNARD, T., AND AFZELIUS, B. A. The matrix granules of mitochondria: a review. *Sub-Cell Biochem.*, **1**, 375–89, 1972.

BELL, P. R. The structure and origin of mitochondria. *Sci. Prog.*, **53**, 33–44, 1965.

BUTLER, W. H., AND JUDAH, J. D. Preparation of isolated rat liver mitochondria for electron microscopy. *J. Cell Biol.*, **44**, 278–89, 1970.

GREEN, D. E. The mitochondrion. *Sci. Am.*, **210**, 67–74, 1964.

PALADE, G. An electron microscope study of mitochondrial structure. *J. Histochem. Cytochem.*, **1**, 188–211, 1953.

PARSONS, D. F. New developments in electron microscopy as applied to mitochondria. *Meth. Enzymol.*, **10**, 655–67, 1967.

SJÖSTRAND, F. S., AND BARAJAS, L. A new model for mitochondrial membranes based on biochemical information. *J. Ultrastruct. Res.*, **32,** 293–306, 1970.

TEWARI, J. P., AND MALHOTRA, S. K. A study of the structure of mitochondrial membranes by freeze-etch and freeze fracture techniques. *Cytobios,* **4,** 97–119, 1971.

7. Golgi Complex

BEAMS, H. W., AND KESSEL, R. G. The Golgi apparatus: structure and function. *Int. Rev. Cytol.*, **23,** 209-76, 1968.

FINERAN, B. A. Organization of the Golgi apparatus in frozen-etched root tips. *Cytobiol.*, **8,** 175–93, 1973.

LOCKE, M., AND SYKES, A. K. The role of the Golgi complex in the isolation and digestion of organelles. *Tissue and Cell,* **7,** 143–58, 1975.

NEUTRA, M., AND LEBLOND, C. P. The Golgi apparatus. *Sci Am.*, **220,** 100–7, 1969.

WHALEY, W. G., AND DAUWALDER, M. The Golgi apparatus, the plasma membrane, and functional integration. *Int. Rev. Cytol.*, **58,**' 199–245, 1978.

8. Endoplasmic Reticulum

CARDELL, R. R., JR. Smooth endoplasmic in rat hepatocytes during glycogen deposition and depletion. *Int. Rev. Cytol.*, **48,** 221–279, 1977.

DALLNER, G., SIEKEVITZ, P., AND PALADE, G. E. Biogenesis of endoplasmic reticulum membranes. *J. Cell Biol.*, **30,** 73–96, 1966.

GARRETT, R. A., AND WITTMANN, H. G. Structure and function of the ribosome. *Endeavour,* **32,** 8–14, 1973.

GOLDBLATT, P. J. Molecular pathology of the endoplasmic reticulum. *Sub-Cell Biochem.*, **1,** 147–69, 1972.

PALADE, G. E., AND PORTER, K. R. Studies on the endoplasmic reticulum. 1. Its identification in cells *in situ*. *J. Exp. Med.*, **100,** 641–56, 1956.

RICH, A. Polyribosomes. *Sci. Am.*, **209,** 44–53, 1963.

9. Centrioles

ALOV, I. A., AND LYUBSKII, S. L. Functional morphology of the kinetochore. *Int. Rev. Cytol., Suppl.*, **6,** 59–74, 1977.

BRINKEY, B. R. Ultrastructure and interaction of the kinetochore and centriole in mitosis and meiosis. *Adv. Cell Biol.*, **1,** 119–85, 1970.

SOROKIN, S. P. Reconstruction of centriole formation and ciliogenesis in mammalian lungs. *J. Cell Sci.*, **3,** 207–30, 1968.

STUBBLEFIELD, E., AND BRINKLEY, B. R. Architecture and function of the mammalian centriole. *Symp. Int. Soc. Cell Biol.*, **6,** 175–218, 1967.

10. Lysosomes and Microbodies

ALLISON, A. C. The lysosome. *Discovery,* **26,** 8–13, 1965.

DEDUVE, C., AND WATTIAUX, R. Function of lysosomes. *Ann. Rev. Physiol.*, **28,** 435–92, 1966.

GOSSRAU, R. Die lysosomen des darmepithels. *Adv. Anat. Embryol. Cell Biol.*, **51,** 1–95, 1975.

MERKLE, M. C. The distribution and function of lysosomes in condylar cartilage. *J. Anat.*, **119,** 85–96, 1975.

WEISSMANN, G. Lysosomes (analytical review). *Blood,* **24,** 594–606, 1964.

Microbodies

DEDUVE, C., AND BAUDHUIN, P. Peroxisomes (micro-bodies and related particles). *Physiol. Rev.*, **46,** 323–57, 1966.

HRUBAN, Z., VIGIL, E. L., SLESERS, A., AND HOPKINS, E. Microbodies. Constituent organelles of animal cells. *Lab. Invest.*, **27,** 1984–91, 1972.
NOVIKOFF, P. M., AND NOVIKOFF, A. B. Peroxisomes in absorptive cells of mammalian small intestine. *J. Cell Biol.*, **53,** 532–60. 1972.
TSUKADA, H., MOCHIZUKI, Y., AND KONISHI, T. Morphogenesis and development of microbodies of hepatocytes of rats during pre- and postnatal growth. *J. Cell Biol.*, **37,** 231–43, 1968.

11. Microtubules and Microfilaments

Microtubules

BEHNKE, O., AND FORER, A. Evidence for four classes of microtubules in some vertebrate and invertebrate cells. *J. Cell Sci.*, **2,** 169–92, 1967.
BLOODGOOD, R. A., AND MILLER, K. R. Freeze-fracture of microtubules and bridges in motile exostyles. *J. Cell Biol.*, **62,** 660–71, 1974.
ERICKSON, H. P. Microtubule surface lattice and subunit structure and observations on reassembly. *J. Cell Biol.*, **60,** 153, 1974.
MCINTOSH, J. R. Bridges between microtubules. *J. Cell Biol.*, **61,** 166–87, 1974.
OLSON, L. W., AND HEATH, I. B. Observations on the ultrastructure of microtubules. *Z. Zellforsch.*, **115,** 338–95, 1971.

Microfilaments

CARR, I. The fine structure of microfibrils and microtubules in macrophages and other lymphoreticular cells in relation to cytoplasmic movement. *J. Anat.*, **112,** 383–89, 1972.
MALECH, H. L., AND LEWTZ, T. L. Microfilaments in epidermal center cells. *J. Cell Biol.*, **60,** 473–82, 1974.
MORGAN, J. Microfilaments from *Amoeba proteus*. *Exp. Cell Res.*, **65,** 7–16, 1971.
REAVEN, E. P., AND AVLINE, S. G. Subplasmalemmal microfilaments and microtubules in resting and phagocytizing cultivated macrophages. *J. Cell Biol.*, **59,** 12–27, 1973.
SPOONER, B. S., YAMADA, K. M., AND WESSELS, N. K. Microfilaments and cell locomotion. *J. Cell. Biol.*, **49,** 596–613, 1971.

12. Annulate Lamellae

KESSEL, R. G. Annulate lamellae. *J. Ultrastructure Res.*, (suppl. 10), 1–82, 1968.
KUMEGAWA, M., CATTONI, M., AND ROSE, G. G. Electron microscopy of oral cells *in vitro*. 1. Annulate lamellae observed in strain KB cells. *J. Cell Biol.*, **34,** 897–901, 1967.
MAUL, G. G. Ultrastructure of pore complexes of annulate lamellae. *J. Cell. Biol.*, **46,** 604–610, 1970.
SWIFT, H. The fine structure of annulate lamellae. *J. Biophys. Biochem. Cytol.*, **2** (suppl.), 415–18, 1956.
WISCHNITZER, S. The annulate lamellae. *Int. Rev. Cytol.*, **27,** 65–100, 1970.

13. Inclusion Bodies

Pigment

BJÖRKERUD, S. Isolation of lipofuscin granules from bovine cardiac muscle. *J. Ultrastructure Res.*, **5** (suppl.), 5–49, 1963.
DROCHMAN, P. Melanin granules. Their fine structure, formation, and degradation in normal and pathological tissues. *Int. Rev. Exper. Pathol.*, **2,** 357–422, 1963.
TURNER, W. A. JR., TAYLOR, J. D., AND TCHEN, T. T. Melanosome formation in the goldfish; the role of multivesicular bodies. *J. Ultrastructure Res.*, **51,** 16–31, 1975.

Glycogen

BIAVA, C. Identification and structural forms of human particulate glycogen. *Lab. Invest.*, **12,** 1179–97, 1963.

DE BRUIJN, W. J. Glycogen, its chemistry and morphologic appearance in the electron microscope. *J. Ultrastructure Res.*, **42,** 29–50, 1973.
REVEL, J. P. Electron microscopy of glycogen. *J. Histochem. Cytochem.*, **12,** 104–14, 1964.

Lipid

NAPOLITANO, L. The differentiation of white adipose cells. An electron microscope study. *J. Cell Biol.*, **18,** 663–79, 1963.
PALAY, S. L., AND REVEL. J. P. The morphology of fat absorption. In *Lipid Transport* (H. C. Meng., Ed.), pp. 33–43. Thomas, Springfield, Ill. 1964.
THEONES, W. Fine structure of lipid granules in proximal tubule cells of mouse kidney. *J. Cell Biol.* **12,** 433–37, 1962.

Crystalline

KARASAKI, S. Studies on amphibian yolk. 1. The ultrastructure of the yolk platelet. *J. Cell Biol.*, **18,** 135–66, 1963.
OLLERICH, D. A. An intramitochondrial crystalloid in element III of rat chorioallantoic placenta. *J. Cell Biol.*, **37,** 188–91, 1968.
RICHTER, W. R., STERN, R. J., RDZOK, E. J., MOIZE, S. M., AND BISCHOFF, M. B. Ultrastructural studies of intranuclear crystalline inclusions in the liver of the dog. *Am. J. Path.*, **47,** 587–99, 1965.

14. Nucleus

BOUTEILLE, M., LAVAL, M., AND DUPUYCOIN, A. M. Localization of nuclear function as revealed by ultrastructural autoradiography and cytochemistry. In *The Cell Nucleus* (H. Busch, Ed.), Vol 1, pp. 5–71. Academic Press, New York, 1974.
LAFONTAINE, J. G. Ultrastructural organization of plant cell nuclei. In *The Cell Nucleus* (H. Busch, Ed.), Vol. 1, pp. 149–85. Academic Press, New York, 1974.
MOSES, M. J. The nucleus and chromosomes: a cytological perspective. In *Cytology and Cell Physiology* (G. Bourne, Ed.), pp. 424–558. Academic Press, New York, 1964.
RIS, H. Interpretation of ultrastructure in the cell nucleus. *Symp. Int. Soc. Cell Biol.*, **1,** 69–87, 1962.
WISCHNITZER, S. The ultrastructure of the nucleus and nucleocytoplasmic relations. *Int. Rev. Cytol.*, **10,** 137–62, 1960.

15. Nuclear Envelope

AARONSON, R. P., AND BLOBEL, G. On the attachment of the nuclear pore complex. *J. Cell Biol.*, **62,** 746–54, 1974.
FRANKE, W. W. Structure, biochemistry, and functions of the nuclear envelope. *Int. Rev. Cytol.*, Supplement 4, 72–236, 1974.
KAY, R. R., AND JOHNSTON, I. R. The nuclear envelope: current problems of structure and of function. *Sub-Cell. Biochem.*, **2,** 127–66, 1973.
KESSEL, R. G. Structure and function of the nuclear envelope and related cytomembranes. *Prog. Surface Membrane Sci.*, **6,** 243–329, 1973.
MAUL, G. The nuclear and cytoplasmic pore complex: structure, dynamics, distribution and evolution. *Int. Rev. Cytol.*, Supplement 6, 75–186, 1977.
WISCHNITZER, S. The nuclear envelope: its ultrastructure and functional significance. *Endeavour,* **33,** 137–42, 1974.

16. Nucleolus

BERNARD, W., AND GRANBOULAN, N. Electron microscopy of the nucleolus in vertebrate cells. In *Ultrastructure in Biological Systems* (A. J. Dalton and F. Haguenau, Eds.), pp. 81–149. Academic Press, New York, 1968.
GHOSH, S. The nucleolar structure. *Int. Rev. Cytol.*, **44,** 1–28, 1976.

HAY, E. D. Structure and Function of the nucleolus in developing cells. In *Ultrastructure in Biological Systems* (A. J. Dalton and F. Haguenau, Eds.), pp. 1–79, Academic Press, New York, 1968.

RECHER, L., WHITESCARVER, J., AND BRIGGS, L. The fine structure of a nucleolar constituent. *J. Ultrastructure Res.,* **29,** 1–14, 1969.

SIDEBOTTOM, E., AND HARRIS, H. The role of the nucleolus in the transfer of RNA from nucleus to cytoplasm. *J. Cell Sci.,* **5,** 351–64, 1969.

17. Chromosomes, Mitosis, and Meiosis

Chromosomes

ABRIELO, J. G., AND MOORE, D. E. The human chromosome. Electron microscopic observations on chromatin fiber organization. *J. Cell. Biol.,* **41,** 73–90, 1969.

BRINKLEY, B. R., AND HITTELMAN, W. N. Ultrastructure of mammalian chromosomes aberrations. *Int. Rev. Cytol.,* **42,** 48–101, 1975.

DASKAL, Y., AND BUSCH, H. Ultrastructure of chomatin and chromosomes as visualized by scanning electron microscopy. In *The Cell Nucleus* (H. Busch, Ed.), Vol. 6, part c, pp. 3–45, Academic Press, New York, 1978.

FRENSTER, J. H. Ultrastructure and function of heterochromatin and euchromatin. In *The Cell Nucleus* (H. Busch, Ed.), Vol. 1, pp. 565–80. Academic Press, New York, 1974.

MOTT, M. R., AND CALLAN, H. G. An electron-microscope study of the lampbrush chromosomes of the newt, *Triturus cristatus. J.Cell Sci.,* **17,** 214–61, 1975.

STUBBLEFIELD, E. Electron microscopy and autoradiography of chromosomes. *Meth. Enzymol.,* **40,** 63–75, 1975.

Mitosis

BAJER, A. W., AND MOLE-BAJER, J. Spindle dynamics and chromosome movements. *Int. Res. Cytol.,* Suppl. 3, 1–280, 1972.

FUGE, H. Ultrastructure of the mitotic spindle. *Int. Rev. Cytol.,* Suppl. 6, 1–58, 1977.

HARRIS, P., AND MAZIA, D. The finer structure of the mitotic apparatus. *Symp. Int. Soc. Cell. Biol.,* **1,** 279–305, 1962.

HUDSON, P. R., AND WAALAND, J. R. Ultrastructure of mitosis and cytokinese in the multinucleate green algae. *Aerosiphonia, J. Cell Biol.,* **62,** 274–94, 1974.

REBHUN, L. I., AND SANDER, G. Ultrastructure and birefringenece of the isolated mitotic apparatus of marine eggs. *J. Cell Biol.,* **34,** 859–83, 1967.

Meiosis

BAKER, T. G., AND FRANCHI, L. L. The fine structure of oogonia and oocytes in human ovaries. *J. Cell Sci.,* **2,** 213–24, 1967.

SOTELO, J. R., AND WETTSTEIN, R. Fine structure of meiotic chromosomes. *Natl. Cancer Inst. Monog.,* **18,** 133–52, 1965.

18. Epithelium and Glands

Epithelium

ADAMS, D. R. Olfactory and non-olfactory epithlia in the nasal cavity of the mouse. *Am. J. Anat.,* **133,** 37–50, 1972.

DUNN, J. S. The fine structure of the absorptive epithelial cells of the developing small intestine of the rat. *J. Anat.,* **101,** 57–68, 1967.

MONAS, B., AND ZAMBRANO, D. Ultrastructure of transitional epithelium of man. *Z. Zellforsch.,* **87,** 101–17, 1968.

PARAAKAL, P. F. An electron microscopic study of esophageal epithelium in the newborn and adult mouse. *Am. J. Anat.,* **121,** 175–96, 1967.

PIEZZI, R. S., SANTOLAYA, R. S., AND BERTINI, F. The fine structure of endothelial cells of toad arteries. *Anat. Rec.,* **165,** 229–36, 1969.

PITTMAN, F. E., AND PITTMAN, J. C. An electron microscopic study of the epithelium of normal human sigmoid colonic mucosa. *Gut,* **7,** 644–61, 1966.

RIVA, A. Fine structure of human seminal vesicle epithelium. *J. Anat.,* **102,** 71–86, 1967.

THOMAS, C. E. The ultrastructure of human amnion epithelium. *J. Ultrastructure Res.,* **13,** 65–84, 1965.

WISCHNITZER, S. The ultrastructure of the germinal epithelium of the mouse ovary. *J. Morph.,* **117,** 387–400, 1965.

Epithelial Glands

FREEMAN, J. A. Goblet cell fine structure. *Anat. Rec.,* **152,** 121–48, 1966.

MOE, H. On goblet cells, especially of the intestine of some mammalian species. *Int. Rev. Cytol.,* **4,** 299–334, 1955.

NEUTRA, M., AND LEBLOND, C. P. Synthesis of the carbohydrate of mucus in the Golgi complex as shown by electron microscope radioautography of goblet cells from rats injected with glucose-H^3. *J. Cell Biol.,* **30,** 119–36, 1966.

19. Connective Tissues

Fibers

GOTTE, L., GIRO, M. G., VOLPIN, D., AND HORNE, R. W. The ultrastructural organization of elastin. *J. Ultrastructure Res.,* **46,** 23–33, 1974.

LOW, F. N. Extracellular connective tissue fibrils in the chick embryo. *Anat. Rec.,* **160,** 93–108, 1967.

ROSS, R., AND BORENSTEIN, P. Elastic fibers in the body. *Sci. Am.,* **224,** 44–59, 1971.

TORRI-TARELLI, L., AND PETRUCCIOLI, M. G. Studies on the ultrastructure with negative staining. *J. Submicr. Cytol.,* **3,** 153–70, 1971.

Ground Substance

BRIGGAMAN, R. A., DALLDORF, F. G., AND WHEELER, C. E., JR. Formation and origin of basal lamina and anchoring fibrils in adult human skin. *J. Cell Biol.,* **51,** 384–95, 1971.

GERSH, I., AND CATCHPOLE, H. R. The nature of ground substance of connective tissue. *Persp. Biol. Med.,* **3,** 282–319, 1969.

PIERCE, G. B., AND NAKANE, P. K. Basement membranes, synthesis and deposition in response to cellular injury. *Lab. Invest.,* **21,** 27–41, 1969.

SCALLETA, L. J., AND MACCALLOM, D. K. A fine structural study of divalent cation-mediated epithelial union with connective tissue in human oral mucosa. *Am. J. Anat.,* **133,** 431–54, 1972.

Fibroblasts

ALPERT, E. N. Developing elastic tissue. *Am. J. Path.,* **69,** 89–102, 1972.

REITH, E. J. Collagen formation in developing molar teeth of rat. *J. Ultrastructure Res.,* **21,** 383–414, 1968.

VAN WINKLE, W. JR. The fibroblast in wound healing. *Surg. Gynec. Obstet.,* **124,** 369–86, 1967.

WEINSTOCK, M. Collagen formation. Observations on its intracellular packaging and transport. *Z. Zellforsch.,* **129,** 455–70, 1972.

Mast Cells

COMBS, J. W. Maturation of rat mast cells. An electron microscope study. *J. Cell Biol.,* **31,** 563–75, 1966.

PADAWER, J. The reaction of rat mast cells to polylysine. *J. Cell Biol.,* **47,** 352–72, 1970.

SMITH, D. E. The tissue mast cell. *Int. Rev. Cytol.,* **14,** 327–86, 1963.

WEINSTOCK, A., AND ALBRIGHT, J. T. The fine structure of mast cells in normal human gingiva. *J. Ultrastructure Res.,* **17,** 245–56, 1967.

Adipose cells

BARNARD, T. The ultrastructural differentiation of brown adipose tissue in the rat. *J. Ultrastructure Res.,* **29,** 311–32, 1969.

CUSHMAN, S. W. Structure-function relationship in the adipose cell. I. Ultrastructure of the isolated adipose cell. *J. Cell Biol.*, **46**, 326–41, 1970.
DYER, R. F. Morphological features of brown adipose tissue cell maturation *in vivo* and *in vitro*. *Am. J. Anat.*, **123**, 255–82, 1968.
SLAVIN, B. G. The cytophysiology of mammalian adipose cells. *Int. Rev. Cytol.*, **33**, 297–334, 1972.

20. Cartilage and Bone

Cartilage

ANDERSON, H. C., AND MATSUZAWA, T. Membranous particles in calcifying cartilage matrix. *Trans. N.Y. Acad. Sci.*, Series II, **22**, 619–30, 1970.
BONUCCI, E. Fine structure of early cartilage calcification. *J. Ultrastructure Res.*, **20**, 33–50, 1967.
GHADIALLY, F. N., THOMAS, I., YONG, N., AND LALONDE, J. M. Ultrastructure of rabbit semilunar cartilage. *J. Anat.*, **125**, 499–517, 1978.
SILBERBERG, R. Ultrastructure of articular cartilage in health and disease. *Clin. Orthop.*, **57**, 233–57, 1968.
THYBERG, J. Ultrastructural localization of aryl sulfatase activity in epiphyseal plate. *J. Ultrastructure Res.*, **38**, 332–42, 1972.

Bone

BAUD, C. G. Submicroscopic structure and functional aspects of the osteocytes. *Clin. Orthop.*, **56**, 227–36, 1968.
JANDE, S. S., AND BELANGER, L. F. Electron microscopy of osteocytes and the pericellular matrix in rat trabecular bone. *Cal. Tiss. Res.*, **6**, 280–89, 1971.
KALLIO, D. M., GARANT, P. R., AND MIMKIN, C. Evidence of coated membranes in the ruffled border of the osteoclast. *J. Ultrastructure Res.*, **37**, 169–77, 1971.
SCHERFT, J. P. The lamina limitans of the organic matrix of calcified cartilage and bone. *J. Ultrastructure Res.*, **38**, 318–31, 1972.
SCOTT, B. L., AND GLIMCHER, M. J. Distribution of glycogen in osteoblasts of the fetal rat. *J. Ultrastructure Res.*, **36**, 565–86, 1971.

Bone Development

ASCENZI, A., BONUCCI, E., AND BOCCIARELLI, D. S. An electron microscope study on the primary periosteal bone. *J. Ultrastructure Res.*, **18**, 605–18, 1967.
BERNARD, G. W., AND PEASE, D. C. An electron microscopic study of initial intramembranous osteogenesis. *Am. J. Anat.*, **125**, 271–90, 1969.
DECKER, J. D. An electron microscopic investigation of osteogenesis in the embryonic chick. *Am. J. Anat.*, **118**, 591–641, 1966.
OWEN, M. The origin of bone cells. *Int. Rev. Cytol.*, **28**, 213–38, 1970.
SMITH, J. W. The disposition of proteinpolysacharide in the epiphysial plate cartilage of the young rabbit. *J. Cell Sci.*, **6**, 843–64, 1970.

21. Nervous Tissues

Neuron

BRUNK, U., AND ERICSSON, J. L. E. Electron microscopical studies on rat brain neurons. Localization of acid phosphatase and mode of formation of lipofuscin bodies. *J. Ultrastructure Res.*, **32**, 1–15, 1972.
GREY, E. G. The fine structure of nerve. *Comp. Biochem. Physiol.*, **36**, 419–48, 1970.
HAMBERGER, A., HANSSON, H., AND SJÖSTRAND, J. Surface structure of isolated neurons. *J. Cell Biol.*, **47**, 319–331, 1970.
WAXMAN, S. G., AND PAPPAS, G. D. Changing concepts of the neuron. *Microstructures*, **3**, 13–17, 1972.
WUERKER, R. B., AND KIRKPATRICK, J. B. Neuronal microtubules, neurofilaments, and microtubules. *Int. Rev. Cytol.*, **33**, 45–75, 1972.

Nerve endings

CHIBA, T., AND YAMAUCHI, A. On the fine structure of the nerve terminals in the human myocardium. *Z. Zellforsch.*, **108,** 324–38, 1970.

COËAS, C. Structure and organization of the myoneural junction. *Int. Rev. Cytol.*, **22,** 239–68, 1967.

HALATA, Z. The mechanoreceptors of the mammalian skin: ultrastructure and morphological classification. *Adv. Anat. Embryol. Cell. Biol.*, **50,** 1–77, 1975.

OVALLE, W. K. JR. Motor nerve terminals on rat intrafusal muscle fibers, a correlated light and electron microscopic study. *J. Anat.*, **111,** 239–52, 1972.

PADYKULA, H. A., AND GAUTHIER, G. F. The ultrastructure of the neuromuscular junctions of mammalian red, white and intermediate skeletal muscle fibers. *J. Cell. Biol.*, **46,** 27–41, 1970.

Non-neural structures: Neuroglia, ependyma, choroid plexus, meninges.

BARON, M., AND GALLEGO, A. The relation of the microglia with the pericytes in the cat cerebral cortex. *Z. Zellforsch.*, **128,** 42–57, 1972.

HIMANGO, W. A., AND LOW, F. N. The fine structure of a lateral recess of the subarachnoid space in rat. *Anat. Rec.*, **171,** 1–20, 1971.

HIRANO, A., AND ZIMMERMAN, H. M. Some new cytological observations of the normal rat ependymal cell. *Anat. Rec.*, **158,** 293–302, 1967.

MILHORAT, T. H. Structure and function of the choroid plexsus and other sites of cerebrospinal fluid formation. *Int. Rev. Cytol.*, **47,** 225–88, 1976.

MILLHOUSE, O. E. Light and electron microscopic studies of the ventricular wall. *Z. Zellforsch.*, **127,** 149–74, 1972.

MORSE, D. E., AND LOW, F. N. The fine structure of the pia mater of the rat. *Am. J. Anat.*, **133,** 349–68, 1972.

22. Musclar Tissue

Skeletal Muscle

HUXLEY, H. E. The mechanism of muscular contraction. *Science,* **164,** 1356–66, 1969.

KELLY, D. E., AND CAHILL, M. A. Filamentous and matrix components of skeletal muscle Z-disks. *Anat. Rec.*, **172,** 623–42, 1972.

MCCALLISTER, L. P., AND HADEK, R. Transmission electron microscopy and stereo-ultrastructure of the T system in frog skeletal muscle. *J. Ultrastructure Res.*, **33,** 360–68, 1970.

ROWE, R. W. Ultrastructure of the Z line of skeletal muscle fibers. *J. Cell Biol.*, **51,** 674–85, 1971.

SCHIAFFINO, S., HANZLIKOVA, V., AND PIEROBON, S. Relations between structure and function in rat skeletal muscle fibers. *J. Cell. Biol.*, **47,** 107–119, 1970.

Smooth Muscle

COOKE, P. H., AND FAY, F. S. Correlation between fiber length, ultrastructure and length-tension relationship of mammalian smooth muscle. *J. Cell Biol.*, **52,** 105–16, 1971.

DEVINE, C. E., SOMLYO, A. V., AND SOMYLO, A. P. Sarcoplasmic reticulum and excitation-contraction in mammalian smooth. *J. Cell Biol.*, **52,** 690–718, 1972.

KELLY, R. E., AND ARNOLD, J. W. Myofilaments of the papillary muscles of the iris fixed in situ. *J. Ultrastructure Res.*, **40,** 532–45, 1972.

RICE, R. V., MOSES, J. A., MCMANUS, G. M., BRADY, A. C., AND BASIK, L. M. The organization of contractile filament in mammalian smooth muscle. *J. Cell Biol.*, **47,** 183–96, 1970.

RHODIN, J. A. G. Fine structure of vascular walls in mammal with special reference to smooth muscle component. *Physiol. Rev.*, **42** (Suppl.), 48–81, 1962.

VEHARA, Y., CAMPBELL, G. R., AND BURNSTOCK, G. Cytoplasmic filaments in developing and adult vertebrate smooth muscle. *J. Cell Biol.*, **50,** 487–97, 1971.

Cardiac muscle

CHALICE, C. E. Microstructure of specialized tissues in the mammalian heart. *Ann. N.Y. Acad. Sci.*, **156,** 14–33, 1969.

HIBBS, R. G., AND FERRANS, V. J. An ultrastructural and histochemical study of rat atrium myocardium. *Am. J. Anat.*, **124,** 251–80, 1969.

SCHERMANN, D. W. Über den Feinbau des Myokards von *Rana temporaria* (L). *Adv. Anat. Embryol. Cell Biol.*, **48,** 1–70, 1974.

SPERELAKIS, N., RUBIO, R., AND RADNICK, J. Sharp discontinuity in sarcomere lengths across intercalated disks of fibrillating cat hearts. *J. Ultrastructure Res.*, **30,** 503–32, 1970.

THAEMERT, J. C. Fine structure of the atrioventricular node as viewed in serial sections. *Am. J. Anat.*, **136,** 43–66, 1973.

23. Blood

Erythrocytes

BESSIS, M., AND WEED, R. I. The structure of normal and pathologic erythrocytes. *Adv. Biol. Med. Phys.*, **14,** 36–91, 1973.

HARRIS, J. R., AND AGUTTER, P. A negative staining study of human erythrocytes ghosts and rat liver nuclear membranes. *J. Ultrastructure Res.*, **33,** 219–32, 1970.

KOEHLER, J. K. Freeze-etching observations on nucleated erythrocytes with special reference to the nuclear and plasma membranes. *Z. Zellforsch.*, **85,** 1–17, 1968.

Granulocytes

DAEMS, W. On the fine structure of human neurophilic leukocyte granules. *J. Ultrastructure Res.*, **24,** 343–48, 1968.

SPICER, S. S., AND HARDIN, J. H. Ultrastructure, cytochemistry and function of neutrophil leukocyte granules. A review. *Lab. Invest.*, **20,** 488–97, 1969.

ZUCKER-FRANKLIN, D. Electron microscopic studies of human granulocytes: Structural variations related to function. *Seminars Hemat.*, **5,** 109–33, 1968.

Agranulocytes

FEDORKO, M. E., AND HIRSCH, J. G. Structure of monocytes and macrophages. *Seminar Hemat.*, **7,** 107–24, 1970.

SMETANA, K. Electron microscopy of lymphocytes. *Meth Cancer Res.*, **5,** 455–78, 1970.

WIVEL, N. A., MANDEL, M. A., AND ASOFSKY, R. M. Ultrastructural study of thoracic duct lymphocytes of mice. *Am. J. Anat.*, **128,** 57–72, 1970.

Platelets

BEHNKE, O. Electron microscopic observations on the surface coating of human blood platelets. *J. Ultrastructure Res.*, **24,** 51–69, 1968.

DAVID-FERREIRA, J. F. The blood platelet: electron microscopic studies. *Int. Rev. Cytol.*, **17,** 99–148, 1964.

ZUCKER-FRANKLIN, D. The submembranous fibrils of human blood platelets. *J. Cell Biol.*, **47,** 293–99, 1970.

24. Hemopoeisis

ACKERMAN, G. A. The human neutrophilic promyelocyte. *Z. Zellforsch.*, **121,** 153–70, 1971.

BAINTON, D. F., ULLYOT, J. L., AND FARQUHAR, M. G. The development of neutrophilic polymorphonuclear leukocytes in human bone marrow. *J. Exp. Med.*, **134,** 907–34, 1971.

BRETON-GORIUS, J., AND REYES, F. Ultrastructure of human bone marrow cell maturation. *Int. Rev. Cytol.*, **46,** 251–321, 1976.

ORLIC, D. Ultrastructural analysis of erythropoiesis. In *Regulation of Hematopoiesis,* (A. S. Gordon, Ed.) Vol. 1, 271–96. Appleton-Century-Crofts, New York, 1976.

ROSSE, C., AND TROTTER, J. A. A cytochemical radioautographic analysis of erythropoiesis at the ultrastructural level. *Am. J. Anat.*, **141,** 41–72, 1974.

25. Digestive System and Glands

Oral mucosa

HASHIMOTO, K. Fine structure of horny cells of the vermilion border of the lip compared with skin. *Arch. Oral Biol.*, **16**, 397–410, 1971.

SCALETTA, L. J., AND MACCALLUM, D. K. A fine structural study of divalent cation-mediated epithelial union with connective tissue in human oral mucosa. *Am. J. Anat.*, **133**, 431–54, 1972.

SILVERMAN, S. JR., BARBOSA, J., AND KEARNS, G. Ultrastructural and histochemical localization of glycogen in human normal and hyperkeratetic oral epithelium. *Arch. Oral Biol.*, **16**, 423–34, 1971.

Taste Buds

BARATZ, B. R., AND FARBMAN, A. I. Morphogenesis of rat lingual filiform papillae. *Am. J. Anat.*, **143**, 283–91, 1975.

BECKERS, H. W., AND EISENWACHER, W. Morphologie der papillae fungiformes. *Adv. Anat. Embryol. Cell. Biol.*, **50**, 1–117, 1975.

FERNANDEZ, B., SUAREZ, I., AND ZAPATA, A. Ultrastructure of the filiform papillae on the tongue of the hamster. *J. Anat.*, **126**, 487–94, 1978.

Salivary Glands

COWLEY, L. H., AND SHACKLEFORD, J. M. An ultrastructural study of the submandibular glands of the squirrel monkey, *Saimiri sciureus*. *J. Morph.*, **132**, 117–36, 1970.

ENOMOTO, S., AND SCOTT, B. L. Intracellular distribution of mucosubstances in the major sublingual gland of the rat. *Anat. Rec.*, **169**, 71–96, 1971.

TANDLER, B., DENNING, C. R., MANDEL, I. D., AND KUTSCHER, A. H. Ultrastructure of human labial salivary glands. III. Myoepithelium and ducts. *J. Morph.*, **130**, 227–46, 1970.

Teeth

MOE, H. Morphological changes in the intranuclear portion of the enamel-producing cells during their life cycle. *J. Anat.*, **108**, 43–62, 1971.

REITH, E. J. The stages of amelogenesis as observed in molar teeth of young rats. *J. Ultrastructure Res.*, **30**, 111–151, 1970.

WARSHAWSKY, H. A light and electron microscopic study of the nearly mature enamel of rat incisors. *Anat. Rec.*, **169**, 559–84, 1971.

WHITTAKER, D. K. The enamel-dentina junction of human and *Macaca irus* teeth: a light and electron microscopic study. *J. Anat.*, **125**, 323–35, 1978.

Stomach

FORSSMANN, W. G., AND ORCI, L. Ultrastructure and secretory cycle of the gastrin-producing cell. *Z. Zellforsch.*, **101**, 419–32, 1969.

JOHNSON, F. R., AND MCMINN: R. M. H. Microscopic structure of pyloric epithelium of the cat. *J. Anat.*, **107**, 67–86, 1970.

LILLIBRIDGE, C. G. Electron microscopy of human gastric mucosa. *Prog. Gasterenterol.*, **1**, 22–33, 1970.

Small intestine

BRUNSER, O., AND LUFT, J. H. Fine structure of the apex of absorptive cells from rat small intestine. *J. Ultrastructure Res.*, **31**, 291–311, 1970.

MUKHERJEE, T. M., AND STAEHELIN, L. A. The fine structural organization of the brush border of intestinal epithelial cells. *J. Cell Sci.*, **8**, 573–99, 1971.

REAVEN, E. P., AND REAVEN, G. M. Distribution and content of microtubules in relation to the transport of lipid. An ultrastructural quantitative study of the absorptive cell of the small intestine. *J. Cell. Biol.*, **75**, 559–72, 1977.

TONER, P. G. Cytology of intestinal epithelial cells. *Int. Rev. Cytol.*, **24**, 233–43, 1966.

Large Intestine

ONO, K. Ultrastructure of the surface principal cells of the large intestine in postnatal developing rats. *Anat. Embryol.*, **149,** 155–71, 1976.

VENKATACHAIAM, M. A., SOLTANI, M. H., AND DARIUSH, F. H. Fine structural localization of peroxidase activity in the epithelium of large intestine of rat. *J. Cell Biol.*, **46,** 168–73, 1970.

WOODING, F. B. P., SMITH, M. W., AND CRAIG, H. The ultrastructure of the neonatal pig colon. *Am. J. Anat.*, **152,** 269–86, 1978.

Liver

CARDELL, R. R., JR. Action of metabolic hormones on the fine structure of rat liver cells. I. Effects of fasting on the ultrastructure of hepatocytes. *Am. J. Anat.*, **131,** 21–54, 1971.

FLAKS, B. Observations on the fine structure of the normal porcine liver. *J. Anat.*, **108,** 563–77, 1971.

MA, M. H., AND BIEMPICA, L. The normal human liver cell. *Am. J. Path.*, **62,** 353–76, 1971.

MOTTA, P. M. The three-dimensional fine structure of the liver as revealed by scanning electron microscopy. *Int. Rev. Cytol.*, Suppl. 6, 347–399, 1977.

ROUILLER, C., AND JEZEQUEL, A. M. Electron microscopy of the liver. In *The Liver, Morphology, Biochemistry, Physiology,* (C. Rouiller, Ed.), Vol. 1, pp. 195–264. Academic Press, New York, 1965.

WISSE, E. An ultrastructural characterization of the endothelial cell in the rat liver sinosoid under normal and various experimental conditions, as a contribution to the distinction between endothelial and Kupffer cells. *J. Ultrastructure Res.*, **38,** 528–62, 1972.

Gall Bladder and Bile Duct

CHAPMAN, G. B., CHIARODO, A. J., COFFEY, R. J., AND WIENEKE, K. The fine structure of mucosal epithelial cells of a pathological human gallbladder. *Anat. Rec.*, **154,** 579–616, 1966.

FOX, H. Ultrastructure of the human gallbladder epithelium in cholelithiasis and chronic cholecystitis. *J. Path.*, **108,** 157–64, 1972.

HAYWOOD, A. F. The structure of gallbladder epithelium. *Int. Rev. Gen. Exp. Zool.*, **3,** 205–39, 1968.

LUCIANO, L. Die Feinstruktur der Gallenblase und der Gallengänge. *Z. Zellforsch.*, **135,** 87–114, 1972.

YAMADA, K. Fine structure of rodent common bile duct epithelium. *J. Anat.*, **105,** 511–23, 1969.

Pancreas

GREIDER, M. H., BENCOSME, S. A., AND LECHAGO, J. The human pancreatic islets cells and their tumors. I. The normal pancreatic islets. *Lab. Invest.*, **22,** 344–54, 1970.

KERN, H. F., AND FERNER, H. Die Feinstrukter des exokrinen Pankreasgewebes von Menschen. *Z. Zellforsch.*, **113,** 322–43, 1971.

KLEIN, C. Ultrastructural and cytochemical bases for the identification of cell types in the endocrine pancreas of teleosts. *Int. Rev. Cytol.*, Suppl. 6, 289–346, 1977.

LIKE, A. A. The ultrastructure of the secretory cells of the islets of Langerhans in men. *Lab. Invest.*, **16,** 937–51, 1967.

PALADE, G. E., SIEKEVITZ, P., AND CARO, L. G. Structure, chemistry and function of the pancreatic exocrine cell. Ciba Foundation Symp. Exocrine Pancreas, pp. 23–55, 1963.

26. Respiratory System

Nasal Cavity

ADAMS, D. R. Olfactory and non-olfactory epithelia in the nasal cavity of the mouse. *Am. J. Anat.*, **133,** 37–50, 1972.

MATULIONIS, D. H., AND PARKS, H. F. Ultrastructural morphology of the normal nasal respiratory epithelium in the mouse. *Anat. Rec.*, **176,** 65–83, 1973.

POLYZONIS, B. M., KAFADARIS, P. M., AND GIGIS, P. I., AND DEMETRIOV, T. An electron microscopic study of human olfactory mucosa. *J. Anat.*, **128,** 77–83, 1979.

Trachea and Bronchi

BASKERVILLE, A. Ultrastructure of the bronchial epithelium of the pig. *Souder. Zentra. Veter.*, **17,** 796–802, 1970.

MEVRICK, B., AND REID, L. Ultrastructure of cells in the human bronchial submucosal glands. *J. Anat.*, **107,** 281–99, 1970.

MIANI, A., PIZZINI, G., AND DEGASPERIS, C. "Special type cells" in human tracheal epithelium. *J. Submicr. Cytol.*, **3,** 81–84, 1971.

Lung

BLUMCKE, S., KESSLER, W. D., NIEDORF, H. R., BECKER, N. H., AND VEITH, F. J. Ultrastructure of lamellar bodies of type II pneumocytes after osmium-zinc impregnation. *J. Ultrastructure Res.*, **42,** 417–33, 1973.

GIL, J. AND WEIBEL, E. R. Extracellar lining of bronchioles after perfusion-fixation of rat lungs for electron microscopy. *Anat. Rec.*, **169,** 185–200, 1971.

KIKKAWA, Y. Morphology of alveolar lining layer. *Anat. Rec.*, **167,** 389–400, 1970.

PETRIK, P. Fine structural identification of peroxisomes in mouse and rat bronchiolar and alveolar epithelium. *J. Histochem. Cytochem.*, **19,** 339–48, 1971.

SMITH, V., SMITH, D. S., AND RYAN, J. W. Tubular myelin assembly in type II alveolar cells: freeze-fracture studies. *Anat. Rec.*, **176,** 125–28, 1973.

27. Circulatory System

Arteries

OSBORNE-PELLEGRIN, M. J. Some ultrastructural characteristics of the renal artery and abdominal aorta in the rat. *J. Anat.*, **125,** 641–52, 1978.

PIEZZI, R. S., SANTOLAYA, R. S., AND BERTINI, F. The fine structure of endothelial cells of toad arteries. *Anat. Rec.*, **165,** 226–36, 1969.

SCHWARTZ, S. M., AND BENDITT, E. P. Studies on aortic intima. I. Structure and permeability of rat thoracic aortic intima. *Am. J. Path.*, **66,** 241–64, 1972.

TAKAYANAGI, T., RENNELS, M. L., AND NELSON, E. An electron microscopic study of intimal cushions in intracranial arteries of the cat. *Am. J. Anat.*, **133,** 415–30, 1972.

Veins

HAMMERSEN, F. Zur Ultrastruktur der Venenwand. *Zbl. Phlebol.*, **6,** 221–37, 1967.

TSAO, C-H., GLAGOV, S., AND KELSEY, B. F. Structure of mammalian portal vein: postnatal establishment of two mutually perpendicular medial muscle zones in the rat. *Anat. Rec.*, **171,** 457–70, 1970.

WOOD, E. The venous system. *Sci. Am.*, **218,** 86–96, 1968.

Capillaries

MAUL, G. G. Structure and formation of pores in fenestrated capillaries. *J. Ultrastructure Res.*, **36,** 768–82, 1971.

RHODIN, J. A. G. Fine structure of capillaries. In *Topics in the Study of Life* (H. Ris, Ed.), pp. 215–24. Harper & Row, New York, 1971.

VENKATACHALAM, M. A., AND KARNOVSKY, M. J. Extravascular protein in the kidney. An ultrastructural study of its relation to renal peritubular capillary permeability using protein tracers. *Lab. Invest.*, **27,** 435–44, 1972.

Lymphatics

CLIFF, W. J., AND NICOLL, P. A. Structure and function of lymphatic vessels of the bat's wing. *Quart. J. Exp. Physiol.*, **55,** 112–21, 1970.

LEAK, L. V. Studies on the permeability of lymphatic capillaries. *J. Cell Biol.*, **50,** 300–23, 1971.

TAKADA, M. The ultrastructure of lymphatic valve in rabbits and mice. *Am. J. Anat.*, **132,** 207–17, 1971.

28. Urinary System

Kidney

BARAJAS, L. The ultrastructure of the juxtaglomerular apparatus as disclosed by three-dimensional reconstructions from serial sections. *J. Ultrastructure Res.*, **33,** 116–47, 1970.

GRIFFITH, L. D., BULGER, R. E., AND TRUMP, B. F. The ultrastructure of the functioning kidney. *Lab. Invest.*, **16,** 220–46, 1967.

JORGENSEN, F., AND BENTZON, M. W. The ultrastructure of the normal human glomerulus. Thickness of glomerular basement membrance. *Lab. Invest.*, **18,** 42–8, 1968.

LATTA, H., MAUNSBACH, A. B., AND OSVALDO, L. The fine structure of renal tubules in cortex and medulla. In *Ultrastructure of the Kidney* (A. J. Dalton and F. Haguenau, Eds.), pp. 2–56. Academic Press, New York, 1967.

TISHER, C. C., BULGER, R. E., AND VALIN, H. Morphology of renal medulla in water diuresis and vasopressin-induced antidiuresis. *Am. J. Physiol.*, **220,** 87–94, 1971.

Ureter and Urinary Bladder

DIBONA, D. R., AND CIVAN, M. M. The effect of smooth muscle on the intracellular spaces in toad urinary bladder. *J. Cell Biol.*, **46,** 235–44, 1970.

DIXON, J. S., AND GOSLING, J. A. Electron microscopic observations of the renal caliceal wall in the rat. *Z. Zellforsch.*, **103,** 328–40, 1970.

LIBERTINO, J. A., AND WEISS, R. H. Ultrastructure of human ureter. *J. Urol.*, **108,** 71–6, 1972.

MONIS, B., AND ZAMBRANO, D. Transitional epithelium of urinary tract in normal and dehydrated rats. *Z. Zellforsch.*, **85,** 165–82, 1968.

NOTLEY, R. G. Electron microscopy of the lower ureter in man. *Brit. J. Urol.*, **42,** 439–45, 1970.

29. Male Genital System

Testis

BURGOS, M. H., VITALE-CALPE, R., AND AOKI, A. Fine structure of the testis and its functional significance. In *The Testis* (A. D. Johnson, W. R. Gomes, and Nil Van Demark Eds.), Vol. 1, pp. 551–649. Academic Press, New York, 1970.

FAWCETT, D. W. The mammalian spermatozoon. *Devel. Biol.*, **44,** 394–436, 1975.

KRETSER, D. M. D. The fine structure of the testicular interstitial cells in men of normal androgenic status. *Z. Zellforsch.*, **80,** 594–609, 1967.

LU, C. C., AND STEINBERGER, A., Effects of estrogen on human seminiferous tubules: Light and electron microscopic analysis. *Am. J. Anat.*, **153,** 1–14, 1978.

RATTNER, J. B., AND BRINKLEY, B. R. Ultrastructure of spermiogenesis. *J. Ultrastructure Res.*, **32,** 316–22, 1970.

YASUZUMI, G. Electron microscope studies on spermiogenesis in various animal species. *Int. Rev. Cytol.*, **37,** 53–119, 1973.

Epididymis

HAMILTON, D. W., JONES, A. L., AND FAWCETT, D. W. Cholesterol biosynthesis in the mouse epididymis and ductus deferens: a biochemical and morphological study. *Biol. Reprod.*, **1,** 167–84, 1969.

LADMAN, A. J. The fine structure of the ductuli efferentes of the opossum. *Anat. Rec.*, **157,** 559–76, 1967.

ORGEBIN-CRIST, M. C. Studies on the function of the epididymis. *Biol. Reprod.*, Suppl. 1, 155–75, 1969.

Prostate, Seminal vesicles and Penis

BRANDES, D. The fine structure and histochemistry of prostatic glands in relation to sex hormones. *Int. Rev. Cytol.*, **20,** 207–76, 1966.

FISHER, E. R., AND JEFFREY, W. Ultrastructure of human normal and neoplastic prostate. *Am. J. Clin. Path.*, **44,** 119–34, 1965.

FLICKINGER, C. J. The fine structure and development of the seminal vesicles and prostate in the fetal rat. *Z. Zellforsch.*, **109**, 1–14, 1970.
LEESON, T. S., AND LEESON, C. R. The fine structure of cavernous tissues in the adult rat penis. *Invest. Urol.*, **3**, 144–54, 1965.
RIVA, A. Fine structure of human seminal vesicle epithelium. *J. Anat.*, **102**, 71–86, 1967.

30. Female Genital System

Ovary

CRISP, T. M., DESSOUKY, A. D., AND DENYS, F. R. The fine structure of the human corpus luteum of early pregnancy and during the progestational phase of the menstrual cycle. *Am. J. Anat.*, **127**, 37–70, 1970.
HADEK, R. The structure of the mammalian egg. *Int. Rev. Cytol.*, **18**, 29–72, 1965.
NØRREVANG, A. Electron microscopic morphology of oogenesis. *Int. Rev. Cytol.*, **23**, 114–86, 1968.
PAPADAKI, L., AND BEILBY, J. O. W. The fine structure of the surface epithelium of the human ovary. *J. Cell Sci.*, **8**, 445–65, 1971.

Oviduct

BJÖRKMAN, N., AND FREDAICSSON, B. Ultrastructural features of the human oviduct epithelium. *Int. J. Fertil.*, **7**, 259–66, 1962.
CLYMAN, M. J. Electron microscopy of the human fallopian tube. *Fertil. Steril.*, **17**, 281–301, 1966.
NILSSON, O. Electron microscopy of the fallopian tube epithelium of rabbit in oestrus. *Exp. Cell Res.*, **14**, 541–54, 1958.

Uterus

BO, W. J., ODOR, D. L., AND ROTHROCK, M. L. Ultrastructure of uterine smooth muscle following progesterone or progesterone-estrogen treatment. *Anat. Rec.*, **163**, 121–32, 1969.
GORDON, M. Cyclic changes in the fine structure of the epithelial cells of human endometrium. *Int. Rev. Cytol.*, **42**, 127–72, 1975.
LARKIN, L. H., AND FLICKINGER, C. J. Ultrastructure of the metrial gland cell in the pregnant rat. *Am. J. Anat.*, **126**, 337–54, 1969.

Vagina

COOPER, R. A., CARDIFF, R. D., AND WELLINGS, S. R. Ultrastructure of vaginal keratinization in estrogen treated immature Balb/cCRGL mice. *Z. Zellforsch.*, **77**, 377–403, 1967.
EDDY, E. M., AND WALKER, B. E. Cytoplasmic fine structure during hormonally controlled differentiation in vaginal epithelium (mouse). *Anat. Rec.*, **164**, 205–18, 1969.
STEGNER, H., AND IWATHA, M. Elektronenmikroskopische Untersuchungen am Scheidenepithel der Ratte. *Mikr. Anat. Forsch.*, **76**, 491–508, 1967.

31. Placenta, Umbilical Cord and Mammary Gland

Placenta

BJÖRKMAN, N. Contributions of electron microscopy in elucidating placental structure and function. *Int. Rev. Gen. Exper. Zool.*, **3**, 309–71, 1968.
BOYD, J. D., AND HAMILTON, W. J. Development and structure of the human placenta from the end of the 3rd month of gestation. *J. Obstet. Gyn.* (Br.), **74**, 161–226, 1967.
BURGESS, S. M., AND TAM, W. H. Ultrastructural changes in the guinea-pig placenta, with special reference to organelles associated with steriodogenesis. *J. Anat.*, **126**, 319–27, 1978.
ENDERS, A. C. Fine structure of anchoring villi of the human placenta. *Am. J. Anat.*, **122**, 419–52, 1968.

Umbilical cord

HOYES, A. D. Ultrastructure of the epithelium of the human umbilical cord. *J. Anat.*, **105,** 149–62, 1969.

LEESON, C. R., AND LEESON, T. S. The fine structure of the rat umbilical cord at various times of gestation. *Anat. Rec.*, **151,** 183–97, 1965.

PARRY, E. W. Some electron microscope observations on the mesenchymal structures of full-term umbilical cord. *J. Anat.*, **107,** 505–18, 1970.

Mammary Gland

HELMINEN, H. J., AND ERICSSON, J. L. E. Studies on the mammary gland involution. I. On the ultrastructure of the lactating mammary gland. *J. Ultrastructure Res.*, **25,** 193–213, 1968.

KUROSUMI, K., KOBAYASHI, Y., AND BABA, N. The fine structure of the mammary glands of lactating rats, with special reference to the apocrine secretion. *Ex. Cell. Res.*, **50,** 177–92, 1968.

WOODING, F. B. P. The structure of the milk fat globule membrane. *J. Ultrastructure Res.*, **37,** 388–400, 1971.

32. Integumentary System

Skin

BRODY, I. An electron microscopic study of the fibrillar density in the normal human stratum corneum. *J. Ultrastructure Res.*, **4,** 264–97, 1960.

MARTINEZ, I. R., AND PETERS, A. Membrane-coating granules and membrane modifications in keratinizing epithelia. *Am. J. Anat.*, **130,** 93–120, 1971.

MENTON, D. N., AND EISEN, A. Z. Structure and organization of mammalian stratum corneum. *J. Ultrastructure Res.*, **35,** 247–64, 1971.

MILLWARD, G. R. The substructure of x-keratin microfibrils, *J. Ultrastructure Res.*, **31,** 349–55, 1970.

WIER, K. A., FOKUYAMA, K., AND EPSTEIN, W. Z. Nuclear changes during keratinization of normal human epidermis. *J. Ultrastructure Res.*, **37,** 138–145, 1971.

Glands

BELL, M. A. A comparative study of sebaceous gland ultrastructure in subhuman primates. *Anat. Rec.*, **170,** 331–42, 1971.

HASHIMOTO, K. Demonstration of the intercellular spaces of the human eccrine sweat gland by lanthanum. *J. Ultrastructure Res.*, **37,** 504–20, 1971.

TERZAKIS, J. A. The ultrastructure of monkey eccrine sweat glands. *Z. Zellforsch.*, **64,** 493–509, 1964.

Hair

BREATHNACH, A. S., AND SMITH, J. Fine structure of the early hair germ and dermal papilla in the human foetus. *J. Anat.*, **102,** 511–26, 1968.

ROBINS, E. J., AND BREATHNACH, A. S. Fine structure of bulbar end of human foetal hair follicle at stage of differentiation of inner root sheath. *J. Anat.*, **107,** 131–46, 1970.

Pigment

HASHIMOTO, K. The ultrastructure of the skin of human embryos III. Melanoblast and intrafollicular melanocyte. *J. Anat.*, **108,** 99–108, 1971.

WISE, G. E. Origin of amphibian premelanosome and their relation to microtubules. *Anat. Rec.*, **165,** 185–96, 1969.

33. Nervous System

Brain

BODIAN, D. An electron microscopic characterization of classes of synaptic vesicles by means of controlled aldehyde fixation. *J. Cell Biol.*, **44,** 115–24, 1970.

COLONNIER, M. The fine structural arrangement of the cortex. *Arch. Neurol.*, **16,** 651–57, 1967.
GOBELS, S. Electron microscopical studies of the cerebellar molecular layer. *J. Ultrastructure Res.*, **21,** 430–58, 1968.
KOBAYASHI, H., MATSUI, T., AND SUSUMA, I. Functional electron microscopy of the hypothalamic median eminence. *Int. Rev. Cytol.*, **29,** 281–381, 1971.
PETERS, A. Stellate cells of the rat perietal cortex. *J. Comp. Neurol.*, **14,** 345–73, 197.

Ganglia

KERR, F. W. L. Correlated light and electron microscopic observations on the normal trigeminal ganglion and sensory root in man. *J. Neurosurg.*, **26,** 132–37, 1967.
PANNESE, E. The histogenesis of the spinal ganglia. *Adv. Anat. Embryol. Cell. Biol.*, **47,** 1–97, 1974.
STEER, J. M. Some observations on the fine structure of rat dorsal spinal nerve roots. *J. Anat.*, **109,** 467–85, 1971.

34. Sensory Organs

Eye

HINDS, J. W., AND HINDS, P. L. Early ganglion cell differentiation in the mouse retina: an electron microscopic analysis utilizing serial sections. *Devel. Biol.*, **37,** 381–416, 1974.
MAGALHAES, M. M., AND COLUMBRA, A. Electron microscope radioautographic study of glycogen synthesis in the rabbit retina. *J. Cell Biol.*, **47,** 263, 1970.
NGUYEN-LEGROS, J. Fine structure of the pigment epithelium in the vertebrata retina. *Int. Rev. Cytol.*, Suppl. 7, 287–328, 1978.
ROSENKRANTZ, J. New aspects of the ultrastructure of frog rod outer segments. *Int. Rev. Cytol.*, **50,** 25–158, 1977.

Ear

KAWABATA, I., AND PAPARELLA, M. M. Ultrastructure of normal human middle ear mucosa. *Ann. Otol. Rhin. Laryngol.*, **78,** 125–137, 1968.
KITMURA, R. S. The ultrastructure of the organ of Corti. *Int. Rev. Cytol.*, **42,** 173–222, 1975.
MCMINN, R. M. H., AND TAYLOR, M. Ultrastructure of fibrils in developing human and guinea pig tympanic membrane. *J. Anat.*, **125,** 107–15, 1978.
SMITH, C. A. Electron microscopy of the inner ear. *Ann. Otol. Rhinol. Laryngol.*, **77,** 629–43, 1968.

Smell

ANDRES, K. H. Der olfaktorische saum der katze. *Z. Zellforsch.*, **96,** 250–75, 1969.
CAUNA, N., HINDERER, K. H., AND WENTGES, R. T. Sensory receptor organs of the human nasal respiratory mucosa. *Am. J. Anat.*, **124,** 187–210, 1969.
OKANO, M., WEBER, A. F., AND FROMMES, S. P. Electron microscopic studies of the distal border of the canine olfactory epithelium. *J. Ultrastructure Res.*, **17,** 487–502, 1967.
POLYZONIS, B. M., KAFADARIS, P. M., GIGIS, P. I., AND DEMETRION, T. An electron microscopic study of human olfactory mucosa. *J. Anat.*, **128,** 77–83, 1979.

Taste. See 25. Digestive System and Glands

35. Lymphoid Organs

Lymph nodes

CARR, I. The fine structure of the mammalian lymphoreticular system. *Int. Rev. Cytol.*, **27,** 283–348, 1970.
NOPAJAROONSRI, C., LUK, S., AND SIMON, G. T. Ultrastructure of the normal lymph node. *Am. J. Path.*, **65,** 1–24, 1971
SAINTE-MARIE, G., AND SIN, Y. M. Structures of the lymph node and their possible function during the immune response. *Rev. Can. Biol.*, **27,** 191–207, 1968.

Spleen

BURKE, J. S., AND SIMON, G. T. Electron microscopy of the spleen. I. Anatomy and microcirculation. *Am. J. Path.*, **58,** 127–155, 1970.

HIRASAWA, Y., AND TOKUHIRO, H. Electron microscopic studies on the normal human spleen: especially on the red pulp and the reticuloendothelial cells. *Blood,* **35,** 201–12, 1970.

PICTET, R., ORCI, L., FORSSMANN, W. G., AND GIRARDIER, L. An electron microscope study of tbe perfusion-fixed spleen. *Z. Zellforsch.*, **96,** 372–417, 1969.

Thymus

CHAPMAN, W. L., AND ALLEN, J. R. The fine structure of the thymus of the fetal and neonatal monkey *(Macaca mulatta)*. *Z. Zellforsch.*, **114,** 220–33, 1971.

EVERETTS, N. B., AND TYLER (CAFFREY), R. W. Lymphopoiesis in the thymus and other tissues: functional implications. *Int. Rev. Cytol.*, **22,** 205–86, 1967.

PINKEL, D. Ultrastructure of human fetal thymus. *Am. J. Dis. Child.*, **115,** 222–38, 1968.

36. Endocrine Organs

Thyroid Gland

KLINCK, G. H., OERTEL, J. E., AND WINSHIP, T. Ultrastructure of normal human thyroid. *Lab. Invest.*, **22,** 2–22, 1970.

ERICSON, L. E. Subcellular localization of 5-hydroxytryptamine in the parafollicular cells of the mouse thyroid gland. An autoradiographic study. *J. Ultrastructure Res.*, **31,** 162–77, 1970.

GERSHON, M. D., BELSHAW, B. E., AND NUNEX, E. H. Biochemical, histochemical and ultrastructural studies of thyroid sertonin, parafollicular and follicular cells during development in the dog. *Am. J. Anat.*, **132,** 5–20, 1971.

FUJITA, H. Fine structure of the thyroid gland. *Int. Rev. Cytol.*, **40,** 197–280, 1975.

Pituitary Gland (Hypophysis)

MIKAMI, S. Light and electron microscopic investigations of six types of glandular cells of the bovine adenohypophysis. *Z. Zellforsch.*, **105,** 457–82, 1970.

PAIZ, C., AND HENNIGAR, G. R. Electron microscopy and histochemical correlation of human anterior pituitary cells. *Am. J. Path.*, **59,** 43–73, 1970.

Parathyroid Gland

FETTER, A. W., AND CAPEN, C. C. Tbe ultrastructure of the parathyroid glands of young pigs. *Acta Anat.*, **75,** 359–72, 1970.

NAKAGIMA, K., YAMAZAKI, Y., AND ISUNODA, Y. An electron microscopic study of the human fetal parathyroid gland. *Z. Zellforsch.*, **75,** 89–95, 1968.

NUNEX, E. A., WHALEN, J. P., AND KROOK, L. An ultrastructural study of the natural secretory cycle of the parathyroid gland of the bat. *Am. J. Anat.*, **134,** 459–80, 1972.

Adrenal (Suprarenal) Gland

BROWN, W. J., BARAJAS, L., AND LATTA, H. The ultrastructure of the human adrenal medulla: with comparative studies of white rat. *Anat. Rec.*, **169,** 173–84, 1971.

IDELMAN, S. Ultrastructure of the mammalian adrenal cortex. *Int. Rev. Cytol.*, **27,** 181–281, 1970.

PENNY, D. P., AND BROWN, G. M. The fine structural morphology of adrenal cortices of normal and stressed squirrel monkeys. *J. Morph.*, **134,** 447–66, 1971.

Islets of Langerhans See Pancreas in 25. Digestive System and glands.

Pineal Gland

CLABOUGH, J. W. Ultrastructural features of the pineal gland in normal and light deprived golden hamsters. *Z. Zellforsch.*, **114,** 151–64, 1971.

LUES, G. Die Feinstruktur der Zirbeldrüse normaler trächtiger und experimentell beeinflusster Meerschweinchen. *Z. Zellforsch.*, **114,** 38–60, 1971.

SHERIDEN, M. N., AND REITER, R. J. Observations on the pineal system in the hamster. II. Fine structure of the deep pineal. *J. Morph.*, **131,** 163–78, 1970.

36. Bacteria, Viruses, and Macromolecules

Bacteria

CHAPMAN, G. B. Electron microscopy of ultrathin sections of bacteria. *Sci. Instr. News,* **8,** 1–5, 1963.

DREWS, G. Submikroskopische Cytologie der Bakterienzelle. *Fortschr. Bot.,* **27,** 36–43, 1965.

GREENALGH, G. N., AND EVANS, L. V. Electron microscopy. In *Methods in Microbiology* (O. Booth, Ed.), Vol. 4, pp. 517–65. Academic Press, New York, 1971.

KELLENBERGER, E. Die Elektronenmikroskopie in molekularbiologie und mikrobiologie. *Path. Microbiol.,* **28,** 540–60, 1965.

VAN ITERSON, W. Symposium on the fine structure and replication of bacteria and their parts. II. Bacteriological cytoplasm. *Bact. Rev.,* **29,** 299–325, 1965.

Viruses

CROWTHER, R. A. Three-dimensional reconstruction and the architecture of spherical viruses. *Endeavour,* **30,** 124–29, 1971.

HEINE, U. I., COTTLER-FOX, M., AND WEBER, G. H. Visualization of tumor virus RNA in the electron microscope. *Meth. Cancer Res.,* **11,** 167–203, 1975.

HORNE, R. W. Electron microscopy of viruses. *Sci. Prog.,* **52,** 525–42, 1964.

HULL, R. The structure of tubular viruses. *Adv. Virus Res.,* **20,** 1–32, 1976.

SEMAN, G., AND DMOCHOWSKI, L. Methods for electron microscopy of viruses. *Meth. Cancer Res.,* **12,** 178–255, 1976.

SMITH, K. O. Identification of viruses by electron microscopy. *Meth. Cancer Res.,* **1,** 545–72, 1967.

Macromolecules

GRIFFITH, J. D. Electron microscopic visualization of DNA in association with cellular components. *Meth. Cell Biol.,* **7,** 129–46, 1973.

KLEINSCHMIDT, A. K. Electron microscopic studies of macromolecules without oppositional contrast. *Phil. Trans. Roy. Soc. Lond., B,* **261,** 143–49, 1971.

KOLLER, T., BEER, M., MULLER, M., AND MUHLETHALER, K. Electron microscopy of selectively stained molecules. *Cytobiol.,* **4,** 369–408, 1971.

LANG, D. Individual macromolecule preparation and recent results with DNA. *Phil. Trans. Roy. Soc. Lond. B.* **261,** 151–58, 1971.

VAINSHTEIN, B. K. Electron microscopical analysis of the the three-dimensional structure of biological macromolecules. *Adv. Opt. Electron Micr.,* **7,** 281–377, 1979.

37. Botany

BELL, P. R. Physical interactions of nucleus and cytoplasm in plant cells. *Endeavour,* **34,** 19–22, 1975.

BUVAT, R. Electron microscopy of plant protoplasm. *Int. Rev. Cytol.,* **14,** 41–155, 1963.

CRONSHAW, J., AND ESAU, K. Tubular and fibrillar components of mature and differentiating sieve elements. *J. Cell. Biol.,* **34,** 801–15, 1967.

GOODCHILD, D. J. The ultrastructure of root nodules in relation to nitrogen fixation. *Int. Rev. Cytol.,* Suppl. 6, 235–85, 1977.

MUHLETHALER, K. Ultrastructure and formation of plant cell walls. *Ann. Rev. Plant. Physiol.,* **18,** 1–24, 1967.

ROBINOW, C. F. Electron microscopy of yeasts. *Meth. Cell. Biol.,* **11,** 1–22, 1975.

39. Tumor Cells

BRINKLEY, B. R., AND CHANG, J. P. Mitosis in tumor cells: methods for light and electron microscopy. *Meth. Cancer Res.,* **11,** 247–91, 1975.

BUSCH, H., AND SMETANA, K. The nucleus of the cancer cell. In *The Molecular Biology of Cancer* (H. Busch, Ed.), pp. 41–80. Academic Press, New York, 1974.

HAGUENAU, F. Ultrastructure of the cancer cell. In *The Biological Basis of Medicine* (E. E. Bittar and N. Bittar, Eds.), pp. 433–86. Academic Press, New York, 1968.

MURAD, T. M. Cytological differentiation of carcinoma of the breast by electron microscopy. *Acta Cytol.*, **15,** 400–9, 1971.

ROSSI, J., AND RODRIGUES, H. A. Application of electron microscopy to the differential diagnosis of tumors. *Am. J. Path.*, **50,** 550–62, 1968.

40. Medicine

ARENBERG, I. K. Scanning electron microscopy in otorhinolaryngologic research. *Arth. Otolaryngol.*, **95,** 477–83, 1972.

CAUFIELD, J. B. Application of the electron microscope to renal diseases. *New Engl. J. Med.*, **270,** 183–94, 1964.

HENDERSON, J. A., CHANDLER, J. A., BLUNDELL, G., GRIFFITHS, C., AND DAVIES, J. The application of analytical electron microscopy to the study of diseased biological tissue. *J. Micr.*, **99,** 183–92, 1973.

HIRANO, A. Electron microscopy in neuropathology. *Prog. Neuropath.*, **1,** 232–54, 1971.

JENKINSON, J. A., AND DAWSON, J. M. P. The value of electron microscope studies in diagnosing malignant change in ulcerative colitis. *Gut,* **12,** 110–18, 1971.

SUZUKI, K., ANDREWS, J. M., WALTZ, J. M., AND TERRY, R. D. Ultrastructural studies of multiple sclerosis. *Lab. Invest.*, **20,** 444–54, 1969.

WATSON, J. H. L. Electron microscopy in medical diagnosis. *The Microscope,* **17,** 63–69, 1969.

GLOSSARY*

ABBE'S EQUATION. A mathematical expression of the magnitude of the effect of diffraction on the resolution of a lens.

ABERRATION. An imperfection of a focused optical system that prevents the formation of an image of a point as a perfect Airy disc.

AIR-LOCK. A chamber in an electron microscope between the outer air and inner vacuum, designed so that the outer part closes before the inner part opens and vice versa. Either the volume of the chamber is sufficiently small so that the air that is admitted provides an insignificant load for the pumps, or alternatively, the chamber is independently pre-pumped before opening the inner part. Such a chamber facilitates rapid changing of specimens or photographic plates.

AIRY DISC. The bright central spot and the system of surrounding diffraction rings formed by a perfectly focused optical system with light from a point source.

ALIGNMENT. The process by which the electron gun and all the lenses and apertures are positioned so as to be electron-optically centered relative to one another.

ANODE. A positive or electron attracting electrode.

APERTURE. An opaque optical stop containing a central opening.

ASTIGMATISM. An axial aberration which results in images of points appearing as line elements.

BIAS. The voltage between the cathode and grid of the electron gun that serves to control the current drawn from the filament.

CHROMATIC ABERRATION. The blurring of the image of a point due to the inability of an electron lens to simultaneously focus electrons of different velocity (i.e., wavelength) coming from the same object point.

COMPENSATED OBJECTIVE. An objective lens whose astigmatism has been compensated for by means of a stigmator.

CONDENSER LENS. The lens closest to the radiation source of an optical system that can focus the emitted electrons onto the specimen.

CROSS-OVER. The axial position of the image of the effective electron source formed by the condenser. At fixed emission current and fixed condenser aperture, when this position is coincident with the specimen plane, illumination is maximum.

DEPTH OF FIELD. The range of distances at the specimen level, measured along the optical axis, which simultaneously provide acceptable image sharpness.

DEPTH OF FOCUS. The range of distances in the region of image formation (behind a lens), measured along the optical axis, throughout which the image has acceptable sharpness.

DIFFRACTION. Modification in direction, amplitude and phase which a wave undergoes in passing by an obstruction.

DIFFUSION PUMP. A vacuum pump for producing high vacuum by diffusing gas into a directed high velocity stream of molecules which impel the gas in the direction of the stream.

ELECTRON GUN. The unit of the electron microscope responsible for the formation and acceleration of the electron beam.

ELECTRON STAIN. A substance that is used to provide increased electron contrast between a part of a biological specimen and its surrounds.

ELECTROSTATIC LENS. An arrangement of electrodes so disposed that the resulting electric fields produce a focusing effect on a beam of charged particles.

EQUIPOTENTIAL SURFACE. A surface in an electric field which is the locus of the points at which a test charge would experience forces of the same magnitude.

FILAMENT. The electron-emitting heated wire of the electron gun. (Syn. cathode, emitter.)

FOCAL LENGTH. The distance from the optical center of a lens to its focal point.

FOCAL POINT. The point at which rays initially parallel to the axis of the optical system intersect after passing through it.

FORE-PUMP. A rotary vacuum pump, auxiliary to a diffusion pump, which removes the gas from the high pressure side of the diffusion pump.

* All terms in Glossary will also be found in the Index.

FREE ELECTRONS. Valence electrons that move freely between atoms in a solid when a potential difference is applied across the solid.

FRESNEL (DIFFRACTION) FRINGES. These are fringes observed as a result of bending of radiation waves near the edges of an obstacle and which appear in planes just beyond the obstacle or its image.

GEOMETRIC ABERRATIONS. Those aberrations of a lens or lens system which are independent of wavelength (electron velocity) but depend in magnitude upon the particular geometric path through the lens system (i.e., depend upon both distance from the optical axis and direction of travel with respect to the optical axis), and/or upon geometrical asymmetries in the lens system.

IMAGE DRIFT. Usually a disturbance resulting in lateral movement of the position of the image.

LENS ASYMMETRY. Inhomogeneity of the field of a lens (usually due to physical asymmetries) with respect to rotation about the optic axis.

MAGNETIC LENS. An arrangement of current-carrying coils, core-containing coils or permanent magnets so positioned that the resulting magnetic fields will focus a point source of electrons.

NUMERICAL APERTURE (N.A.). Refers to light-gathering and resolving capacity of a lens. More specifically it is the quantity resulting from multiplying the index of refraction of the medium between the specimen and the lens by the sine of the angle formed by the optical axis and the most oblique ray which enters the lens.

OBJECTIVE LENS. The lens of a microscope that forms the initial image of the specimen.

OVER-FOCUS. Adjustment of the focus of the objective lens to a level between the plane of the specimen and the objective.

POLE PIECE. A piece of soft iron at a pole of an electromagnet shaped to concentrate or direct the magnetic flux in some desired path.

PROJECTION LENS. The last lens which magnifies the image initially formed by the objective lens.

REPLICA. Reproduction, usually with the aid of the shadow-casting technique, of a facsimile of the surface of an object.

RESOLUTION. A measure of the limit of the capability of an optical system to produce separate images of adjacent objects.

SHADOW CASTING. The production of exaggerated contrast by "irradiating" the specimen obliquely with a beam of metal atoms (in vacuum) which covers the specimen with electron opaque film on the side of the specimen facing the source and leaves electron transparent shadows on the opposite side.

SHIELD. The negative electrode that is shaped as a shield around the filament. The bias voltage is applied between this element and the filament. (Syn. grid, grid cap, Wehnelt cylinder, focusing electrode.)

SPHERICAL ABERRATION. The absence of a common focal point for rays which pass axially and off-axis through a lens, due to the fact that the zones of the lens farther from the axis have an inappropriate refractive power.

STIGMATOR. A device placed in or near an objective lens so as to make possible correction for its astigmatism.

THERMIONIC EMISSION. The "boiling off" of some of the free electrons of a metal as a result of heating.

THROUGH-FOCUS SERIES. A series of photographs of a specimen, taken with it in closely spaced under-focus, in-focus and over-focus positions.

UNDER-FOCUS. The adjustment of the focus of the objective lens to a level between the plane of the specimen and the condenser.

VECTOR. A quantity which has magnitude and direction.

WAVE FRONT. A surface composed at any instant of all the points just reached by a vibrational disturbance in its propagation through a medium.

WAVELENGTH. The distance between the crests of adjacent waves.

WORK FUNCTION. The energy needed to remove an electron from the surface of a solid to a point an infinite distance away.

AUTHOR INDEX*

Abbe, E. 10
Airy, G. 10
Aldridge, W.G. 119
Amer, A. 3rd. **200**
Ampere, A.M. 20
Andrews, P. **271**
Ardenne, M. von, 5, 216, 254
Axelrod, J. 119

Bachmann, L. 119
Bahr, G.F. 117, 207
Baker, R.F. 117, **118**
Balcerzak, S.P. 264
Barber, V.C. 264
Barrnett, R.J. 119
Bartle, P. 119
Bensch, K. 119
Bernhard, W. 119, 143
Birch-Anderson, A. 117, 147
Blum, J. 117, 157
Borries, B. von, 5
Borysko, E. 117
Boyde, A. 264, **268**
Bradbury, S. 131
Bradfield, J.R.G. 117
Bragg, W.H. 323
Brandes, D. 117
Brenner, S. 117
Broers, A.N. 224
Brüche, E. 5
Bullivant, S. **200**
Burton, E.F. 5
Busch, H. 4, 31, 32, 216

Camp, A.S. 264
Caro, L.G. 119, 186
Caufield, J.B. 127
Castaing, R. 254
Claude, A. 117, 163
Cosslett, V.E. 254
Crewe, A.V. 253
Crowell, J. 119, **198**

Dalton, A.J. 127
Davis, B.J. 119
Davis, R. **178**
Davison, C.J. 15
de Broglie, L. 4, 14, 216, 294, 295
de Forest, L. 299
Devlin, M. **175**
Dronzek, B. 264
Duncumb, P. 254
Driest, E. 5

Echlen, P. 264
Einstein, A. 293, 294
Ellis, E. **200**
Essner, E. 117, 174
Everhart, T.E. 216

Fahimi, H.D. **120**
Faraday, M. 24
Farenbach, W.H. 157
Farquhar, M.G. **180**
Fernández-Morán, H. 117, 119, **172**
190
Flood, P.R. 264
Freeman, J.A. 152
Fresnel, A.J. 90, 237
Frisch, B. **281**
Fujita, T. **287**

Gabor, D. 32, 33
Germer, L.H. 15
Gettner, M.E. 117, 164
Gibbons, I.R. 117
Glauert, A.M. 117, 151
Glauert, E.H. 11
Goldfischer, S. 177
Guinier, A. 254

Haggis, G.H. 264
Hall, C.E. 117, **190**, 191
Hanker, J.S. 179
Hartmann, J.F. 117

* Numbers in **bold** refer to electron micrograph contributions.

Hattori, A. 284
Hay, E.D. 187
Hayes, T.L. 264
Heidenreich, R.D. 195
Hellstrom, B. 158
Hertz, H.R. 293
Hicks, R.M. 119
Hillier, J. 5, 117, 164
Holt, S.J. 119
Horne, R.W. 117, 173
Humphreys, W.H. **279**
Huygens, C. 291

Jaques, W.E. 264
Jamieson, J.D. **188**
Johannson, H. 5

Karnovsky, M. 117, 119, 179, 181, **182**, **183**, 187, 269
Kellenberger, E. 117
Kleinschmidt, A.K. 11, 204, **206**
Knoll, M. 4, 33, 216
Koelle, G.B. **178**

Laane, M.M. **214**
Lafferty, J.M. 224
Langmiur, D.B. 221
Laue, M. von 323
Latta, H. 117
Leduc, E.H. 119, 143, **144**
Lehrer, G.M. 119
Lesko, J. 119
Lewis, S.M. **281**
LoBuglio, A.F. 264
Luft, J.H. 117, 119, 129, 131, 151

Maalϕe, O. 117, 147
Mahl, H. 5
Malhotra, S.K. 119, **199**
Mautner, W. 119
Maswell, J.C. 293
McMullen, D. 216
Meek, G.A. 131
Miller, M.M. **267**
Millonig, G. 117
Miyoshi, M. **287**
Molday, R.S. **282**, 283
Mollenhauer, H.H. **130**, **196**
Moor, H. 117
Moore, D.H. 117
Morgan, C. 117, **184**
Morre, D.J. **196**
Moseley, H.G.J. 254
Moses, M.J. **174**
Muhlethaler, K. 117
Muller, H.O. 5

Nakane, P.K. 119, 155
Nemanic, M.K. 283
Newman, S.B. 117, 139
Newton, I. 293, 305
Novikoff, A. 117, **176**

Oatley, C.W. 216
Oersted, H.C. 20, 21
Ornstein, L. 119

Paden, R. 264
Palade, G.E. 117, **188**
Palay, S.L. 119, 121
Parducz, B. 266
Pease, D.C. 117, **118**, 119, 135, **136**, 137
Pease, R.F.W. 216, 264
Phegan, W. 174
Planck, M. 293
Porter, K.R. 117, **128**, 134, 157
Potter, L.J. 119
Prebus, A. 5

Ramberg, E.G. 5
Rasch, E. 117
Rayleigh, Lord 10
Rebhun, L. 119, 199
Reid, A. 15
Revel, J.P. 119, 181, 187, **267**
Reynolds, E.A. 117
Richardson, H.C. 119
Richardson, O.W. 221
Rinehart, J.J. 264
Robertson, J.D. **132**
Roentgen, W.K. 323
Rogers, G.E. 117
Rose, H.M. 117
Rosenberg, M. 119
Rudenberg, R. 4
Ruska, E. 4, 5, 32, 33, 34
Ryter, A. 117

Sabatini, D.D. 119, 177
Salpeter, M.M. 119, 189
Scheid, W. 264

Schlegel, R.A. 202
Schwab, W. 117
Sheldon, H. 117
Siekevitz, P. 117
Singer, J.J. 119, 181
Sjöstrand, F.S. 117
Smith, K.C.A. 216
Spurlock, B.O. 152
Steere, R.L. 117
Stewart, A.D.G. 264
Steinberger, L.A. 181
Swerdlow, M. 117
Swift, H. 117, 177

Thomson, G.D. 15, 25
Thomson, J.J. 216
Thornley, R.M.F. 216
Tokunaga, J. **287**
Traut, H. 264

Valantine, R.C. 173
Vance, A.W. 5
Van Harreveld, A. 119, **198**, 199

Wachtel, A.W. 119, 157, 164, 166, 177, **214**
Watson, M. 117, 119, 170
Wayte, R. 264
Wissig, S.L. 117, 180
Wolf, D.E. 119
Wood, C. 264

Young, T. 291

Zahn, R.E. 119, 204
Zeitler, E. 117, 207
Zetterquist, H. 117, 127
Zworkin, V.K. 216

SUBJECT INDEX

Abbe's equation, 11
Aberrations, 53, 228
  astigmatism, 89, 230
  correction, 58
  chromatic, 54,229,313
  coma, 53
  curvature of field, 53
  distortion, 64
  field, 53
  geometric, 53
  spherical 53, 229, 311
Absorbtion
  of electrons, 59
  method, 283
Accelerating potential
  SEM, 254
  TEM, 15, 80
Accessories
  electron diffraction, 104
  electron microscope, 102
Air lock, 79
Airy disc, 10
Aldehyde fixatives, 129
Alignment, 81
  adjustment of illumination tilt, 85
  electron gun—condenser, 84
  illumination—objective lens, 85
  objective—intermediate lens, 87
  objective lens—projector lens, 87
  objective aperture, 75, 87
  preset, 83
Ampere-turns, 33
Amplifier, 331
Analogy
  light and electron lenses, 28
  light and electron microscope, 37
Angles
  aperture, 48
  clearance, 155
  knife, 155
  rake, 155
Anode, 40
Antibody—latex sphere coupling, 283
Antibody markers, 284
Antigen—sandwich labeling, 284
Aperture
  angle, of condenser, 48, 305
  angle, of objective, 58
  condenser, 50
  final condenser, 242
  heated, 74
  material, 74
  objective, 73
  opening
    SEM, 242, 255
    TEM, 74
  optimal, 58
  ultrathin metal, 74
Aquon, 141
Araldite, 151
Astigmatism
  correction
    SEM, 230, 255
    TEM, 89
  detection, 89
  quantitative determination, 320
Asymmetrical lens, 89
Autoradiography, 18
Barrel distortion, 65
Beam
  brightness, 45
  electron
    focusing of, 46
    intensity level of, 70
  specimen interaction, 234, 251
  wobbler, 94, 101
Bias, 300
  fixed, 302
  self, 302
Bragg angle, 329
Bragg's equation, 329, 330
Brightness of source, 45
Buffering media, 126
  Osmium tetroxide, 123
Calibration
  diffraction, 330
  magnification, 101
Cambridge-Huxley microtome, 160

Camera, 75
  constant, 330
Carbon particles, 89, 318
Cardolite, 152
Cathode, 39
Cell
  fractions, 195
  surface labeling, 280
Chromatic aberration, 54, 313
  constant of, 55
  correction of, 55
Clearance angle, 155
Coating
  materials, 275
  methods, 275
Coil
  current, 48
  design, 32
  of magnetic lens, 31
Collisions, electron, 59
Coma, 53
Comparative valve of SEM, 258
Compensated objective, 91
Compound microscope, 2
Condenser aperture angle, 48
  derivation of equation of, 305
Condenser lens, 46
  alignment of, 84
  aperature angle of, 48, 305
  double, 2
  final, 231
  operation of, 49
Constructive interference,
Contamination, 95, 21, 99
  objective aperature, 73
  of specimen, 97
Contrast,
  SEM, 246
  TEM, 59, 164
Converging lens, 35
Core, 22
Corpuscular theory of light, 6
Covalent linkage method, 233
Critical point drying, 273
Cross-over, 43
Cryofixation, 197
Cryogenic pump, 342
Cylindrical lens, 89, 315
Cytochemistry, 177
Dark field microscopy, 4
Dehydration, 137
  duration of, 137
  for SEM, 272
  for TEM, 133
  media, 13
Depth of field, 61, 307
Depth of focus, 63, 307
D.E.R.—334, 152
Destructive interference, 319
Diamond knoves, 162
Diffraction, 8
  effective magnitude of, 11
  fringes, Fresnel, 90, 317
  grating, 102
  of edge, 8
  of light, 8
  of water waves, 8
Diffusion pump, 77
Diode, 299
Dispersion methods, 203
Distortion
  barrel, 65
  pincusion, 64
Disturbances, 93
  contanimation, 95
  electrical, 94
  magnetic fields, 93
  methanical, 93
  specimen instability, 94
Double
  condenser lens, 102
  projector lens, 66
  staining, 167
Drift
  image, 93
  specimen, 72
Droplet method, 203
Durcupan, 143
Drying,
  air, 272
  critical point, 273
Einstein equation, 294
Elastically scattered electrons, 59
Electrical disturbances, 94
Electromagnetic theory, 293
Electromagnet, 22
Electron
  absorption, 59
  analogy with light, 211
  autoradiography, 186
  ballistics, 17
  collisions, 59

emission, 15
free, 16
image, 66
micrograph, 70
scattering, 59
valence, 16
velocity, 15
wavelength, 7, 294
Electron ballistics, 17
Electron beam
alignment, 81
contamination, 95, 97
formation, 39
nature of, 14
Electron collection, 239
Electron diffraction, 323
accessories, 104
Electron emission, 15
Electron gun
centration, 45, 83
components, 38
prerequisites, 39
temperature, 40
SEM, 219
TEM, 38
Electron microscope, SEM and TEM
accessories for, 102
alignment, 81
analogy with light microscope, 37
analysis of cell fractions, 195
contrast, 246
differences from light microscope, 111
electrostatic, 105
focusing, 28
history, 4, 216
image
demagnification in, 227
formation in, 59
magnetic, 38
operation of, 100
other (non-lens) components of, 72
other (non-magnetic) types of, 105
power supply for, 80
resolution of, 15, 58, 189, 248
scanning, 105, 209
transmission, 38
vacuum system of, 76, 244
Electron microscopy, SEM and TEM
basic theory, 6, 217
high resolution, 189
historical review, 4, 216
quantitative, 207
Electron optical column, SEM, 219
Electron optics, 19
analogy with light optics, 28
history, 31
Electronics, 299
Electrons
atomic, 15
in electrostatic field, 34
in magnetic field, 24
scattering of, 60
thermionic emission of, 18
trajectory, 24
Electrostatic electron microscope, 105
Electrostatic field
equipotential, 34
non-uniform, 34
uniform, 34
Electrostatic lens, 34
converging, 35
diverging, 35
Embedding, 139
aquon, 143
araldite, 151
D.E.R.—152, 334
durcupan, 143
epoxy resins, 147
glycol methacrylate, 143
Maraglas, 152
media, 139
methacrylate, 139
polyester resins, 145
Vestopal W, 145
water-soluble, 143
Emitter
non-biased and biased, 40
Emission
current density, 221
electron, 15
thermionic, 18
Enhancement in final image, 248
Enzyme cytochemistry, 177
Epon, 151
Epoxy resins, 147
Equation
Abbe's, 11
Bragg's, 323, 324
chromatic aberration, 55, 313
condenser aperature angle, 48, 305
contrast, 160
de Broglie's, 14

Equation *(Continued)*
  depth of field, 62, 307
  depth of focus, 63, 307
  mass energy equivalence, 294
  maximal useful magnification, 67
  Newton's, 48, 305
  quantum, 294
  spherical aberration, 54, 311
  total magnification, 66
Equipotential surfaces, 34
Exposure,
  to electron beam, 97
  to hydrocarbons, 97
  to non-hydrocarbon residual gases, 99
  to vacuum, 96
  photographic, 70
Ferritin labeled antibody, 284
Field
  aberrations, 53
  curvature of, 53
  depth of, 61
  electrostatic, 34
  emission gun, 224
  intensity, 20
  magnetic, 20
Filament, 39
  centration, 40
  circuit, 299
  current supply, 80
  life, 39
  material, 40
  pointed, 46
  temperature, 40
  tungsten, 219
Films, selection of, 70
Final
  condenser lens, 231
  image formation, 65
Fixation, 125, 269
  aldehyde, 129
  buffering media for, 125
  cryo-, 197
  dripping, 271
  duration, of, 12
  for SEM, 269
  immersion, 271
  injection, 271
  in situ, 121
  osmium tetroxide, 125
  perfusion, 121, 271
  permanganate, 131
  *p*H of, 126
  for TEM, 121
  temperature of, 126
  tonicity during, 126
Fixed bias, 303
Fluorescence, 69
Fluorescent observation screen, 69
Focal length, 28
Focal point, 19
Focusing
  electrode, 40
  of condenser, 46
  of electrons, 28
  of light, 19
  of magnetic lenses, 30
Fore-pump, 77
Formalin fixation, 129
Fracture ridge, 161
Free electrons, 16
Freeze
  drying, 273
  etching, 195, 279
  fracturing, 277
  substitution, 197
Fresnel diffraction fringes, 23, 90
Fringes, Fresnel, 23, 90
Freon, 272
Geometric aberrations, 53
Glass knives, 161
Glutaraldehyde fixation, 129
Glycol methacrylate, 143
Grating, diffraction, 102
Grid cap, 40
Grid circuit, 299
Gun, electron, 38, 219
  alignment, 84
  self-biased, 42
Handling
  of human material, 124
  of small specimens, 123
High voltage
  electron microscopy, 107
  supply, 80
High resolution electron microscopy, 189
Historical Review
  SEM
    instrumentation, 216
    methodology, 261
  TEM
    instrumentation, 4
    methodology, 112

Human material, 124
Hysteresis, 23
Illumination
  alignment, 85
  aperture angle of, 48
  intensity of, 48
Image
  bright phase contrast, 23
  characteristics, 246
  dark field, 19
  dark phase contrast, 321
  defects (*see* Aberrations)
  demagnification, 229
  distance, 28
  drift, 93
  final electron, 66
  formation, 54
    final, 65
  intensifiers, 103
  phase contrast, 196
  system, 38
Image translating system
  SEM, 233
  TEM, 68
Imaging system, 50
Immersion oil, 13
Immuno-electron microscopy, 181
Immunouranium technique, 181
Index of refraction, 11
Indium staining, 175
Inelastically scattered electrons, 59
In situ fixation, 121
Intensity of illumination, 48
Interference, 211
  colors, 166
  constructive, 219
  destructive, 319
  microscopy, 4
Intermediate lens, 66
Ion beam etching, 279
Knife angle, 15
Knives, 160
  diamond, 152
  fracture ridge of, 161
  glass, 161
Kohler illumination, 36
Langmuir
  equation, 221
  trough, 201
Lanthanum Rexaboride cathode, 223
Lead salts, as stains, 171
Lens
  aberrations, 53, 228
  asymmetry, 89
  condenser, 46
  converging, 35
  current supply, 81
  cylindrical, 89
  electromagnetic, 24
  electrostatic, 34
  final condenser, 231
  focusing, 30
  intermediate, 170
  objective, 50
  permanent magnet as, 31
  projector, 62
  scanning electron microscope, 226
Light
  corpuscular theory of, 6
  diffraction of, 8
  electromagnetic theory of, 6
  nature of, 6
  quantum theory of, 293
  wave theory of, 6
Light microscope
  analogy with electron microscope, 37
  differences from electron
    microscope, 111
  resolution of, 13
Limit of resolution, 10
Lines of force, 20, 34
Living material, 285
LKB ultratome, 15
Load resistor, 301
Low voltage microscopy, 252
Macromolecules, 2, 3
Magnetic center, 81
Magnetic density, 20
Magnetic field, 20
  disturbance, 93
  intensity, 23
  lens action of, 24
  motion of electron in, 25, 27
  strength of, 30
  uniform, 25
Magnetic field intensity, 20
Magnetic focusing, 28
Magnetic lens
  aberrations, 53
  action, 24
  alignment, 81
  coil design, 31

Magnetic lens *(Continued)*
  condenser, 46
  design, 31, 51
  double
    condenser, 46
    projector, 62
  evolution of, 31
  focal length, 30
  objective, 50
  permanent, 31
  pole pieces, 34, 51
  projector, 62
  rotation
    of electrons in, 26
    of image in, 27
Magnetic electron microscope, 38
Magnetic shielding, 243
Magnification
  and final image formation, 65
  and objective lens, 51
  and projector lens, 66
  determination of, 105
  range of, 66
Magnifier, 68
Maraglas, 152
Markers, 284
Mass density, 60
Mechanical disturbances, 93
Methacrylate embedding, 139
  problems with, 142
Methodology
  SEM, 261
  TEM, 115
Mica substrate technique, 189
Micrograph, electron, 70
Microprobes, 109
Microtomy, 154
  disturbances of, 166
  for SEM, 277
  for TEM, 150
Mild pretreatment, 265
Momentum, 295
Motion of electrons, 21
Negative-feedback, 303
Negative staining, 171
Newton's equation, 48, 305
Non-uniform field, 34
Nucleoprotein localization, 175
Numerical aperture, 11
Objective aperture, 73
  alignment of, 87
  cleaning, 74
  diameter, 74
  optimal, 58
Objective lens, 50
  aberrations of, 53
    limits of, 57
  aperture angle, 58
  coil design, 32
  compensated, 91
  contrast, 59
  magnification, 51
  operation, 62
  pole pieces, 34, 51
Observation of electron image, 69
Operation of electron microscope,
  SEM, 249
  TEM, 100
Operational requirements
  SEM, 255
  TEM, 81
Optimal aperture angle, 58
Orbitron pump, 339
Osmium tetroxide fixation, 125
  modification of, 127
Over-focus, 320
Paraxial focal plane, 311
Perfusion, 121
Permanent magnet, 31
Permanganate fixation, 131
*p*H of buffering media, 126
*p*H of fixative, 126
Phase contrast
  image, 195, 319, 321
  microscopy, 4
Photoelectrid effect, 291
Photoelectrons, 291
Photographic
  apparatus, 75
  emulsion, choice of, 70
  recording, 69
Photography, 104
Photon, 293
Pincushion distortion, 64
Planck's constant, 14, 293
Plate circuit, 299
Pointed filaments, 46
Pole pieces, 34, 51
Polyester resins, 145
Polystyrene latex spheres, 102
Positive electrode, 40
Positive staining, 170

Postfixation handling, 272
Potential
  difference, 34
  energy, 17
  gradient, 34
Power supply, 80
Prefixation handling, 269
Projector lens, 62
  distortion, 64
  double, 66
  magnification, 66
Quanta, 293
Quantitative electron microscopy, 207
Quantum
  equation, 293
  theory, 293
Rake angle, 155
Range of magnification, 66
Recording of electron image, 69
Refraction, 19
Refractive index
  in electron imcroscope, 15
  in light microscope, 11
Regulator, electrical, 80
Replica technique
  single-stage, 204
  two-stage, 204
Replication, 281
Resins
  epoxy, 147
  polyester, 145
Resolution
  definition of, 8
  electron microscope
    SEM, 249
    TEM, 15, 58, 189
  eye, 67
  high, 185
  light microscope, 13
  limit of, 10, 191
    on fluorescent screen, 69
    on photographic plate, 70
  maximum, 13
  meaning of, 13
  of points, 13
  of SEM,
  test of, 102
Richardson's equation, 221
Rotation of image, 27
Saturation, 43
Scan rates, 256
Scanning electron microscope, 105, 211
Scanning transmission electron microscopy, 253
  basic theory, 217
  column, 219
  comparative value of, 258
  current developments, 252
  design, 219
Scanning systems, 233
Scanning electron microprobe analysis, 254
Scattering of electrons
  elastic, 59
  inelastic, 59
Screen, fluorescent observation, 69
Secondary electron emission mode, 246
Section
  interference color of, 166
  thickness, 164
Sectioning (see Microtomy)
Self-bias gun, 42
  operation of, 44
Shadow casting, 327
Shadowing
  apparatus, 326, 328
  applications, 329
  metals, 328
Shield, 40
Signal
  amplification 239
  display, 240
  monitor, 257
  processing, 234
  selection, 241
Silica sphere labeling, 283
Small particles, surface examination of, 203
Small specimens, 123
Soft-iron, 33, 51
Solenoid, 22
Sorvall
  MT-1, 157
  MT-2, 158
Source, electron, 37
Space
  charge, 42
  limited, 42
Specimen
  chamber, 72
  coating, 274
  contamination, 97

Specimen *(Continued)*
damage, 96
drift, 72
grid, 73
handling of human, 124
handling of small, 123
holder, 72
instability, 94
mounting, 274
obtaining of, 120
preparation of excised, 120
stage
SEM, 243
TEM, 72
thickness, 152, 164
Spherical aberration, 53, 311
constant of, 54
correction of, 58
Spray method, 203
Sputter
coating, 275
ion pump, 337
Staining, 159
negative, 171
of nucleic acids, 175
positive, 170
principles of, 169
Staining media
ammonium molybdate, 171
bismuth, 175
lead ammonium acetate, 171
lead cacodylate, 171
lead citrate, 171
lead tartate, 171
phosphomolybdic acid, 170
phosphotungstic acid, 170
sodium uranate, 170
uranyl acetate, 170
uranyl magnesium acetate, 171
Stereoscopic accessories, 104
Stigmator, 91
Surface spreading, 202
Technique
cryofixation, 197
freeze-etching, 201
freeze-substitution, 201
immunouranium, 181
mica substrate, 189
replica, 204
surface spreading, 204
Temperature of fixative, 126
Theory
corpuscular, 6, 291
de Broglie's, 4, 14, 293
Einstein, 293
electromagnetic, 6, 293
electron ballistics, 17
Fresnel Fringes, 317
Huygens', 291
Maxwell's, 293
Movement of charged particles, 17
Planck, 293
quantum, 14, 293
wave, 9, 291
Thermionic
emission, 18
materials, 40
Three-stage unit, 66
Through-focus series, 101
Titanium sublimation pump, 341
Tonicity of fixative, 126
Tracers, 180
ferritin, 180
gold, 180
horse radish peroxidase, 181
lanthanum, 181
thorotrast, 180
Trajectories, electron, 27
Transmission, electron microscope, 38
Triode, 299
Trough, 163
Tungsten filaments, 40, 46, 219

Ultramicrotomes, 157
Cambridge-Huxley, 160
LKB, 158
Sorvall MT-1, 157
Sorvall MT-2, 158
Ultrastructural
enzyme cytochemistry, 177
immuno-electron microscopy, 181
nucleoprotein localization, 175
Ultrathin metal apertures, 74
Under-focus, 320
Uniform magnetic field, 25
Vacuum
evaporation, 275
gauges, 74
leaks, 95
pumps, 244, 335
requirements, 76
system,

SEM, 244
TEM, 76
Valence electron, 16
Variable self-bias gun, 43
Vector quantities, 297
Vectors
components, 26
resolution of, 297
summation of, 297
Velocity of electrons, 14
Vestopal W, 145
Vibrations, 93
Viewing screen, 69
Viruses, 2, 3
Voltage
accelerating, 15, 80
centration, 81
regulation, 184
Water-soluble embedding media, 143
Wave front, 8
Wavelength, 7
electron, 14, 294
light, 7
Wave mechanics, 294
Wave theory of light, 6, 291
Wehnelt cylinder, 40
Work function, 17